AF615043

CHLORINATED DIOXINS AND DIBENZOFURANS IN THE TOTAL ENVIRONMENT

Edited by

Gangadhar Choudhary

Lawrence H. Keith

Christoffer Rappe

BUTTERWORTH PUBLISHERS
Boston • London
Sydney • Wellington • Durban • Toronto

An Ann Arbor Science Book

ISBN 0-250-40604-7
Library of Congress Catalog Card Number 83-14472

Butterworth Publishers
10 Tower Office Park
Woburn, MA 01801

Printed in the United States of America

CONTENTS

Lawrence H. Keith

Gangadhar Choudhary

Christoffer Rappe

Lawrence H. Keith is Chemistry Development Coordinator at Radian Corporation in Austin, Texas. Before joining Radian, Dr. Keith was a research chemist with the U.S. Environmental Protection Agency in Athens, Georgia. He received his Ph.D. in chemistry from the University of Georgia in 1966. He is a member of the Executive Committee of the American Chemical Society Division of Environmental Chemistry, having served as secretary, chairman, and currently as alternate councilor and program chairman. Dr. Keith is an Advisory Board Member of *Environmental Science and Technology* and a delegate to the U.S. National Committee to the International Association for Water Pollution Research. An editor or co-editor of nine books and over fifty publications and chapters in various journals and books, Dr. Keith's technical interests center on methods for analysis of organic pollutants in the environment and on the safe handling of hazardous chemicals.

Gangadhar Choudhary is a chemist in the Measurement Research Support Branch of the National Institute of Occupational Safety and Health in Cincinnati, Ohio. He has previously worked in the Reduction Research Division of Reynolds Metals Company in Alabama and the Center for Disease Control in Atlanta, Georgia, serving as a senior research scientist, GC-MS chemist, organic-analytical group leader, project director and project officer for governmental contracts in special trace organic projects. Dr. Choudhary has been involved with chlorinated dioxins and dibenzofurans for the past seven years and is the author or editor of three books and twenty-two technical articles involving multidisciplinary chemistry. A native of Uchhati, Bihar State, India, Dr. Choudhary is a naturalized U.S. citizen and received his Ph.D. in physical-organic chemistry from the University of Ottawa, Canada, in 1965.

Christoffer Rappe is Professor and Chairman of Organic Chemistry at the University of Umeå, Sweden. He was previously an associate professor at the University of Uppsala, Sweden, where he also received his Ph.D. in 1965. Professor Rappe is Rapporteur in ad hoc groups on dioxin problems of the World Health Organization (WHO) and of the International Agency for Research on Cancer (IARC). An internationally recognized authority on chlorinated dioxins and dibenzofurans, Professor Rappe has been involved with their synthesis, occurence, formation, degradation, and analysis for over twelve years. Other interests include the synthesis, analysis and occupational exposure to nitrosamines, and synthesis, rearrangements and mechanistic aspects of halogenated ketones. Professor Rappe has authored over 150 papers in various fields of synthetic and analytical organic chemistry.

CONTRIBUTING AUTHORS

Donald G. Barnes
U.S. Environmental Protection Agency
Washington, DC 20460

Robert A. Bell
General Electric Co.—CRD
1 River Road, K-1, 3B15
Schenectady, NY 12301

P.-A. Bergqvist
Department of Organic Chemistry
University of Umeå
S-901 87 Umeå
Sweden

Joseph J. Brooks
Dayton Laboratory
Monsanto Research Corporation
1515 Nicholas Rd.
Dayton, OH 45418

William W. Bunn
Region VII Laboratory
U.S. Environmental Protection Agency
25 Funston Rd.
Kansas City, KS 66115

Maria Antonia Canegrati
Istituto di Ricerche Farmacologiche Mario Negri
Via Eritrea, 62
20157 Milan
Italy

Gangadhar Choudhary
Division of Physical and Engineering Centers for Disease Control
National Institute for Occupational Safety and Health
Cincinnati, OH 45226

R.R. Delongchamp
National Center for Toxicological Research
Food and Drug Administration
Department of Health and Human Services
Jefferson, AR 72079

D.L. Foerst
Environmental Monitoring and Support Laboratory
U.S. Environmental Protection Agency
Cincinnati, OH 45268

A. Garå
Department of Organic Chemistry
University of Umeå
S-901 87 Umeå
Sweden

Silvio Garattini
Istituto di Ricerche Farmacologiche Mario Negri
Via Eritrea, 62
20157 Milan
Italy

John H. Garrett
Brehm Laboratory
Wright State University
Dayton, OH 45435

G. Gustafsson
Department of Organic Chemistry
University of Umeå
S-901 87 Umeå
Sweden

Michael D. Hale
Dayton Laboratory
Monsanto Research Corporation
1515 Nicholas Rd.
Dayton, OH 45418

R.C. Hall
Radian Corporation
8501 MoPac Blvd.
Austin, TX 78758

R.C. Hanisch
Radian Corporation
8501 MoPac Blvd.
Austin, TX 78758

M. Hansson
Department of Organic Chemistry
University of Umeå
S-901 87 Umeå
Sweden

Lennart Hardell
Department of Oncology
University Hospital
S-901 85 Umeå
Sweden

Daniel J. Harris
Region VII Laboratory
U.S. Environmental Protection
Agency
25 Funston Rd.
Kansas City, KS 66115

Fred D. Hileman
Dayton Laboratory
Monsanto Research Corporation
1515 Nicholas Rd.
Dayton, OH 45418

J.L. Johnson
Columbia National Fisheries
Research Laboratory
U.S. Fish and Wildlife Service
Route 1
Columbia, MO 65201

A.E. Jones
Radian Corporation
8501 MoPac Blvd.
Austin, TX 78758

L.H. Keith
Radian Corporation
8501 MoPac Blvd.
Austin, TX 78758

David E. Kirk
Dayton Laboratory
Monsanto Research Corporation
1515 Nicholas Rd.
Dayton, OH 45418

Robert D. Kleopfer
Region VII Laboratory
U.S. Environmental Protection
Agency
25 Funston Rd.
Kansas City, KS 66115

W.A. Korfmacher
National Center for Toxicological
Research
Food and Drug Administration
Department of Health and
Human Services
:fferson, AR 72079

Hiroaki Kuroki
Daiichi College of Pharmaceutical Sciences Minami-Ku
Fukuoka, 815
Japan

P.-Y. Lau
Food Division, Health Protection Branch
Health and Welfare Canada
Ottawa, Ontario K1A OL2
Canada

D. Lewis
Food Division, Health Protection Branch
Health and Welfare Canada
Ottawa, Ontario K1A OL2
Canada

K. Linnainmaa
Institute of Occupational Health
Department of Industrial Hygiene and Toxicology
Haartmaninkatu 1
00290 Helsinki 29
Finland

J.E. Longbottom
Environmental Monitoring and Support Laboratory
U.S. Environmental Protection Agency
Cincinnati, OH 45268

C.R. McMillin
Dayton Laboratory
Monsanto Research Corporation
1515 Nicholas Rd.
Dayton, OH 45418

S. Marklund
Department of Organic Chemistry
University of Umeå
S-901 87 Umeå
Sweden

Yoshito Masuda
Daiichi College of Pharmaceutical Sciences Minami-Ku
Fukuoka, 815
Japan

Thomas Mazer
Monsanto Research Corporation
Dayton Laboratory
1515 Nicholas Rd.
Dayton, OH 45418

R.K. Mitchum
National Center for Toxicological Research
Food and Drug Administration
Department of Health and Human Services
Jefferson, AR 72079

G.F. Moler
National Center for Toxicological Research
Food and Drug Administration
Department of Health and Human Services
Jefferson, AR 72079

Junya Nagayama
Department of Public Health
Faculty of Medicine
Kyushu University
Higashi-Ku
Fukuoka, 812
Japan

M. Nygren
Department of Organic Chemistry
University of Umeå
S-901 87 Umeå
Sweden

Roy W. Noble
Dayton Laboratory
Monsanto Research Corporation
1515 Nicholas Rd.
Dayton, OH 45418

L.D. Ogle
Radian Corporation
8501 MoPac Blvd.
Austin, TX 78758

M.W. Orbanosky
California Analytical
Laboratories, Inc.
Sacramento, CA 95824

B.A. Pearce
National Center for Toxicological
Research
Food and Drug Administration
Department of Health and
Human Services
Jefferson, AR 72079

J.D. Petty
Columbia National Fisheries
Research Laboratory
U.S. Fish and Wildlife Service
Route 1
Columbia, MO 65201

J.C. Pilon
Food Division, Health Protection
Branch
Health and Welfare Canada
Ottawa, Ontario K1A OL2
Canada

C. Rappe
Department of Organic Chemistry
University of Umeå
S-901 87 Umeå
Sweden

J.J. Ryan
Health Protection Branch, Food
Division
Health and Welfare Canada
Ottawa, Ontario K1A OL2
Canada

Marina Sironi
Istituto di Ricerche Farmacologiche
Mario Negri
Via Eritrea, 62
20157 Milan
Italy

L.M. Smith
Columbia National Fisheries
Research Laboratory
U.S. Fish and Wildlife Service
Route 1
Columbia, MO 65201

Joseph G. Solch
Brehm Laboratory
Wright State University
Dayton, OH 45435

D.L. Stalling
Columbia National Fisheries
Research Laboratory
U.S. Fish and Wildlife Service
Route 1
Columbia, MO 65201

Michael L. Taylor
Brehm Laboratory and Department
of Pharmacology and Toxicology
Wright State University
Dayton, OH 45435

P.A. Taylor
California Analytical
Laboratories, Inc.
Sacramento, CA 95824

Thomas O. Tiernan
Brehm Laboratory and Department
of Chemistry
Wright State University
Dayton, OH 45435

L.J. Thibodeaux
Department of Chemical Engineering
University of Arkansas
Fayetteville, AR 72701

John A. Todhunter
U.S. Environmental Protection Agency
Washington, DC 20460

Helle Tosine
Ontario Ministry of the Environment
P.O. Box 213
Resources Rd.
Rexdale, Ontario M9W 5L1
Canada

Garrett F. VanNess
Brehm Laboratory
Wright State University
Dayton, OH 45435

Annunciata Vecchi
Istituto di Ricerche Farmacologiche Mario Negri
Via Eritrea, 62
20157 Milan
Italy

Beverly Warner
Dayton Laboratory
Monsanto Research Corporation
1515 Nicholas Rd.
Dayton, OH 45418

R.J. Wesselman
Environmental Monitoring and Support Laboratory
U.S. Environmental Protection Agency
Cincinnati, OH 45268

A.S. Wong
California Analytical Laboratories, Inc.
Sacramento, CA 95824

D. Wood
Dayton Research Corporation
Monsanto Research Corporation
1515 Nicholas Rd.
Dayton, OH 45418

Kenneth T. Yue
Region VII Laboratory
U.S. Environmental Protection Agency
25 Funston Rd.
Kansas City, KS 66115

PREFACE

Since the days of the "Agent Orange" herbicide spraying in Vietnam in the middle and late 1960s, the "Yusho" disease in Japan in 1968, the "Horse arena" accident in the United States in 1971, and the "Seveso" explosion in Italy in 1976, there has been increasing concern regarding environmental exposure to chlorinated dioxins and dibenzofurans. In the United States recent concerns were manifested in the Love Canal (New York), Times Beach (Missouri), and similar incidents during the past few years. Exposure to such chemicals is particularly crucial because some members (particularly 2,3,7,8-tetrachlorodibenzo-*p*-dioxin and 2,3,7,8-tetrachlorodibenzofuran) of these classes of compounds are supertoxins to humans. In recent years, these concerns have spurred worldwide attention in industry, academia, and government resulting in several topical meetings on the subject, occasional polemical discussions, and numerous publications.

The human environment includes the environment within—that is, occupational—and the environment without—that is, ecological. Because of this fact, the total environment must be considered in relation to chlorinated dioxin and dibenzofuran exposure for any health assessment purpose. All facts must be fully explored before any meaningful legislative and regulatory action can be developed. The growing concern for the health hazards of exposure to chlorinated dioxins and dibenzofurans in the total environment led to the planning of two symposia in conjunction with the national meetings of the American Chemical Society. The purpose of the symposia was to provide state-of-the-art technology with a view to present new information, problems identified in sampling and analysis of the analyte, toxicology, collection and interpretation of epidemiological data, and the need for further research in the field. The 1982 symposia brought together experts, both national and international, to a common forum.

This volume is developed from the proceedings of the first part of these symposia, held in Kansas City, Missouri, in September 1982. An Excellent cross-section of the arena of chlorinated dioxins and dibenzofurans in the total environment were presented at this symposium in the form of reviews as well as original research papers. The subdivisions of the book encompass general perspective, synthesis, environmental, analytical, and occupational topics, which are based on the logical categorization used to develop various sessions of the symposium.

It is hoped that this volume will become a valuable reference in literature as well as open the avenues to further research in the field, particularly in relation to human health risk conditions. The second symposium of this series, to be held in Washington, DC., in 1983, and its intended proceedings, is expected to be of greater value in this regard.

The editors wish to acknowledge, with sincere appreciation, the publication of the contributions of the authors and reviewers without whose time and effort this book would not have been possible. The encouragement and support given by the Division of Environmental Chemistry of the American Chemical Society is gratefully acknowledged.

Gangadhar Choudhary
Lawrence H. Keith
Christoffer Rappe

I

Perspectives

1

Dioxins: A Canadian Perspective

Helle Tosine

Most scientists are aware of the 1976 plant explosion in Seveso, Italy, which released, among other chemicals, 2,3,7,8-tetrachlorodibenzo-*p*-dioxins (tetra-CDD) into the area, contaminating a fairly extensive land mass adjoining the plant. In Canada, the incident initially received scant media coverage, and those of the public who read of the incident sympathized with the people as they would have with anyone affected by any industrial accident.

Canadian and Ontario officials, however, had already been well aware of "dioxin," specifically in 2,4,5-T, as a problem in Canada, several years before the Seveso incident. In 1970 the Canadian federal government limited tetra-CDD levels to 0.5 ppm in 2,4,5-T. In the same year, the Ontario government had limited 2,4,5-T to "schedule 2" use, which removed it from sale to farmers and home gardeners and required that it be applied only by licensed exterminators. Thereafter, the chief use of 2,4,5-T was on rights-of-way rather than on agricultural land. At that time, the label on 2,4,5-T read "Not for use around homes or recreational areas. Do not expose pregnant women and children to this chemical."

As is frequently the case, only the discovery of dioxins on the very doorstep of Toronto finally made "dioxin" a household word. The Canadian public became aware of the seriousness of dioxins, indeed the seriousness of industrial chemicals, when on April 25, 1979, two million Torontonians read the news: "Dioxin, one of the most toxic of chemicals, has been found in two fish taken from Lake Ontario. N.Y. State Dept. of Health confirmed that tests on fish from the lake contained 6.5 ppt and 4.6 ppt TCDD." Immediately, the public began to ask questions. "What is dioxin?" "Are all the fish contaminated?" "What about our drinking water?" "What is the government doing about it?"

Both federal and provincial government agencies (in this case the province of Ontario) had already taken steps to limit public exposure to tetra-CDD from herbicidal use of 2,4,5-T. In this particular situation of a contaminated environment in boundary waters, both the Canadian federal and provincial governments along with their U.S. counterparts share responsibility for the Great Lakes and boundary waters such as the Niagara River, through the Canada–Ontario and

International Joint Commission (IJC) agreements. The two Canadian levels of government cooperate in the investigation of problems that involve international issues as well as health and environment issues specific to a province or region. Since the initial dioxin findings were made in fish from boundary waters, the responsibility for investigating the problem became a joint one.

Thus, in fall 1979 a Canadian ad hoc dioxin committee was formed to include federal and Ontario provincial analytical chemists and biologists to review analytical problems and future monitoring programs for various environmental matrices in Lake Ontario and the Niagara River for tetra-CDD. Discussions were initiated between the Canadian committee and representatives from the New York state Department of Health about the most efficient and effective manner in which to approach dioxin monitoring and analysis.

In December 1980 and July 1981, meetings were held in Washington, D.C., and Ottawa, Ontario, respectively, between U.S. and Canadian committees regarding the significance and implications of any Canadian or U.S. tetra-CDD data. The groups also met to set guidelines for 2,3,7,8-tetra-CDD levels in fish.

In Canada, action arising from this initial ad hoc Canadian dioxin committee has occurred on many fronts. Provincial-federal cooperation has resulted in committees both of an advisory nature (giving advice to senior management regarding toxicity and limits of exposure) and of a task-oriented nature (inter-analyst discussions regarding method development and research plans). One of these committees was formed in 1980 by the National Research Council of Canada (NRCC), comprising thirteen individuals from government, university and industrial organizations, and was directed as follows:

1. to identify the key questions needed to address the issue of dioxin in the environment;
2. to develop and assess the adequacy of scientific criteria for polychlorinated dibenzodioxins (PCDD) needed to provide the answers; and
3. to determine any gaps in the present scientific information and recommend actions necessary to improve the data base.

By December 1981 the NRCC panel completed its task by issuing a two-part report [1,2]. Another group, called the Dioxin Working Group, was established in 1982 under the Federal Interdepartmental Committee on Toxic Chemicals to review dioxin program status within various agencies in Canada and to develop a plan of action [3].

As it is the intent of this chapter to discuss the Canadian view of the impact of dioxins on the environment, the findings of the two mentioned committee reports will be used to cover:

1. Canadian sources of PCDD,
2. input of PCDD into the Canadian environment, and
3. Canadian exposure to PCDD.

SOURCES

Two major sources of PCDD into the Canadian environment have been identified as combustion and use of chlorophenols (CP) and their derivatives. Another major, but regional, contributing source is chemical waste disposal sites associated with the production of CP and their derivatives.

Combustion

The role of combustion as a source of PCDD has been investigated internationally since 1977. Since that time, two schools of thought have formed regarding PCDD formation during combustion:

1. The "trace chemistries of fire" hypothesis [4] maintains that the formation of PCDD is a general phenomenon in all types of combustion (e.g., wood stoves and cigarettes); i.e., PCDD and polychlorinated dibenzofurans (PCDF) are ubiquitous.
2. PCDD are formed as a result of burning municipal and industrial garbage, which contains aromatics and organics containing chlorine, and which acts as a direct PCDD/PCDF precursor.

In Canada, the only documented combustion source has been municipal incineration [5].

Chlorophenols

CP are a major group of chemicals used for a variety of biocidal and industrial purposes and as precursors for other pesticides, for example, 2,4-D, 2,4,5-T and hexachlorophene. They are frequently marketed as complex mixtures of CP, not as single chemicals. For example technical-grade penta-CP is predominantly penta-CP but can contain up to 12% tetra-CP. Currently, there is only a single Canadian manufacturer producing 2,4-di-, 2,3,4,6-tetra- and penta-CP.

The major concern with CP has been the contamination of CP mixtures with various isomers of PCDD. The CP that have been shown to contain various PCDD homologs are:

1. tetra- and penta-CP;
2. 2,4-di-CP as a precursor of 2,4-D; and
3. 2,4,5-tri-CP as a precursor of 2,4,5-T.

Briefly, the PCDD detected in these CP homologs are:

- tetra-CDD in tri-CP;
- hexa-, hepta- and octa-CDD in tetra- and penta-CP;

- di-, tri- and tetra-DCC in 2,4,D; and
- tetra-CDD in 2,4,5-T.

In Canada, the main use of technical tetra- and penta-CP has been in the wood industry as wood preservatives. The amount of penta-CP used for pressure treatment of wood could equal around 80% of the total known Canadian consumption of penta-CP [5]. Tetra-CP is used primarily as the sodium salt formulation for surface treatment of lumber. Penta- and tetra-CP use is regional. It is limited primarily to Pacific and Atlantic timbering regions, and to a lesser extent, to Quebec.

An example of the extent of contamination that exists due to the wide use of chlorophenols is an incident of PCDD-contaminated chickens. Penta-CP–treated wood shavings had been used as bedding for the chickens, which resulted in accumulations of PCDD in the chickens. Restrictions under the federal Pest Control Products Act to the use of penta-CP in the textile, leather and agriculturally related industries may have reduced the potential for contamination from these sources.

Around 1970 the use of tetra- and penta-CP in Ontario pulp and paper mills was restricted from paper products used in connection with food and was phased out entirely around 1976–1977.

As was previously mentioned, PCDD have been detected in 2,4-D, 2,4,5-T and hexachlorophene. Besides details on the extent of PCDD contamination of these chemicals, lists [7] have been compiled of the hypothetical or potentially contaminated chemicals yet to be investigated, for example, the pesticide dicamba. Most chemical products such as 2,4,5-T and 2,4-D can be purified of their PCDD contaminants. This has been the case with 2,4,5-T and penta-CP, although purified penta-CP has not been used extensively in Canada.

2,4-D is used extensively in the agricultural industry in Canada. The latest Canadian federal regulations concerning 2,4-D require that less than 10 ppb of any given PCDD isomer be present. This has resulted in the removal from the market of high-volatility 2,4-D ester formulations and the phasing out of the 2,4-D low-volatility esters, leaving only the 2,4-D amine (low-dioxin-containing) formulations on the market.

In 1979 use of 2,4,5-T in Ontario was discounted by government policy. No permits were issued for use; although the federal government did not deregister 2,4,5-T, it was essentially "banned" in Ontario. Nova Scotia and Alberta are debating whether to impose similar restrictions. The concern with the use of CP and derivatives is not only with the occupational and environmental exposure, but also with the disposal of the wastes associated with their production.

Chemical Dumps

As a result of the New York state Department of Health discovery of 2,3,7,8,-tetra-CDD contamination in fish from Lake Ontario, one of the major suspected sources became landfill leachates containing wastes from 2,4,5-tri-CP production. The only major producers of 2,4,5-tri-CP in the Great Lakes Basin have been Hooker Chemical in Niagara Falls, New York (along the Niagara River), and Dow Chemicals at Midland, Michigan (on the Titabawassi–Saginaw River system, which drains into Saginaw Bay, Lake Huron).

Hooker closed down its tri-CP reactors in the Niagara area in 1974 and has not since produced tri-CP in the Niagara frontier. Dow stopped production in 1978 but has since installed new equipment to produce "dioxin-free" 2,4,5-T. The New York state Department of Environmental Conservation supplied the Ontario Ministry of the Environment with a list of potential landfill sources of dioxins leaching into the Niagara River. The four major landfills used by Hooker affecting Canadians are:

- Hyde Park,
- Love Canal,
- S. Area, and
- 102nd Street.

INPUT INTO THE ENVIRONMENT

An important but perhaps unresolvable question concerning PCDD is "How much PCDD are contributed to the Canadian environment from the sources listed above?" The NRCC committee report on dioxins [1,2] provided the following estimates of inputs and sources.

Combustion

As was previously mentioned, the only documented combustion source of PCDD in Canada to date is municipal incinerators. This particular study reported concentrations of tetra- to octa-CDD found on fly ash. The distribution of the PCDD homologs on the fly ash over time was in good agreement with the study of Dutch incinerators by Lustenhower [8]. In the Dutch study, Lustenhower also found that particulate matter forming emissions had a higher PCDD concentration than the precipitated material. He suggested a tenfold multiplier be used for comparative purposes.

Thus, to estimate the amount of PCDD entering the Canadian environment from combustion sources, the following assumptions were made by Environment Canada:

1. The average PCDD content for Ontario municipal incinerators (the only data currently available) was applied to total national incinerator fly ash figures.

2. The amount of precipitated fly ash was based on the estimated efficiency of burning for each type of incinerator.
3. The Lustenhower tenfold multiplier was applied to the figures for particulate emissions (although no evidence exists in Canadian facilities to support the use of this multiplier).

Table 1.1 lists the estimated environmental input from incinerators for the various regions of Canada. The total estimated PCDD input is 13.5 kg/yr. In comparison to the figure of 1000 kg estimated for the U.S. [9], the estimated PCDD discharge for Canada appears to be low. It must be pointed out that this figure is only an estimate, and may be low for the following reasons:

1. To date, lower PCDD levels have been reported in emissions from Canadian incinerators compared to U.S. incinerators.
2. No attempts were made to include the burning of wood that had been treated with CP.
3. The amount of particulate emissions from wigwam burners is comparable to that from refuse incineration. Although no data are available regarding the PCDD content of these emissions, wigwam burners could conceivably contribute as much PCDD to the environment as municipal incinerators.

At present, two Canadian agencies are involved with monitoring incineration emissions: the Atmospheric Environment Services of Environment Canada, and the Ontario Ministries of Energy and Environment. Complete analyses of the total input and emissions of chlorinated organics [chlorobenzenes, chlorophenols, polychlorinated biphenyls (PCB), PCDD and PCDF] from a municipal waste incinerator, a municipal sludge incinerator and an industrial waste incinerator are being conducted. Analyses of the first two incinerators have been completed. When this study has been completed, it is hoped that a more detailed outline of PCDD/PCDF discharges from these types of waste disposal units than exists at present will be available.

Chlorophenols

Recent Canadian federal and provincial reports [10-13] have estimated that 6 million lb (13.2×10^6 kg) of CP and 8 million lb (17.6×10^6 kg) of 2,4-D are used in Canada per year. Both the penta-CP and the 2,4-D used in Canada have been analyzed by Agriculture Canada for PCDD content. The average concentrations of hexa-, hepta- and octa-CDD were 10, 140 and 370 mg/kg, respectively, in penta-CP. The average concentrations of di-, tri- and tetra-CDD were 2400, 670 and 840 μg/kg, respectively, in the ester formulation for 2,4-D, and 45, 120 and 44 μg/kg, respectively, in the amine formulation for 2,4-D. Thus, by applying

Table 1.1. Estimated Input of PCDD to the Canadian Environment from Combustion Sources [1]

	Metric Tons	*Tetra-CDD*	*Penta-CDD*	*Hexa-CDD*	*Hepta-CDD*	*Octa-CDD*	*Total*
Precipitated Fly Ash from a Municipal Incinerator							
Estimated PCDD Content ($\mu g/kg$)		14	23	26	15	6	
Estimated PCDD Input (g)							
Maritimes	1,000	14	23	26	15	6	100
Quebec	60,000	840	1,400	1,600	900	360	5,100
Ontario	19,000	270	440	490	285	114	1,600
Prairies	980	14	23	26	15	6	100
British Columbia	33	<1	<1	<1	<1	<1	100
Total	81,000	1,100	1,900	2,100	1,200	490	6,800
Emitted Particles from a Municipal Incinerator							
Estimated PCDD Content ($\mu g/kg$)		140	230	260	150	60	
Estimated PCDD Input (g)							
Maritimes	700	89	160	180	110	42	600
Quebec	2,600	360	600	680	390	160	2,200
Ontario	2,600	360	600	680	390	160	2,200
Prairies	1,800	250	410	470	270	110	1,500
British Columbia	1	<1	<1	<1	<1	<1	100
Total	7,700	1,100	1,800	2,000	1,200	470	6,600

these average concentrations of PCDD to the total loading of CP into the Canadian environment, estimates of the PCDD input from CP sources can be made (Table 1.2). The total gross estimate of PCDD input to the Canadian environment is 1.5 metric ton/yr. These limited data suggest that, compared to the total input from combustion, CP and related materials could be the major source of total PCDD input to the Canadian environment. This statement, however, requires some qualification:

1. Since penta-CP is used primarily as a wood preservative, the major portion of applied penta-CP will end up in preserved wood, although exposure to wood preservers and point-source contamination from preservation plants remain significant.
2. Of the preserved wood, 75% is exported, and therefore these residues do not remain in the Canadian environment.
3. Although British Columbia uses tetra-CP containing 40% penta-CP, total tetra-CP was included in the estimate of total penta-CP used.

Penta-CP as a major source needs confirmation before regulatory action can be taken.

Chemical Dump Sites

The Ontario Ministry of the Environment (MOE), as part of its fish contaminant monitoring program, has sampled and analyzed spottail shiners young-of-the-year (i.e., fish born during the sampling year). These fish have not yet migrated from their spawning location and thus become excellent point-source indicators of pollution. During the 1981–1982 field year, MOE found 2,3,7,8-tetra-CDD concentrations as high as 57 ppt in young-of-the-year taken from the mouth of Cayuga Creek, which drains from the Love Canal landfill site. This is the only Canadian evidence to date implicating 2,3,7,8-tetra-CDD leaching from a chemical dump site.

A total of 130 adult fish from various locations in Ontario were also analyzed for 2,3,7,8-tetra-CDD during the same field year. A large percentage of the lake trout sampled from western Lake Ontario were found to contain 2,3,7,8-tetra-CDD concentrations above the 20-ppt guideline. Ontario's concern over this continuing source of 2,3,7,8-tetra-CDD to the Niagara River and Lake Ontario from these landfills, resulting in contamination of western Lake Ontario lake trout, was made known to New York state and U.S. federal officials. In July 1982 the U.S. Environmental Protection Agency (EPA) announced plans to spend an initial $7.5 million in further cleanup, leachate control and contaminated sediment removal at the Love Canal. In Ontario, extensive investigations are underway at a former 2,4,5-T production facility. Control actions on this plant's waste disposal system date back to the late 1960s.

Table 1.2. Estimated Input of PCDD to the Canadian Environment from Chemical Sources [1]

		PCDD Homolog						
	Metric Tons Used	*Di-CDD*	*Tri-CDD*	*Tetra-CDD*	*Hexa-CDD*	*Hepta-CDD*	*Octa-CDD*	*Total*
2,4-D								
Estimated PCDD Content ($\mu g/kg$)								
Ester		2,400	670	840				
Amine		45	120	44				
Estimated PCDD Input (g)								
Prairies (Ester)	2,300	5,500	1,500	1,900				8,900
British Columbia (Amine)	1,500	68	180	66				300
Total	3,800	5,600	1,700	2,000				9,200
P_5CP								
Estimated PCDD Content (mg/kg)					10	140	370	
Estimated PCDD Input (g)								
Maritimes	240				2,400	34,000	89,000	130,000
Quebec	210				2,100	29,000	78,000	110,000
Ontario	670				6,700	94,000	250,000	350,000
Prairies	690				6,900	97,000	260,000	360,000
British Columbia	1,100				11,000	150,000	410,000	570,000
Total	2,900				29,000	400,000	1,100,000	1,500,000

EXPOSURE

Exposure is possible through dermal contact, respiration or dietary intake. It is not possible to cover in detail, in the short space available here, the exposure vectors related to metabolism and bioaccumulation patterns. One can refer to the detailed discussions and extrapolations of "worst-case scenarios" as documented in the NRCC report [1,2]. However, it can be said that no data exist at present in Canada to make a case for exposure through dermal contact or respiration. Certainly, the exposure encountered by, for example, workers in the wood industry needs to be closely documented to determine if a risk exists due to skin contact or inhalation of CP vapors. In western Canada, the British Columbia government has been doing a considerable amount of work in this area and has established guidelines for handling CP.

Based on the behavior of other chlorinated compounds, such as PCB, fish are considered to be a likely high accumulator. Thus, concentrations of PCDD in fish form the largest data base for the dietary vector from which a risk level or extent of exposure can be established. To define a "virtually safe dose" (VSD), (i.e., a dose for which there would be an increase of 1 cancer in 1 million individuals, extrapolations were made as follows. The dose of 2,3,7,8-tetra-CDD observed to have no detectable carcinogenic effect in rats was 1 ng 2,3,7,8-tetra-CDD/kg body weight per day. While it is argued that extrapolations from rats to humans provide a conservative basis for criteria, their application to the human situation remains unproven. Nevertheless, these extrapolative techniques were used to define the level of risk, the VSD, as being between 2 and 6 pg/day 2,3,7,8-tetra-CDD over the life of a 70-kg individual.

Given a meal of 120 g (about 0.25 lb) of fish, intakes of 60 ng would be encountered when fish containing 500 ng 2,3,7,8-TCDD/kg are consumed daily. This assumes that no other contaminated food items are consumed. Using these criteria, Health and Welfare Canada set a limit of 20 ppt as the VSD for fish. Fish containing more than the VSD are not recommended for consumption.

This VSD applies if the only dietary dioxin contribution comes from fish. A few samples of Canadian foods other than fish have been examined for concentrations of PCDD. Specifically, PCDD have been reported in the livers of chickens raised on penta-CP–contaminated litter, and in pork liver and milk from animals raised in penta-CP–treated pens. However, it is not known whether these residues are accurate representations of problems in the food industry or simply isolated cases. Although they provide no solid basis for developing risk assessments of the dietary vector, these residues indicate that there is a need to monitor foods other than fish if the Canadian situation is to be delineated more thoroughly. Other foodstuffs also deserve attention: vegetables grown or stored in CP-treated materials, and meat from animals raised in treated pens and on contaminated litter. Although the use of contaminated materials (e.g., CP-treated wood) as an agricultural practice has been restricted, the effectiveness of this control should be determined through monitoring.

The need for more extensive monitoring programs to increase the data base for environmental matrices is a problem not unique to Canada. The problem of limited residue data exists primarily due to the complexity of the analytical procedure required to generate the data:

1. For valid dioxin analysis, ultrarefined sensitive separation techniques are needed both to separate interferences from PCDD and to distinguish individual isomers for toxicity evaluation.
2. An ultralow (ppt) level of detection and confirmation is needed.
3. The time and expense involved for analysis is considerable. This limits the extent of interlab quality assurance programs for both analysis and sampling, and this may compromise the level of data compatibility.
4. Appropriate PCDD standards with which to confirm isomer identity are unavailable.

Considering the complexity and cost of each analysis, good communication among scientists regarding the extent of individual future monitoring programs is a key factor in optimizing the consolidation of an appropriately sized data base.

A recent publication of the Canadian federal interagency dioxin committee listed all Canadian facilities involved in dioxin investigations, together with their research and monitoring plans as a means of keeping abreast of future plans in the dioxin issue and further emphasizing the need for good communication to avoid wasteful duplication of effort. The Ontario MOE Laboratory Services Branch publishes *Dioxin News*. This newsletter provides scientists with a forum for discussion on developments in analysis toxicology and regulation of dioxin.

The present and future regional, national and international problems involved with acquiring valid scientific data by which to assess chemical toxicology and to provide valid risk assessment permit us to consider "dioxins" as a prototype in the field of hazardous contaminants, which:

- are highly toxic to man at ultratrace levels,
- are highly stable environmentally, and
- present highly complex problems for analytical and biological research due to the many homologs demanding separation and cleanup requirements, presence of ultratrace levels, ubiquity in many diverse environmental compartments.

In Canada, from the dioxin issue we have seen the need for cooperation (at the scientific level) in the fields of analytical chemistry, toxicology and risk assessment. Cooperation has occurred within the various levels of government and among various levels of government. Information exchange has also occurred among government, industry and academia through nongovernment agencies such as NRCC.

Through these cooperative efforts we have developed a framework for addressing future issues in the field of ultrahazardous contaminants. The challenges and problems, however, will continue to center on sparse and uncertain data bases, inadequate human health and toxicology data, and difficulties in isolating sources and introducing the necessary corrective measures.

REFERENCES

1. "Polychlorinated Dibenzo-*p*-dioxins: Criteria for Their Effects on Man and His Environment," NRCC 18574, National Research Council, Canada (1981).
2. "Limitations to Current Analytical Techniques," NRCC 18576, National Research Council, Canada (1981).
3. "Management Strategy for PCDDs and PCDFs," interim report, Federal Interdepartmental Committee on Toxic Chemicals, Environment Canada (1982).
4. Bumb, R.R., W.B. Crummett, S.S. Cutie, J.R. Gledhill, R.H. Hummel, R.O. Kage, L.L. Lamparski, E.V. Luoma, D.L. Miller, T.J. Nestrick, L.A. Shadoff, R.H. Stehl and J.S. Woods. "Trace chemistries of fire: A source of chlorinated dioxins," *Science* 210:385 (1980).
5. Eicemen, G.A., R.A. Clement and F.W. Karasek. *Anal. Chem.* 51:2343 (1979).
6. Jones, P. "Chlorinated Phenols in the Environment," EPS-3-EC-81-2, EPS, Environment Canada (1981).
7. Esposito, M.P., T.O. Tiernan and F.E. Dryden. "Dioxins," EPA-600/2-80-197, U.S. EPA (1980).
8. Lustenhower, J.W.A., K. Olie and O. Hutzinger. *Chemosphere* 9:501 (1980).
9. Crummett, W.B., R.R. Bumb, L.L. Lamparski, N.H. Mahle, T.J. Nestrick and L.W. Whiting. "Environmental Chlorinated Dioxins from Combustion—The Trace Chemistries of Fire Hypothesis," in *Impact of Chlorinated Dioxins and Related Compounds in the Environment,* O. Hutzinger, R.W. Frei, E. Merian and F. Pocchian, Eds. (Oxford: Pergamon Press, Ltd., 1981).
10. Bacon, G. B. "Bioaccumulation of Toxic Compounds in Pulpmill Effluents by Aquatic Organisms in Receiving Waters," M/79-76, annual report, CPAR project 657, prepared for CPAR Secretariat, Environment Canada, Ottawa (1978).
11. Cain, R.T., M.J.R. Clark and N.R. Zorkin "Fraser River Estuary Study—Water Quality (Trace Organic Constituents in Discharges)," report to the government of Canada and province of British Columbia.
12. "Monitoring Environmental Contamination from Chlorophenol Contaminated Wastes Generated in the Wood Preservation Industry," Progress Report 79-24, EPS, Environment Canada (1979).
13. Chittim, B.G. "Dioxin Levels in Products and Environmental Samples in the Vicinity of Two Wood Preserving Operations Using Pentachlorophenol," EPS/Atlantic Region, Environment Canada (1980).

2

Dioxins and the Regulatory Process at the U.S. Environmental Protection Agency

Donald G. Barnes and John A. Todhunter

Two decades ago the group of chemicals referred to popularly—if not accurately—as "dioxins" were thought to reside primarily in the relative obscurity of chemical laboratories, making only an occasional appearance in unfortunate industrial accidents. Today, the word "dioxins" competes favorably for space in newspaper headlines, the evening newscasts and Congressional speeches. Because of this recent coverage, the public has come to associate the term with a variety of images: soldiers in Vietnam, aerial application of herbicides, leachate from abandoned dump sites, smoke emitted from incinerators and home fireplaces, and the phrase, "the most toxic man-made chemical."

The U.S. Environmental Protection Agency (EPA) has gained a special perspective of this problem through a long, and sometimes rocky, relationship with "dioxins." Over the years, EPA has remained in the vanguard of those dealing with dioxin-related issues. Given the multifaceted nature of these problems, virtually all parts of EPA have addressed these issues at one time or another. For example, the EPA Office of Research and Development's Dioxin Monitoring Program played a pioneering role in the development of the gas chromatography/mass spectrometry (GC/MS) analytical techniques used to measure dioxins in the parts-per-trillion range. Further, EPA has used its authority under various statutes and regulations to address environmental contamination by dioxins. For example, there are the ongoing 2,4,5-T/silvex cancellation proceedings under the Federal Insecticide, Fungicide, and Rodenticide Act (FIFRA), a draft Water Quality Criteria document under the Clean Water Act (CWA) and consideration of waste disposal requirements under the Resource Conservation and Recovery Act (RCRA). To coordinate these and other related activities, EPA established a work group, the Chlorinated Dioxins Work Group (CDWG).

In an atmosphere of rising public awareness of and sensitivity to the problems posed by toxic materials in the environment, EPA is treating dioxin-related problems with sophisticated techniques of analytical chemistry, an appreciation of the toxicological effects involved, and an array of Congressionally mandated legislation designed to protect human health and the environment from unreasonable risks of injury. In applying our best scientific approach and judgment to these cases, we are aware of heightened public concern about the "dioxin" issue. Therefore, our work has included efforts to clarify points of fact and to explain the risk/benefit balancing tools provided by Congress.

This chapter will address two of the points EPA is attempting to clarify in the public's mind. In addition, some specific examples will illustrate how EPA has used risk/benefit analysis to reach decisions on dioxin-related problems.

TWO BASIC POINTS

The first problem area is, in many respects, a child of our own creation. Here the "our" refers to society, including the scientific community, and the "child" refers to the public image conjured up by the mention of the word "dioxin." Almost invariably, whenever "dioxin" is used in the popular press, it is coupled with the toxicological properties of 2,3,7,8-tetrachlorodibenzo-*p*-dioxin (2,3,7,8-tetra-CDD), e.g., delayed acute effects, the "wasting away" syndrome, notable species variation, and reproductive and carcinogenic effects at lower doses. Sometimes this association is correct; sometimes it is not. Nevertheless, the public perception of all dioxins appears to be defined in terms of what is known about 2,3,7,8-tetra-CDD. Unfortunately, at least a part of this public misconception may be due to our own imprecision of language.

While there may have once been a time when the only "dioxin" of concern was 2,3,7,8-tetra-CDD and it was understandable and perhaps even excusable to speak simply of tetra-CDD or "dioxin," that time is behind us. As we are aware, a number of other tetra-CDD isomers and other polychlorinated dibenzo-*p*-dioxin (PCDD) congeners have now been encountered in the environment. In fact, environmental samples have been found that contain virtually all of the 22 tetra-CDD isomers and most of the other 75 PCDD congeners. More recently, we have been confronted with the presence of polychlorinated dibenzofurans (PCDF), the 135 chemical cousins of the more familiar PCDD.

It is important to recognize that, although detailed toxicological data do not exist on most of these other PCDD and PCDF, there is sufficient information to suggest that they are not likely to be as toxic as 2,3,7,8-tetra-CDD. Yet, we often find ourselves speaking and reading about "dioxins" (and "furans") in such a way that the less well tutored among us gain the impression that any of these 210 "dioxin" species is as toxic as 2,3,7,8-tetra-CDD.

The foregoing pedantic discussion of basic nomenclature is an appeal for clarity in our discussion of this important topic. The price of imprecision in our language is confusion in the public's mind, which can lead to unfortunate misconceptions regarding adverse effects on public health and the environment.

The second basic point involves the EPA risk/benefit analysis approach to decision-making. EPA is charged with protecting public health and the environment from unreasonable risk. Operationally, the risk associated with any chemical becomes unreasonable when that risk outweighs the benefits associated with the use of that chemical, taking into consideration the availability of acceptable alternatives. In practical terms, this means that risk is just one factor, albeit an important one, in the EPA decision-making process; other factors must also be considered. Public perception often appears to be that any plausible risk should trigger EPA action, regardless of other considerations. This seems to be particularly true in issues related to chemicals that have achieved the level of notoriety of PCDD, where all PCDD are publicly equated with 2,3,7,8-tetra-CDD.

In the EPA decision-making process, however, the administrator weighs risks and benefits. If the determination is made that risks outweigh the benefits, EPA takes action to reduce the risks to an acceptable level, a process for which a number of legislatively established tools exist. The risk, in turn, is a function of both its hazard (inherent toxicity) and exposure; i.e., as either toxicity or exposure goes toward zero, the risk goes toward zero. Exposure has a tendency to get lost in the discussion of risk. As you will see in a moment, however, as analytical procedures are devised to detect PCDD/PCDF at lower and lower levels, the importance of exposure assessments increases.

Having made these general points, the last part of this chapter will discuss some specific examples of EPA decisions that illustrate the use of risk/benefit analysis in cases involving PCDD and PCDF.

ILLUSTRATIONS OF UNREASONABLE RISK CONSIDERATIONS

Manufacturing Wastes

Wastes contaminated by 2,3,7,8-tetra-CDD are generated as a consequence of the manufacture of 2,4,5-trichlorophenol (2,4,5-tri-CP). In the past, some of this contaminated waste material has been handled in such a way that the 2,3,7,8-tetra-CDD was released into the environment. In 1979 EPA moved to mitigate these problems by promulgating regulations under the Toxic Substances Control Act (TSCA) to control the movement of wastes resulting from the production of 2,4,5-tri-CP and its pesticide derivatives [1]. All movement of such wastes is strictly prohibited without prior notification to EPA. Following notification, EPA has 60 days to determine if the proposed action represents an unreasonable risk and to take appropriate action. The EPA regulation is based on its determination that indiscriminate handling and disposal of wastes contaminated by 2,3,7,8-tetra-CDD are an unreasonable risk. The thrust of this EPA action is to ensure that when the wastes are eventually moved and/or treated, the procedures and processes used will be appropriate.

EPA is also using its authority under TSCA and the Superfund legislation to address other PCDD-contaminated waste sites. In these situations, including those in the Niagara Falls area of New York and various sites in Missouri, EPA is assessing the environmental impacts on a site-by-site basis, using risk-assessment procedures in which estimated exposure levels play an essential role. The risk-assessment approach being used in these instances is similar to the one developed for other situations and discussed below.

Products of Combustion

In Section 6 of TSCA, Congress directs EPA to eliminate any significant entry of polychlorinated biphenyls (PCB) into the environment and to approve means of disposal for PCB currently in use. EPA has determined that incineration is an appropriate means of disposal. Therefore, through regional offices, EPA has been involved in permitting facilities for incineration of PCB.

Aware of data indicating that the products of PCB combustion may include PCDD and PCDF, EPA has moved to assess the risk associated with such emissions during PCB incineration. Accordingly, in 1979 EPA carefully examined the emission products from two commercial incinerators used to burn PCB, to determine the levels of tetra-CDD and/or tetra-CDF emitted in the process. In each case, detectable levels of these compounds were discovered.

The next step was to assess the potential risks associated with the tetra-CDD and tetra-CDF emissions. In the absence of definitive data on toxicity and exposure, it was necessary to make certain conservative assumptions to generate risk estimates:

1. The carcinogenic properties and reproductive effects of all tetra-CDD and tetra-CDF are the same as those of 2,3,7,8-tetra-CDD.
2. The air dispersion model and the factor used to convert to the annual concentration adequately represent the transport of the emissions to ground level.
3. The composition of emission products found at ground level is identical to to the composition (but not the concentration) in the stack.
4. All of the exposed population is subjected to the concentration found at the point of the maximum average annual concentration.
5. All of the particulate emissions from the stack are respirable and are retained in the body.
6. All of the tetra-CDD and tetra-CDF on the particles are biologically available to the people who inhale them.
7. Combustion will proceed for 70 years, 24 hours per day.
8. Human sensitivity to tetra-CDD/tetra-CDF is comparable to that of animals, taking body surface areas into account.

The assumption that all the tetra-CDD and tetra-CDF are as toxic as 2,3,7,8-tetra-CDD is conservative in light of our knowledge of the relative LD_{50}

values of the various tetra-CDD and tetra-CDF and the results from enzyme induction studies using these chemicals. In assessing exposure intensity, an air dispersion model was used, employing worst-case atmospheric conditions to estimate ground-level concentrations, and it was assumed that all of the exposure would take place at the point of maximum average annual concentration. Because nearly all of the tetra-CDD and tetra-CDF were found on particulate matter in the emission stream, we assumed that all of these particulates were in the respirable range and were retained in the body. We further assumed that the tetra-CDD and tetra-CDF in the particles were biologically available. Finally, for exposure duration, we assumed a 70-year lifetime exposure.

To estimate the potential risk to persons in the area surrounding the PCB incineration facilities, the resulting estimated exposures were compared to the dose levels used in animal experiments designed to investigate the reproductive and carcinogenic effects of exposure to 2,3,7,8-tetra-CDD. For the cancer endpoint, the comparison was made using the linearized multistage risk extrapolation model, which generated an estimated upper limit to the associated cancer risk for humans. For the reproductive endpoint, the estimated human dose was compared to the lowest dose used in the animal tests. In each case, EPA concluded that the risks of PCB incineration were not unreasonable in light of the conservative assumptions made in the risk estimates and the potential risks of alternative means of disposing of the PCB [2].

A similar approach was used in an interim estimate of the significance of the emission of tetra-CDD from municipal waste combustion [3]. In this case, five different plants were sampled and analyzed. A range of values for tetra-CDD was reported, from "not detected" to some upper value. Air dispersion models were run for each location and estimates of the maximum average annual ground level concentrations were generated. Again, using the assumptions above in lieu of definitive data (considering tetra-CDD, not tetra-CDF), upper bounds of cancer risk estimates were derived, this time using a variety of low-dose extrapolation models. The models used and the upper limits of cancer risk were:

- linearized multistage: up to 8×10^{-6};
- probit: up to 15×10^{-6};
- logit: up to 15×10^{-6};
- Weibull: up to 14×10^{-6}; and
- gamma multihit: up to 19×10^{-6}.

Risk is a probability, and, hence, is unitless. These values can be interpreted as the excess risk of cancer that will not exceed x chances out of one million, where x is the integer above from 8 to 19.

Based on these estimates, EPA determined that the tetra-CDD emissions from these municipal waste combustors did not present a significant health risk for residents in the immediate vicinity and that as long as the emission levels of tetra-CDD at other locations did not greatly exceed those seen at these plants, then there should be no reason for concern.

EPA is continuing its assessment of tetra-CDD emissions resulting from the combustion of municipal wastes. It is interesting to note that since the November 1981 report by EPA, [3], reports from other locations have essentially confirmed these analytical findings [4,5].

CONCLUSION

EPA has pioneered in detecting low levels of PCDD and PCDF in environmental samples. It has supplemented existing toxicological data with conservative assumptions about species that have not been subjected to thorough testing. In addition, EPA used computer models, coupled with further conservative assumptions, to estimate possible exposure levels to which humans might be subjected. These data have led to estimates of health effects risks. Finally, EPA has gained experience in assessing the significance of such findings and in making decisions based on this information. In this way EPA has been able to deal with complex and potentially dangerous situations that call for appropriate measures to protect human health and the environment from unreasonable risks.

REFERENCES

1. *Federal Register* 45:15547 (1980).
2. "PCD Disposal by Thermal Destruction," PB 82-241860, EPA-906/9-82-003, National Technical Information Service (1982).
3. "Interim Evaluation of Health Risks Associated with Emissions of TCDD from Municipal Waste Resource Recovery Facilities," U.S. EPA (1981).
4. Cassitto, L., P. Magani, S. Crescenti and P. Gugole. "Report on Micropollutant Detections on Emissions of Municipal Solid Waste Incineration—National and International Results," Milan, Italy (1982).
5. Cassitto, L., P. Magani, S. Crescenti and P. Gugole. "Environmental Pollution Due to Dioxins and Furans from Communal Rubbish Incineration Plants," *Schrift. Umwelts.* (5):19 (1982).

Synthesis

3

Characterization of Tetrachlorodibenzofurans

Thomas Mazer, Fred D. Hileman, Roy W. Noble, Michael D. Hale and Joseph J. Brooks

Polychlorinated dibenzofurans (PCDF) have been the subject of a considerable research effort, because of their structural similarity to the polychlorinated dibenzo-*p*-dioxins (PCDD) [1]. Much of this research has been directed toward the tetrachlorodibenzofurans (tetra-CDF), for which there are 38 positional isomers, making analysis of specific tetra-CDF isomers a difficult task.

This chapter reports the preparation of the 38 tetra-CDF isomers by pyrolysis of specific polychlorinated biphenyl (PCB) congeners [2,3], by ultraviolet (UV) photolysis of pentachlorodibenzofuran (penta-CDF) isomers and by chlorination using electrophilic substitution of specific trichlorodibenzofuran (tri-CDF) isomers. Characterization of the tetra-CDF isomers was accomplished by high-resolution gas chromatography/mass spectrometry (GC/MS) and UV photolysis [4]. The GC elution patterns of these tetra-CDF were characterized and compared to the elution patterns of the tetrachlorodibenzo-*p*-dioxins (tetra-CDD) as reported by Buser and Rappe [5]. The UV photolysis characteristics of selected TCDF have also been studied, and a set of guidelines has been established to predict the relative rate of chlorine loss of various configurations.

EXPERIMENTAL

Persons attempting the synthesis of PCDF should be familiar with the appropriate techniques for their safe handling and disposal.

PCB Congeners

2,2′,3,4′- and 2,3,3′,4′-Tetrachlorobiphenyl, 2,2′,3,3′,6-pentachlorobiphenyl, and 2,2′,3,3′,4,6- and 2,2′,3,3′,5,6′-hexachlorobiphenyl were obtained from C. A.

Wachtmeister, Wallenberglaboratoriet, Stockholms Universitet, Stockholm; 2,3,4,4′-tetrachlorobiphenyl and 2,3,3′,4,5-pentachlorobiphenyl were obtained from J. Pyle, Miami University, Oxford, Ohio; 2,2′,3,4′,5,6-hexachlorobiphenyl was obtained from M. D. Mullins, U.S. Environmental Protection Agency (EPA), Grosse Ile, Michigan. The remaining PCB congeners used in this study were obtained from Ultra Scientific, Inc., Hope, Rhode Island.

All PCB congeners were analyzed by GC/flame ionization detection (FID) before pyrolysis to assess gross purity. Purity was generally $\geq 97\%$, assuming an equivalent FID response for any PCB congener analyzed. If a given PCB congener was contaminated with another PCB congener, pyrolysis would yield minute amounts of undesired PCDF, which were readily identifiable as a contaminant.

PCB Pyrolysis

The PCB samples were pyrolyzed in glass mini ampules prepared from disposable borosilicate glass pasteur pipettes (Model 13-678-20C, Fisher Scientific Co., Cincinnati, Ohio). The specific PCB congener in hexane (10 μL of a 10-μg/mL solution) was placed in an ampule; when the solvent was completely evaporated, the ampule was carefully flame-sealed. The ampule was then inserted into a large vial along with a cold junction referenced chromel-alumel thermocouple. This thermocouple was connected to a digital voltmeter to allow accurate temperature measurements of the ampule. The ampule and thermocouple were placed into a muffle furnace (Type 1500, Thermolyne Corp., Dubuque, Iowa) maintained at 600 °C. When the temperature of the ampule reached 550 °C, pyrolysis of the PCB was allowed to continue for an additional 5 sec. The ampule was removed from the furnace and allowed to cool to ambient temperature, at which time it was opened and the contents were thoroughly rinsed out with 1 mL of hexane.

After some pyrolyses it was desirable to remove any unreacted PCB that might be present in the hexane wash. The unreacted PCB was removed by subjecting the pyrolysate to chromatographic separation on a mini-column of Woelm Basic Alumina (ICN Pharmaceuticals, Cleveland, Ohio) by eluting the PCB with 10 mL of 2% methylene chloride in hexane, and then eluting the retained PCDF with 15 mL of 50% methylene chloride in hexane. The PCDF isomers were often separated from one another using the high-performance liquid chromatography (HPLC) techniques described below before being subjected to either chlorination or dechlorination processes. This was done to eliminate any ambiguities that might occur in isomer identification when mixtures of PCDF were subjected to these processes.

High-Performance Liquid Chromatography

HPLC was used in both the normal- (NH_2) and reverse-phase (C_{18}) modes to isolate individual PCDF isomers. In the normal-phase mode a Zorbax NH_2 column

(25 cm × 4.6 mm) obtained from Du Pont, Analytical Instrument Division, Wilmington, Delaware, was used at a flowrate of 1.5 mL/min with hexane as the eluant. For reverse phase operation a μ-Bondapak C_{18} column (25 cm × 4.6 mm), available from Waters Associates, Inc., Milford, Massachusetts, was used with 75% acetonitrile/25% water (v/v) as the eluant at a flowrate of 0.75 mL/min. In both modes of operation, Altex Model 110 pumps (Beckman Instruments, Inc., Fullerton, California) were used with a Valco Model CV-6-UHPA-N60 injector (Valco Instruments Co., Houston, Texas). Detection was carried out using a Beckman Model 100-10 UV spectrophotometer operated at 235 nm.

In special cases it was necessary to obtain larger quantities of specific PCDF for further chlorination or dechlorination. In this case, a mixture of PCDF obtained by extensive chlorination of dibenzofuran [6] was separated by reverse phase HPLC. In most cases, further separation by normal phase HPLC was necessary to isolate an individual isomer. These isomers were collected by repeated separation of the mixture with collection of the individual fractions on a Gilson Model FC-80 fraction collector. Samples obtained from the reverse phase eluant were extracted out of the eluant by the addition of 8 mL of 2% sodium bicarbonate solution to every 20 mL of eluant. This aqueous mix was then extracted with 5 mL of hexane; this hexane extract was removed and concentrated to obtain the desired product. Samples were obtained from the normal-phase eluant by simple evaporation of the hexane to obtain the desired PCDF isomer.

GC/MS Analyses

Capillary separations were conducted on a Hewlett-Packard 5985B GC/MS system using the splitless injection technique to introduce the sample in tetradecane onto the capillary column. Two capillary columns were necessary to conduct this work, as will be explained in the Results and Discussion section. The first was a wide-bore fused-silica column coated with SE-54 (30 m × 0.25 mm i.d.) obtained from J&W Scientific, Rancho Cordova, California. The second column was a glass capillary coated with SP-2330 (60 m × 0.25 mm i.d.) available from Supelco, Inc., Bellefonte, Pennsylvania.

Helium carrier gas at 7 psi (μ = 38.5 cm/sec) was used for the SE-54 separations with temperature programming starting from an initial temperature of 200°C with a 1-min hold followed by a 10°C/min program to 275°C. Hydrogen carrier gas at 15 psi (μ = 40 cm/sec) was used for the SP-2330 separations with temperature programming starting from an initial temperature of 200°C with a 1-min hold and then programmed at 8°C/min to 250°C.

The PCDF were detected by selected ion monitoring MS in the electron-impact ionization mode. The capillary column was connected directly to the ion source via a 45-cm × 0.20-mm (i.d.) SE-30-coated length of fused silica tubing. Ions characteristic of all classes of PCDF possible in a synthesis were monitored including the molecular ions for di-, tri-, tetra- and penta-CDF, as well as the $(M\text{-}COCl)^+$ ions for tetra- and penta-CDF. Additional ions that were monitored included the molecular ion of the $^{37}Cl_4$-2,3,7,8-tetra-CDF internal standard (Kor Isotopes, Cambridge, Massachusetts), which was used as a chromatographic reference point, and the molecular ions of polychlorinated diphenyl ethers, which were monitored to ensure that these compounds were not interfering in the analyses.

Chlorination

Selected PCDF congeners were chlorinated on a micro-scale (4 mL vial) by adding 5–8 drops of antimony pentachloride (Matheson, Coleman & Bell, Cincinnati, Ohio) to 200 μL of a solution containing 10–50 μg of the PCDF isomer in CCl_4. After swirling and allowing the solution to stand for 5 min, an additional 750 μL of CCl_4 was added. Following 30 sec of shaking, 2 mL of 0.1 *N* HC1 was added to quench the reaction. The CCl_4 layer was washed twice with 2-mL portions of water, dried with 2 g of Na_2SO_4 and quantitatively transferred to a clean vial for subsequent analysis.

Dechlorination

Dechlorination by UV photolysis was conducted by placing a hexane solution containing approximately 50 μg of a specific PCDF into a 10-mm-path-length quartz cuvette and irradiating the sample at 253.7 nm with a UV light source (Gelman Instrument Co., Model 51438, Ann Arbor, Michigan) at a distance of 2.0 cm. Typically, penta-CDF were irradiated for 3–4 hr and tetra-CDF were irradiated for 1–2 hr, since these irradiation times generally gave a large yield of the PCDF resulting from the loss of a single chlorine.

RESULTS AND DISCUSSION

Structural Assignments of Tetra-CDF by Cross-Correlation

When a PCB congener is pyrolyzed, PCDF are generated by the four mechanisms defined by Buser and Rappe [2], shown in Figure 3.1. Since many of the PCB congeners when pyrolyzed yielded several different tetra-CDF isomers, cross-correlation charts were prepared to organize the pyrolysis data. For example, Figure 3.2 shows the PCB congeners pyrolyzed on one axis and the tetra-CDF isomers that were formed on the other axis. From this chart, one can see which PCB will yield a given tetra-CDF on pyrolysis. Further, since the chlorination and dechlorination of a given PCDF will yield several possible products, charts were also prepared (Figures 3.3 and 3.4) showing the reactants and products on the axes. Figure 3.3 shows all the possible tri-CDF of one axis and all the possible tetra-CDF formed by chlorination on the other axis. From this chart one can imme-

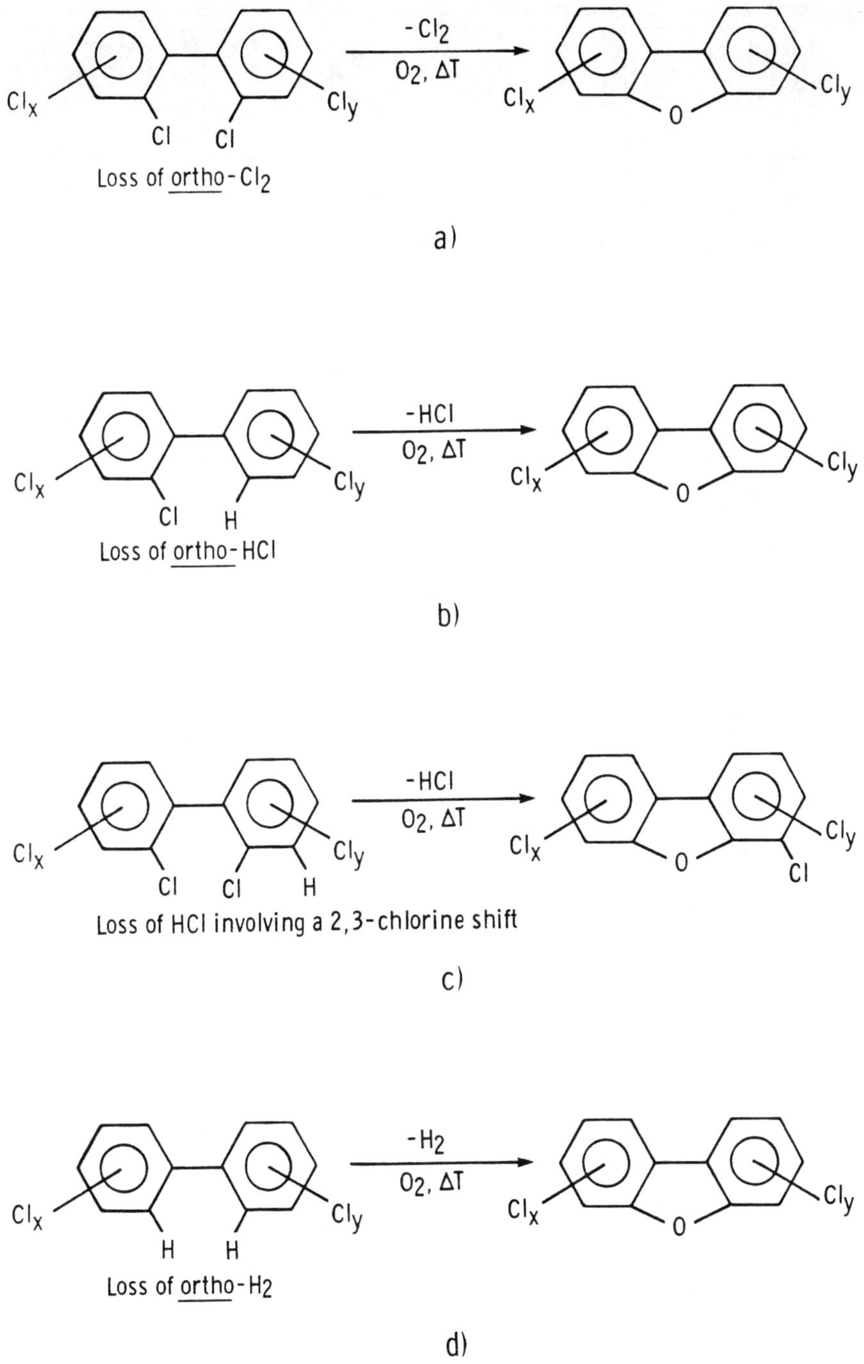

Figure 3.1. The Four Reaction Routes from which PCDF Are Produced from Pyrolysis of PCB Congeners.

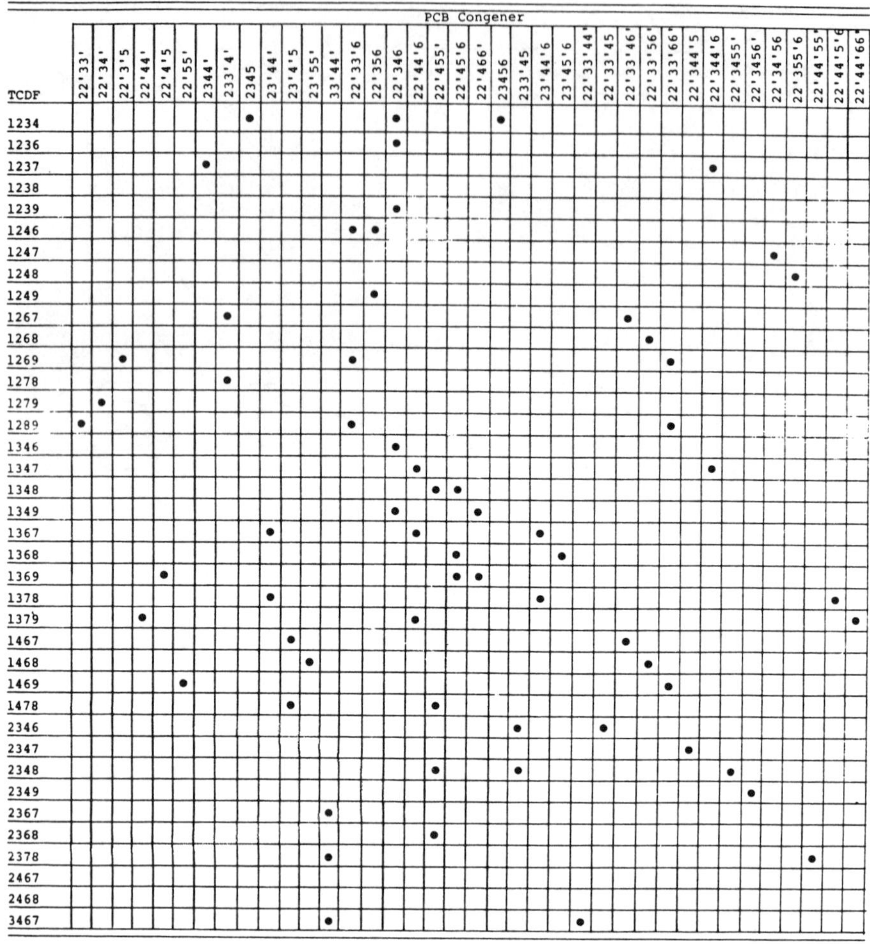

Figure 3.2. Cross-Correlation Chart of PCB Congeners and Tetra-CDF Isomers Generated.

diately see which tetra-CDF will be formed from chlorinating a given tri-CDF or, conversely, which tri-CDF will be generated by the dechlorination of a given tetra-CDF. Figure 3.4 gives the same type of information for the tetra-CDF vs penta-CDF.

When making structural assignments for the tetra-CDF, pyrolysis was considered to be valid only if the expected PCDF congeners were formed. It was often the case that more than one tetra-CDF isomer was formed in a pyrolysis and the ambiguous tetra-CDF had to be identified by comparing them to tetra-CDF formed in other pyrolyses. Assignment of structural identity was then made by matching the chromatographic retention characteristics of the tetra-CDF from the

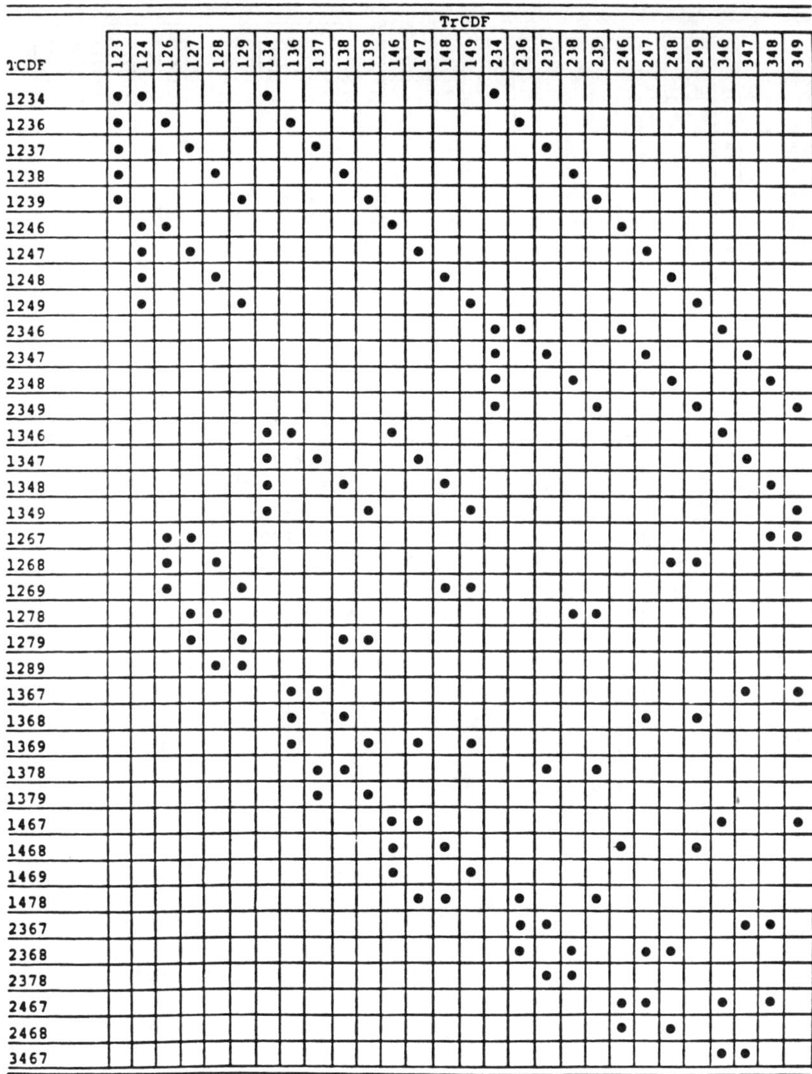

TCDF	123	124	126	127	128	129	134	136	137	138	139	146	147	148	149	234	236	237	238	239	246	247	248	249	346	347	348	349
	TrCDF																											
1234	•	•					•									•												
1236	•		•					•									•											
1237	•			•					•									•										
1238	•				•					•									•									
1239	•					•					•									•								
1246		•	•									•									•							
1247		•		•									•									•						
1248		•			•									•									•					
1249		•				•									•									•				
2346																•	•				•				•			
2347																•		•				•				•		
2348																•			•				•				•	
2349																•				•				•				•
1346							•	•				•													•			
1347							•		•				•													•		
1348							•			•				•													•	
1349							•				•				•													•
1267			•	•																							•	•
1268			•		•																		•	•				
1269			•			•								•	•													
1278				•	•														•	•								
1279				•		•				•	•																	
1289					•	•																						
1367								•	•																	•		•
1368								•		•												•		•				
1369								•			•		•		•													
1378									•	•								•		•								
1379									•		•																	
1467												•	•												•			•
1468												•		•							•			•				
1469												•			•													
1478													•	•			•			•								
2367																	•	•								•	•	
2368																	•		•			•	•					
2378																		•	•									
2467																					•	•			•		•	
2468																					•		•					
3467																									•	•		

Figure 3.3. Cross-Correlation Chart of Tri-CDF Isomers vs Tetra-CDF Isomers.

various pyrolyses. An example (Figure 3.5) illustrates this process for the identification of 2,3,6,7-tetra-CDF. The pyrolysis of 3,3′,4,4′-tetrachlorobiphenyl yields 3,4,6,7-, 2,3,7,8- and 2,3,6,7-tetra-CDF (see Figure 3.2). Since 2,3,7,8-tetra-CDF is the only tetra-CDF formed from the pyrolysis of 2,2′,4,4′,5,5′-hexachlorobiphenyl and 3,4,6,7-tetra-CDF was uniquely formed from 2,2′,3,3′,4,4′-hexachlorobiphenyl (see Figure 3.2), the third tetra-CDF formed from the pyrolysis of 3,3′,4,4′-tetrachlorobiphenyl was, by elimination, 2,3,6,7-tetra-CDF.

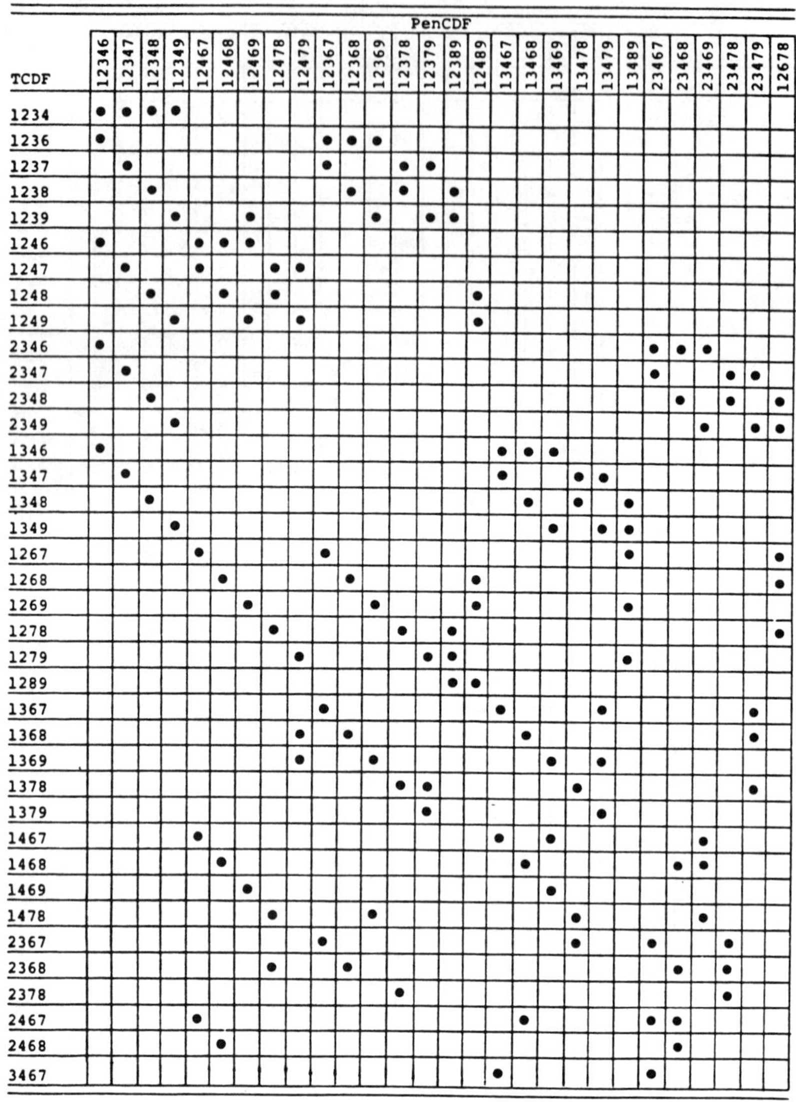

TCDF	12346	12347	12348	12349	12467	12468	12469	12478	12479	12367	12368	12369	12378	12379	12389	12489	13467	13468	13469	13478	13479	13489	23467	23468	23469	23478	23479	12678
	PenCDF																											
1234	●	●	●	●																								
1236	●									●	●	●																
1237		●								●			●	●														
1238			●								●		●		●													
1239				●			●					●		●	●													
1246	●				●	●	●																					
1247		●			●			●	●																			
1248			●			●		●								●												
1249				●			●		●							●												
2346	●																						●	●	●			
2347		●																					●			●	●	
2348			●																					●		●		●
2349				●																					●		●	●
1346	●																●	●	●									
1347		●															●			●	●							
1348			●															●		●		●						
1349				●															●		●	●						
1267					●					●												●						●
1268						●					●					●												●
1269							●					●				●						●						
1278								●					●		●													●
1279									●					●	●							●						
1289															●	●												
1367										●							●				●						●	
1368									●		●							●									●	
1369									●			●							●		●							
1378													●	●						●							●	
1379														●							●							
1467					●												●		●						●			
1468						●												●						●	●			
1469							●												●									
1478								●				●								●					●			
2367										●										●			●			●		
2368								●			●													●		●		
2378													●													●		
2467					●													●					●	●				
2468						●																		●				
3467																	●						●					

Figure 3.4. Cross-Correlation Chart of Penta-CDF Isomers vs Tetra-CDF Isomers.

Since another PCB congener was not available that would generate 2,3,6,7-tetra-CDF in order to confirm the original assessment, an alternative source was necessary. This source was the photolytic dechlorination of 1,2,3,6,7-penta-CDF, which formed 2,3,6,7-tetra-CDF along with 1,2,3,6-, 1,2,3,7-, 1,2,6,7- and 1,3,6,7-tetra-CDF (see Figure 3.4).

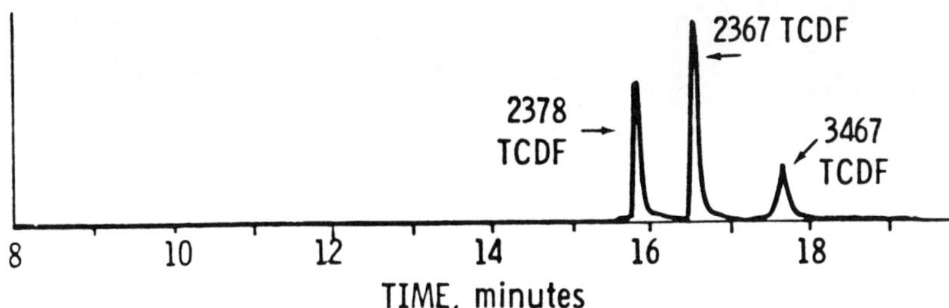

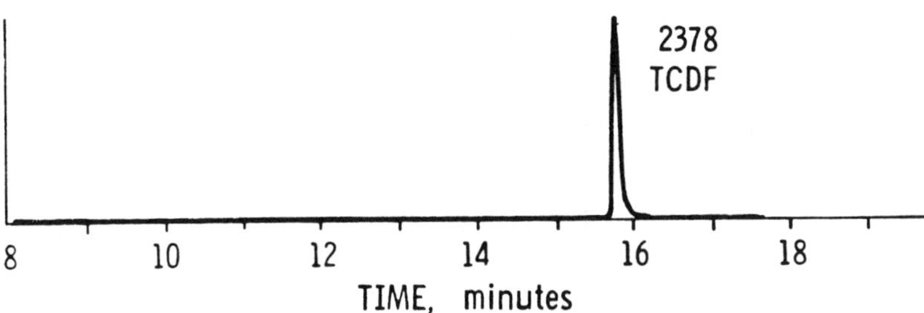

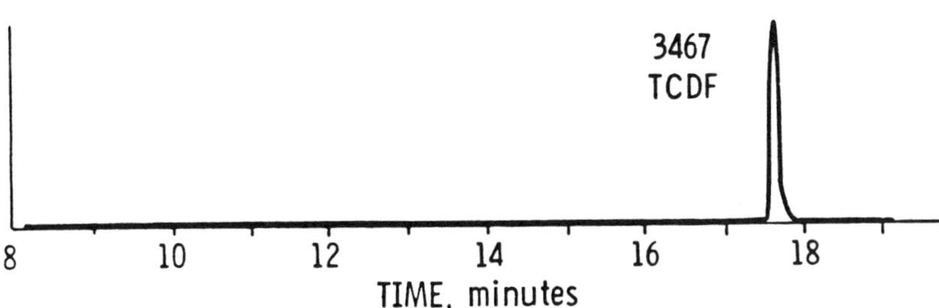

Figure 3.5. GC/MS Analysis of Tetra-CDF Formed from Pyrolysis of Various PCB Congeners (SP-2330 Capillary Column).

A list of the 38 tetra-CDF isomers and their routes of synthesis and confirmation is presented in Table 3.1. PCB congeners that generated a single tetra-CDF isomer are noted. Other pyrolyses as well as dechlorinations and chlorinations that were performed to verify tetra-CDF identities are also indicated in Table 3.1.

Table 3.1. Synthesis Routes for the 38 Tetra-CDF

Tetra-CDF	*Primary PCB Congener*	*Secondary PCB Congener*	*PCDF*		*Confirmed by Photolysis*
			Dechlorinated	*Chlorinated*	
1,2,3,4	2,3,4,5,6[a]; 2,3,4,5[a]	2,2′,3,4,6			
1,2,3,6	2,2′,3,4,6		1,2,3,6,7		
1,2,3,7	2,3,4,4′[a]	2,2′,3,4,4′,6	1,2,3,6,7	1,2,3	
1,2,3,8			1,2,3,4,8	2,3,8; 1,2,3; 1,2,8	Yes
1,2,3,9	2,2′,3,4,6			1,2,3	
1,2,4,6	2,2′,3,3′,6	2,2′,3,5,6		1,2,4	
1,2,4,7	2,2′,3,4′,5,6[a]			1,2,4	
1,2,4,8	2,2′,3,5,5′,6[a,b]			1,2,4; 1,2,8	Yes[c]
1,2,4,9	2,2′,3,5,6			1,2,4	
1,2,6,7	2,3,3′,4′	2,2′,3,3′,4,6′	1,2,3,6,7		
1,2,6,8	2,2′,3,3′,5,6′		1,2,4,6,8		
1,2,6,9	2,2′,3′,5[a]	2,2′,3,3′,6; 2,2′,3,3,′,6,6′			
1,2,7,8	2,3,3′,4′		2,3,4,6,9	2,3,8; 1,2,8	Yes[c]
1,2,7,9	2,2′,3,4′[a,d]	2,2′,3,4′[d]			
1,2,8,9	2,2′,3,3′	2,2′,3,3′,6; 2,2′,3,3′,6,6′			
1,3,4,6	2,2′,3,4,6		1,3,4,6,7		
1,3,4,7	2,2′,3,4,4′,6	2,2′,4,4′,6	1,3,4,6,7		
1,3,4,8	2,2′,4,5,5′	2,2′,4,5′,6			

1,3,4,9	2,2′,4,6,6′	2,2′,3,4,6			
1,3,6,7	2,3′,4,4′	2,2′,4,4′,6; 2,3′,4,4′,6	1,3,4,6,7; 1,2,3,6,7		
1,3,6,8	2,3′,4,5′,6[a]	2,2′,4,5′,6			
1,3,6,9	2,2′,4′,5[d]	2,2′,4,6,6′; 2,2′,4,5′,6			
1,3,7,8	2,2′,4,4′,5′,6[a]	2,3′,4,4′; 2,3′,4,4′,6			
1,3,7,9	2,2′,4,4′[a]	2,2,4,4′,6			
1,4,6,7	2,3′,4′,5	2,2′,3,3′,4,6′	1,3,4,6,7; 2,3,4,6,9		
1,4,6,8	2,3′,5,5′[a]	2,2′,3,3′,5,6	2,3,4,6,9		Yes[c]
1,4,6,9	2,2′,5,5′[a]	2,2′,3,3′,6,6′			
1,4,7,8	2,3′,4′,5′	2,2′,4,5,5′	2,3,4,6,9		
2,3,4,6	2,2′,3,3′,4,5	2,3,3′,4,5			
2,3,4,7	2,2′,3,4,4′,5[a]				
2,3,4,8	2,2′,3,4,5,5′[a]	2,2′,4,5,5′; 2,3,3′,4,5		2,3,8	
2,3,4,9	2,2′,3,4,5,6′[a]				
2,3,6,7	3,3′,4,4′		1,2,3,6,7		
2,3,6,8	2,2′,4,5,5,′		2,3,4,6,8	2,3,8	Yes[c]
2,3,7,8	2,2′,4,4′,5,5′[a]	3,3′,4,4′			Yes
2,4,6,7	[b]		2,3,4,6,8		Yes[c]
2,4,6,8	[b]		2,3,4,6,8		Yes[c]
3,4,6,7	2,2′,3,3′,4,4′[a]	3,3′,4,4′	1,3,4,6,7		

[a]Represents those cases in which a single tetra-CDF isomer was formed from the pyrolyses.

[b]The relative retention times of these isomers were identical with the same isomer synthesized by the palladium acetate cyclization of the appropriate diphenyl ethers [7,8].

[c]These tetra-CDF were also received from D. Firestone and checked against our own syntheses.

[d]These are identical PCB congeners from different sources. Pyrolyses of both gave identical products.

Since some of the tetra-CDF isomers were synthesized and/or confirmed by either chlorination of a tri-CDF isomer or by dechlorination of a penta-CDF isomer, these PCDF had to be of confirmed identity. The assignment of structure and the subsequent confirmation of that assigned structure were accomplished in a manner similar to that already described for the tetra-CDF. A list of the penta-CDF that were used in dechlorination syntheses is presented in Table 3.2. Again, as in Table 3.1, PCB isomers that generated a single penta-CDF are noted. In similar fashion, Table 3.3 lists the tri-CDF used in chlorination syntheses and their routes of confirmation.

It should be emphasized here that, when using chlorinations or dechlorinations as a synthesis route, the starting material must be a pure PCDF isomer or the reaction will yield a multitude of undesired tetra-CDF. Pure starting materials were obtained by using HPLC in combination with fraction collection as described earlier, with the purity of the collected PCDF isomers being assessed by GC/MS.

Table 3.2. Synthesis Routes for the Penta-CDF Isomers Used in This Study

Penta-CDF Isomer	*Primary PCB Congener*	*Secondary PCB Congener*
1,2,3,6,7	2,2′,3,3′,4,4[a]	2,2′,3,4,4′,6; 2,2′,3,3′,4,4′,6
1,2,3,4,8	2,2′,3,4,5,5′,6[a]	2,3,3′,4,5; 2,2′,3,4,5,5′
2,3,4,6,9	2,2′,3,4,5,6′	2,2′,3,4,5,5′
1,3,4,6,7	2,2′,3,4,5,5′	2,2′,3,4,4′,5′
2,3,4,6,8	2,2′,3,4,5,5′	
2,3,4,6,7	2,3,3′,4′,4,5	2,2′,3,4,4′,5

[a]Represents those cases in which a single penta-CDF isomer was formed from the pyrolysis.

Table 3.3. Synthesis Routes for the Tri-CDF Isomers Used in This Study

Tri-CDF Isomer	*Primary PCB Congener*	*Secondary PCB Congener*	*Dechlorination*
1,2,3	2,3,4[a]	2,2′,3,4,6	
1,2,4	2,3,5,6[a]	2,2′,3,5,6	
1,2,8	2,2′,3′,5		1,2,7,8
2,3,8	2,2′,4,5,5,′	2,3′,4′,5	

[a]Represents those cases in which a single tri-CDF isomer was formed from the pyrolysis.

Structural Confirmation of Tetra-CDF by Photolytic Dechlorination

Photolytic (UV) dechlorination of penta-CDF served as an alternative synthetic method for the formation of tetra-CDF. During these studies it became apparent that as an extension of this method, the photolytic dechlorination of a tetra-CDF to known tri-CDF could be used to confirm the structure of the tetra-CDF even on a micro-scale once the tetra-CDF had been isolated by the methods previously described. It was also apparent that if information dealing with the relative rates of formation of the tri-CDF were available, then the relative amounts of the tri-CDF formed would be useful in confirming the identity of a tetra-CDF. Following preparation of the tri-CDF necessary for structural identification of tetra-CDF photolysis products, a study was conducted to try to understand the photolytic dechlorination of tetra-CDF. Some of the tri-CDF isomers used in this study are listed in Table 3.3 along with the PCB isomers used to synthesize them as well as the confirming PCB pyrolysis, if available.

When a tetra-CDF is subjected to UV irradiation, it loses a single atom of chlorine to form a tri-CDF:

Cl_4 (tetra-CDF) $\xrightarrow{UV}$ Cl_3 (tri-CDF)

The loss of any one of the four chlorines may occur, thus resulting in four tri-CDF being formed, except in the cases of the six symmetrical tetra-CDF, for which only two tri-CDF form. An example of this is the UV dechlorination of 1,4,6,9-tetra-CDF:

1469-TCDF —UV→ loss of #4 or #6 Cl → 149-TrCDF
1469-TCDF —UV→ loss of #1 or #9 Cl → 146-TrCDF

Once a tetra-CDF was dechlorinated, the resulting mixture was analyzed by GC/MS using the selected ion mode to generate mass fragmentograms. Masses were monitored for the original tetra-CDF isomer (m/z 303, 306), for the tri-CDF (m/z 270, 272) and finally for di-CDF (m/z 236, 238). The masses for di-CDF were monitored to ensure that the dechlorination of the tetra-CDF was stopped before dechlorination of tri-CDF to di-CDF. This enabled the relative ratios of the tri-CDF formed to be determined. A sample mass fragmentogram is shown in Figure 3.6 for the photolysis of 2,3,6,8-tetra-CDF to yield 2,4,7-, 2,4,8-, 2,3,8- and 2,3,6-tri-CDF. The tetra-CDF for which dechlorination was studied are listed in Table 3.4, along with the tri-CDF formed from each in the observed order of decreasing yield, assuming an equivalent mass spectrometer response for all tri-CDF. The relative ratios of the tri-CDF formed are also given based on a value of 1.00 for the major tri-CDF formed in each case. The reproducibility of the quantification for three photolyses was ±10%.

The compound chosen for the initial photolysis study was 1,2,3,4-tetra-CDF, since all four unique positions on the molecule are chlorinated. Table 3.4 shows that the 3- chlorine is the most easily lost, resulting in 1,2,4-tri-CDF being the major tri-CDF formed. The loss of chlorine involves a homolytic bond cleavage in which the intermediate free radical apparently gains stabilization from the presence of the two vicinal chlorines at positions 2- and 4-, and perhaps some added stabilization from the biphenyl bond located *para* to it.

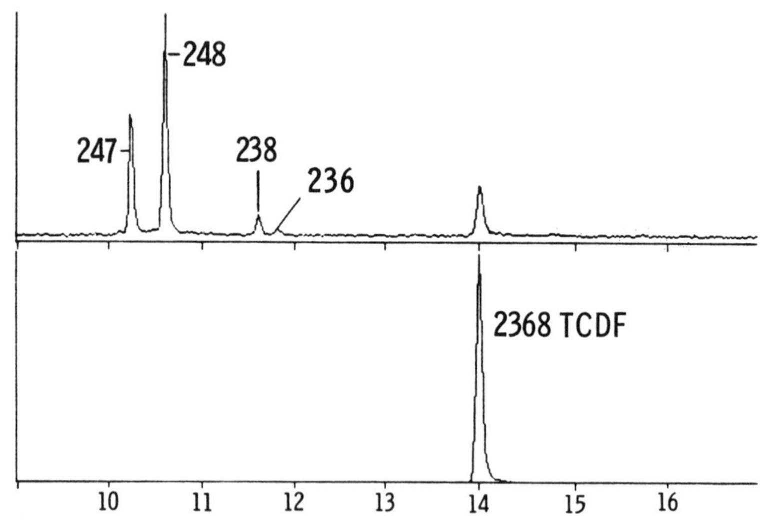

Figure 3.6. Mass Fragmentograms Showing Tri-CDF Formed from UV Photolysis of 2,3,6,8-Tetra-CDF.

Table 3.4. Tri-CDF Formed from Photolysis of Tetra-CDF

	Tri-CDF and Their Relative Ratios Formed by Photolysis			
Tetra-CDF	*Major*	*Second*	*Third*	*Fourth*
4:0[a]/1,2,3,4-	1,2,4-/1.0	1,3,4-/0.38	2,3,4-/0.13	1,2,3-/0.12
3:1/1,2,4,8-	2,4,8-/1.0	1,4,8-/0.4	1,2,8-/0.1	1,2,4-/ND[b]
2:2/2,3,6,8-	2,4,8-/1.0	2,4,7-/0.59	2,3,8-/0.09	2,3,6-/0.03
2:2/1,4,6,8-	2,4,6-/1.0	1,4,6-/0.70	2,4,9-/0.60	1,4,8-/0.21
2:2/2,4,6,8-	2,4,8-/1.0	2,4,6-/0.40		
2:2/2,4,6,7-	2,4,7-/1.0	2,4,6-/0.45	3,4,8-/0.1	3,4,6-/ND
2:2/2,3,7,8-	2,3,8-/1.0	2,3,7-/0.52		
2:2/1,2,7,8-	1,2,8-/1.0	2,3,8-/0.94	1,2,7-/0.75	2,3,9-/0.37

[a]4:0 substitution indicated 4 chlorines on one aromatic ring and no chlorines on the other aromatic ring.

[b]ND = none detected at the point at which di-CDF began to form.

The second most abundant isomer formed is 1,3,4-tri-CDF, resulting from the loss of the 2- chlorine. Note that the 2- chlorine also has two vicinal chlorines and is located *para* to the ether linkage. No justification was found to allow accurate prediction of the order of loss of the 1- or 4- chlorines. Both have one vicinal chlorine and are *ortho* to either the biphenyl bond or the ether linkage. There is evidently little or no stabilization associated with either of these positions, since the 2,3,4- and 1,2,3-tri-CDF are both formed in relatively small amounts and in approximately equivalent abundances.

This same line of reasoning was then applied to the other members of the test set of tetra-CDF. For 1,2,4,8-tetra-CDF, representing the class of 3:1 substitution (i.e., three chlorines on one ring and one chlorine on the other), 2,4,8-tri-CDF is the major product, followed by 1,4,8-tri-CDF. These results are not unexpected, since the 1- and 2- chlorines are the only ones in the molecule that have a vicinal chlorine (in this case they are vicinal to each other). The other two possible losses (4- and 8- chlorine) are both very minor components of the mixture, due primarily to the fact that neither has a vicinal chlorine to lend additional stabilization. According to Hey [9], any substituent on the aromatic ring tends to stabilize homolytic bond cleavage. This is consistent with the observation that 1,2,8-tri-CDF is more abundant than 1,2,4-tri-CDF. In other words, the 4- chlorine is lost preferentially over the 8- chlorine because the 4- chlorine was on the ring having the most substituents. This leads to the generalization that the aromatic ring with the most chlorines will be the first to lose a chlorine in photolytic degradation.

The structures of the 2:2-substituted tetra-CDF (2,3,6,8-, 1,4,6,8-,

2,4,6,8- and 2,4,6,7-tetra-CDF) are shown in Figure 3.7 along with the tri-CDF formed during photolysis. Examination of Table 3.4 reveals that chlorines having a vicinal chlorine are the most readily lost, while the lone chlorines are lost to only a slight extent. For example, 2,4,6,7-tetra-CDF photodegrades to yield 2,4,7- and 2,4,8-tri-CDF as the major tri-CDF. In each case, the chlorine that was lost had one vicinal chlorine. The other possible tri-CDF are formed to only a slight extent, again indicating the absence of any factor strongly influencing the preferential loss of either lone chlorine.

The absence of any factor directing a particular chlorine loss in tetra-CDF without vicinal chlorines is evident in the case of the 1,4,6,8-tetra-CDF, since the three major products here are formed in nearly equal amounts.

Finally, 2,3,7,8- and 1,2,7,8-tetra-CDF were studied. Here each molecule has two pairs of vicinal chlorines with at least one of the pairs located in the 2,3-position as shown in Figure 3.8. Referring again to Table 3.4, it is seen that in both cases the 3- chlorine (7- chlorine in the case of 1,2,7,8-tetra-CDF) is lost preferentially, but not to a major extent, since all chlorines on the molecule are vicinal to one chlorine. In other words, all four of the chlorines are likely to be lost due to the presence of the vicinal chlorine in each case. Therefore, the actual order of losses is directed by other factors, which are not as strongly influencing as the presence of vicinal chlorines.

Consideration of these data led to the following guidelines for predicting the order of loss of chlorine from a TCDF during photolysis:

1. Chlorines will be lost preferentially from the ring containing the most chlorines.
2. Vicinal chlorines promote the loss of a particular chlorine (i.e., the greater the number of vicinal chlorines about a given chlorine, the greater the probability of initially losing that particular chlorine).
3. Given an equal number of vicinal chlorines, the 3- chlorine will be lost before the 2- chlorine.

The techniques of photolytic dechlorination and the preceding guidelines are intended to be used as an aid in the synthesis and the structural confirmation of a particular tetra-CDF. To test the utility of these guidelines, they were applied to the case of an isomer believed to be 1,2,3,8-tetra-CDF. The initial synthesis of 1,2,3,8-tetra-CDF was carried out via dechlorination of 1,2,3,4,8-penta-CDF. The other tetra-CDF formed were known, so the remaining tetra-CDF was considered to be 1,2,3,8-tetra-CDF; however, no PCB congeners were available that gave 1,2,3,8-tetra-CDF when pyrolyzed. This isomer was then isolated using HPLC techniques and photolyzed, and the tri-CDF were analyzed. The results of the analysis are given in Table 3.5 and are consistent with the established guidelines. Referring to Table 3.2, it should be noted that the final column in this table shows the tetra-CDF whose structural assignments were confirmed using this technique.

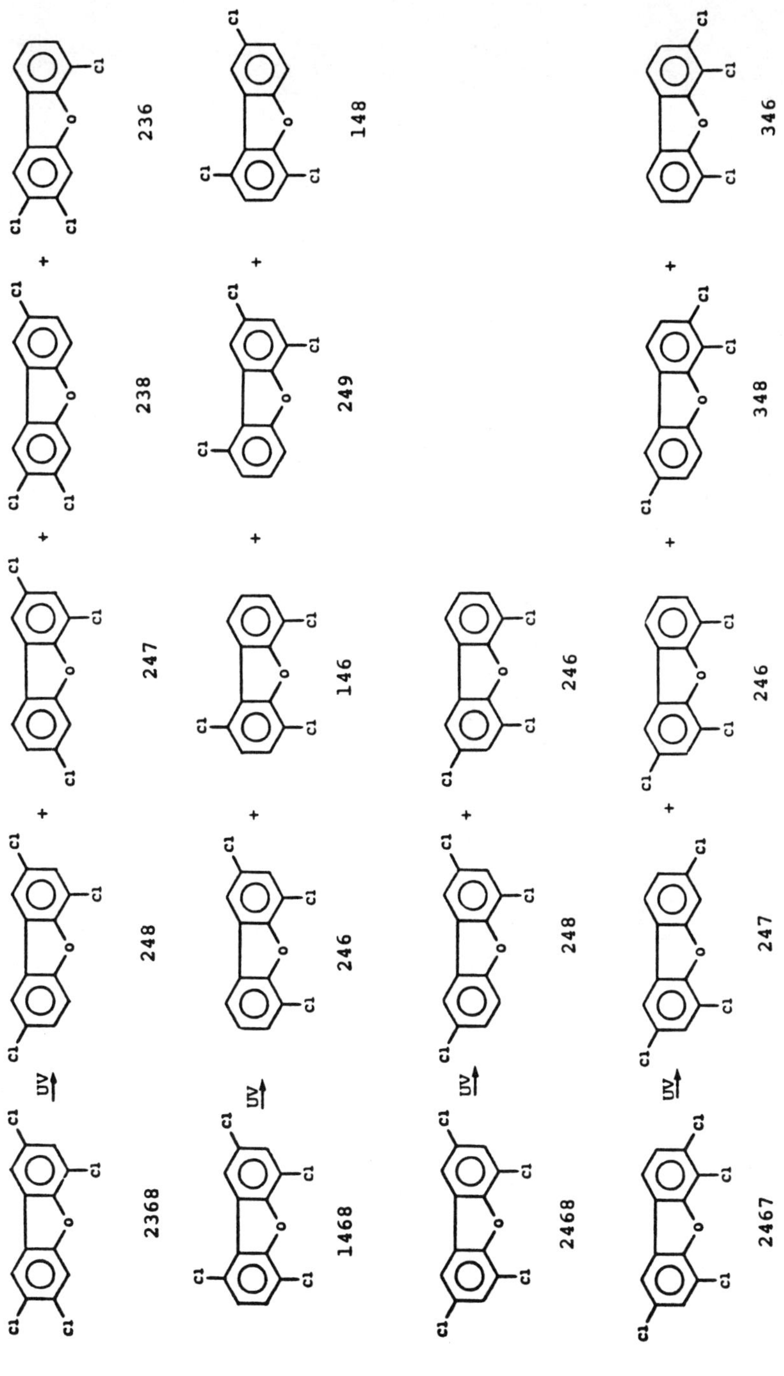

Figure 3.7. Tri-CDF Formed from Photolysis of 2:2 Tetra-CDF Having Chlorines in the 2,4-Positions.

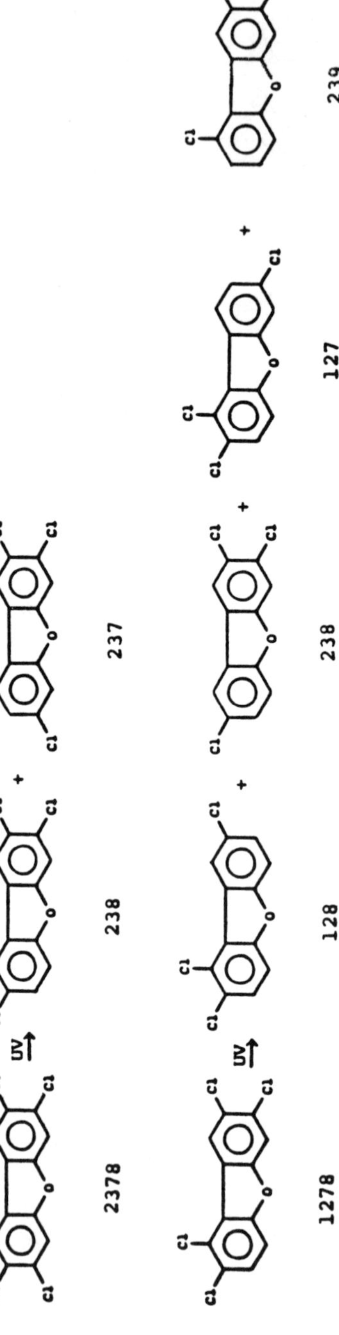

Figure 3.8. Tri-CDF Formed from Photolysis of Tetra-CDF Having Chlorines in the 2,3-Positions.

Table 3.5. Tri-CDF Formed from the Photolysis of 1,2,3,8-Tetra-CDF

Tri-CDF and Their Relative Ratios Formed by Photolysis			
Major	*Second*	*Third*	*Fourth*
1,3,8-/1.0	1,2,8-/0.48	2,3,8-/0.39	1,2,3-/ND[a]

[a]ND means none detected at the point at which di-CDF began to form.

GC Elution Behavior of Tetra—CDF

One of the primary reasons for preparing tetra-CDF and determining their elution characteristics by capillary column GC was to be able to perform an unambigous identification of 2,3,7,8-tetra-CDF. The retention data for the 38 tetra-CDF are depicted in Figure 3.9 for the SP-2330 capillary column and in Figure 3.10 for SE-54 capillary column. The GC was operated under the conditions previously described. For isomer-specific analysis of 2,3,7,8-tetra-CDF, both capillary columns were required. On both columns one or more isomers coeluted with 2,3,7,8-tetra-CDF. However, the differences in the liquid phases were such that the coeluting isomers were separated on the other column. On SE-54, seven isomers coelute with 2,3,7,8-tetra-CDF, and on SP-2330, only 1,2,6,9-tetra-CDF coelutes with 2,3,7,8-tetra-CDF. Tetra-CDF analyses were normally conducted on the SE-54 column first due to its ease of handling and higher temperature limit, allowing a broader range of PCDF to be analyzed. For all analyses, an internal standard of $^{37}Cl_4$-2,3,7,8-tetra-CDF was coinjected to allow calculation of relative retention times (RRT), where $^{37}Cl_4$-2,3,7,8-tetra-CDF = 1.000. When a peak eluted on the SE-54 column within the RRT window of 2,3,7,8-tetra-CDF (RRT = 1.000 ± 0.01), it was necessary to quantify the peaks eluting at RRT = 1.035 ± 0.01). The peaks within this window were identified as 2,3,6,7-, 3,4,6,7-and 1,2,6,9-tetra-CDF. Repeating the analysis on SP-2330 then allowed quantification of the coeluting 1,2,6,9- and 2,3,7,8-tetra-CDF pair as well as the now individually resolved 2,3,6,7-tetra-CDF and 3,4,6,7-tetra-CDF peaks. This information allowed a determination to be made of the amount of 2,3,7,8-tetra-CDF present in the sample.

An additional point should be made about the analysis of 2,3,7,8-tetra-CDF on these two capillary columns. One tetra-CDF isomer elutes very near the 2,3,7,8-tetra-CDF on both columns: 2,3,4,8-tetra-CDF. This isomer can interfere in the analysis of 2,3,7,8-tetra-CDF, particularly if an isotropically labeled internal standard tetra-CDF is not used. The importance of the possible interference of this isomer in analyses can be seen in Figure 3.11, which depicts the mechanism

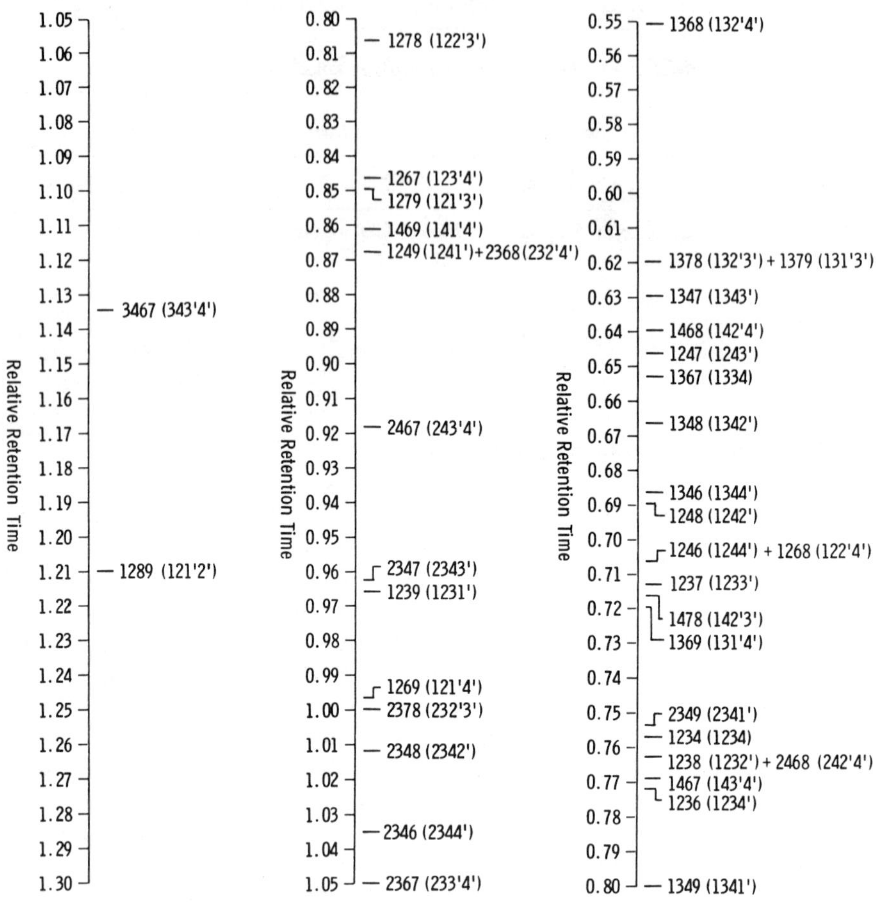

Figure 3.9. Relative Retention Times for Tetra-CDF on SP-2330.

of formation of PCDF in chlorophenols (CP) as reported by Rappe et al. [10]. In the chlorination of the phenol to form penta-CP, 2,4-di-CP and 2,3,4,6-tetra-CP are two of the predominant intermediate products. Dimerization-cyclization of these two intermediates, shown in Figure 3.11, could result in the formation of 2,3,4,8-tetra-CDF, This isomer could, according to the results shown in Figures 3.9 and 3.10, possibly cause errors in the identification and quantification of 2,3,7,8-tetra-CDF in CP samples and environmental samples contaminated with CP.

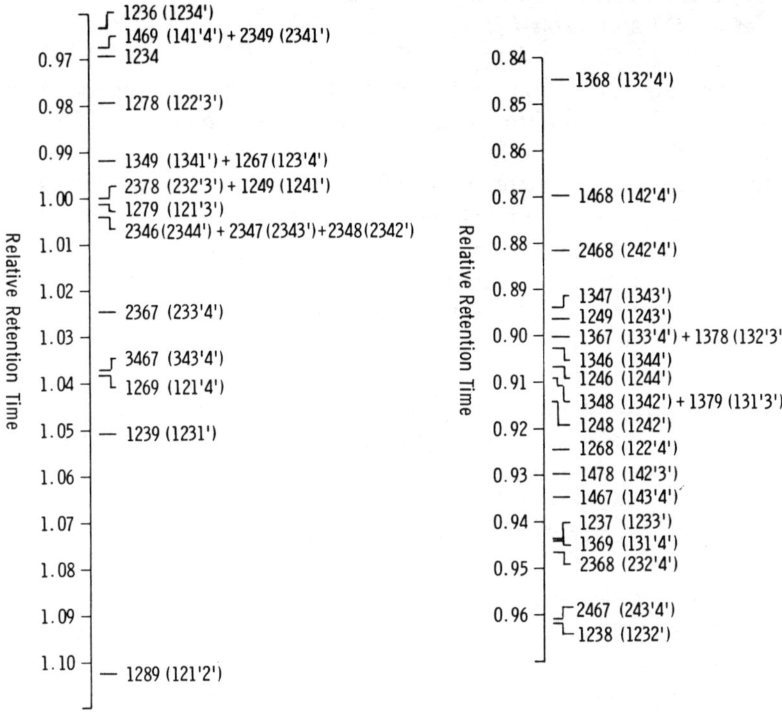

Figure 3.10. Relative Retention Times for Tetra-CDF on SE-54.

A

B

2346-TCP

24-DCP

2348-TCDF

Figure 3.11. Dimerization-cyclization of 2,4-di-CP and 2,3,4,6-tri-CP to Form 2,3,4,8-Tetra-CDF via the Mechanism of Rappe et al. [10].

Structure/Retention Relationships for Tetra-CDF and Tetra-CDD

At the beginning of this chapter it was stated that there are structural similarities between tetra-CDD and tetra-CDF. A direct comparison of tetra-CDD and tetra-CDF is complicated by the fact that dibenzo-*p*-dioxin is a highly symmetrical molecule with two equivalent ether linkages joining the aromatic rings. In contrast, dibenzofuran has a biphenyl and an ether linkage joining the two rings, which reduces this symmetry. However, in reviewing the retention data for tetra-CDF shown in the previous section, the similarities were evident in the gross elution behavior of tetra-CDD and tetra-CDF isomers with similar arrangement of chlorines about the aromatic rings. This led to detailed examination of the GC elution behavior of the tetra-CDD and the formulation of trends that were applied to understanding the elution behavior of the tetra-CDF. The tetra-CDF data were then evaluated to determine whether they were consistent with the trends drawn from the tetra-CDD.

To simplify matters, only the qualitative elution orders for tetra-CDD and tetra-CDF have been considered. In other words, trends were established for the fact that isomer A elutes before isomer B, which elutes before isomer C without being too concerned that A and B elute closely to each other while C elutes considerably later. Furthermore, the discussions were limited to situations where the chlorine substitution pattern on one ring was held constant while the elution order was examined for various substitutions on the second ring. The consistency in the elution orders was evaluated as a means of providing a further check on the structural assignments given to the tetra-CDF in the previous section.

Eventually, the relationships found in this study could be used to devise or calculate molecular descriptions that may someday be used in quantitative structural/retention relationships. These relationships may then allow quantitative prediction of the retention characteristics for all tetra-CDF isomers and possibly for all PCDF.

To facilitate these studies, an alternative numbering scheme has been adopted, in which one ring is numbered in the conventional manner (i.e., 1 through 4) and the other ring is numbered in an identical fashion but these numbers are primed (as in PCB) to distinguish them. As an example for the tetra-CDD, 1,3,6,8-tetra-CDD becomes 1,3,2′,4′-tetra-CDD:

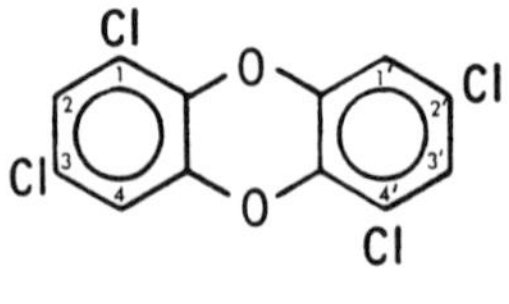

1368- or 132'4'-TCDD

For convenience (and completeness) the tetra-CDD can sometimes have two sets of numbers given to them. For example, 1,2,6,8-tetra-CDD is most properly named 1,2,2′,4′-tetra-CDD; however, it may alternatively be named 1,3,3′,4′-tetra-CDD.

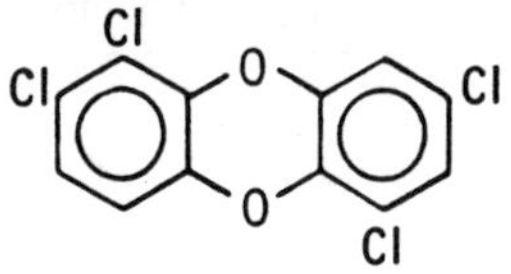

1268- or 122'4'- or 133'4'- TCDD

As a convenient starting point for comparing the tetra-CDD and tetra-CDF, the GC elution parameters for the 2:2-substituted (two chlorines on each ring) tetra-CDD on OV-101 as reported by Buser et al. [5] are given in Table 3.6. This new numbering scheme is illustrated in Figure 3.12, showing the elution order for the 2:2-substituted tetra-CDD, with one ring receiving the unprimed numbers and the other ring the primed numbers. A substitution was selected for ring 1; holding this constant, the elution order for the various substitutions on ring 2 were listed in ascending elution order. For ring 1 there were four possible substitutions and they are listed across the top of Figure 3.12. Once the substitution on ring 1 was fixed, ring 2 could have six possible substitutions. As an example, fixing ring 1 as a 1,3-substitution permits six possible isomers to be formed with varying substitutions on ring 2:

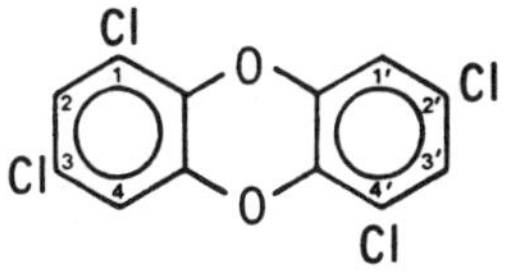

1368 or 132'4'- TCDD

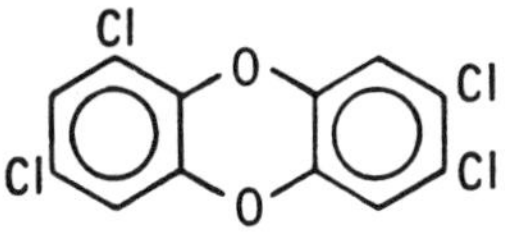

1378 or 132'3'- TCDD

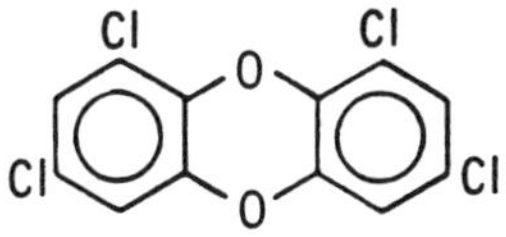

1379 or 131'3'- TCDD

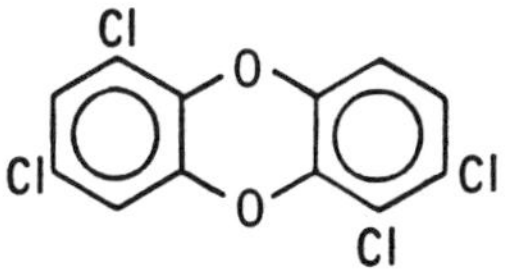

1268 or 133'4'- TCDD

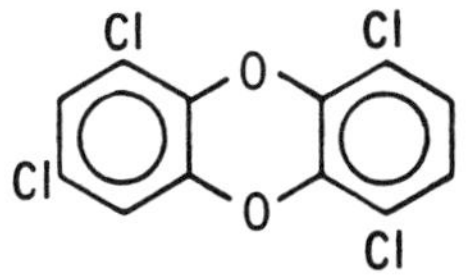

1369 or 131'4'- TCDD

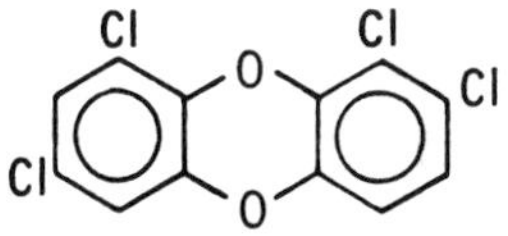

1279 or 131'2'- TCDD

An exception to this was when the substitution on ring 1 was either 2,3- or 1,4-. In these cases the 1,3 and 2,4- as well as 1,2 and 3,4- substitutions on ring 2 were

Table 3.6. Elution Temperatures for the Thirteen 2:2-Substituted Tetra-CDD:OV-101

Name	*Alternative Name*	*Elution Temp. (°C)*
1,3,6,8	1,3,2′,4′	215.2
1,3,7,9	1,3,1′,3′; 2,4,2′,4′	215.7
1,3,6,9	1,4,1′,3′; 1,4,2′,4′	216.5
1,4,6,9	1,4,1′,4′	218.4
1,3,7,8	1,3,2′,3′; 2,3,2′,4′	218.4
1,2,6,8	1,2,2′,4′; 1,3,3′,4′	219.0
1,4,7,8	1,4,2′,3′	219.4
1,2,7,9	1,2,1′,3′; 2,4,3′,4′	219.8
1,2,6,9	1,2,1′,4′; 1,4,3′,4′	221.0
2,3,7,8	2,3,2′,3′	221.5
1,2,7,8	1,2,2′,3′; 2,3,3′,4′	222.5
1,2,6,7	1,2,3′,4′	223.0
1,2,8,9	1,2,1′,2′; 3,4,3′,4′	224.6

OV-101

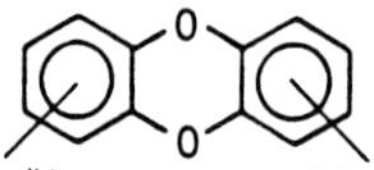

Ring #1 Ring #2

SUBSTITUTION PATTERN—RING #1

SUBSTITUTION PATTERN RING #2 ASCENDING ORDER OF ELUTION

13	14	23	12
24			24
	24 = 13	24 = 13	
13			13
14	14	14	14
23	23	23	23
34			34
	34 = 12	34 = 12	
12			12

Figure 3.12. Elution Order for the 13 2:2-Substituted Tetra-CDD: OV-101.

equivalent, as shown in Figure 3.12. An example shows this equivalency by simple rotation of the TCDD about the longitudinal axis of the molecule:

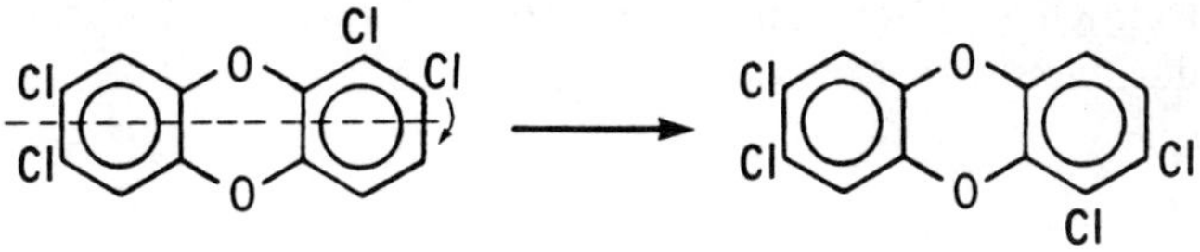

1278 or 122'3' or 233'4'- TCDD

The one fact that stands out from Figure 3.12 is that the elution orders for substitutions on ring 2 are always the same and are independent of the substitution on ring 1. From the elution orders observed in Figure 3.12 the following observations can be made:

1. Substitutions having no vicinal hydrogens or vicinal chlorines (1,3- or 2,4-) result in early elution. Substitutions with only vicinal hydrogens (1,4-) elute next, followed by those substitutions with only vicinal chlorines (2,3-). Substitutions with both vicinal hydrogens and vicinal chlorines (1,2- or 3,4-) elute last.
2. For tetra-CDD with comparable substitution on both rings (in regard to the number of vicinal hydrogens and vicinal chlorines), the later-eluting isomer will be the one with a mirror plane through the oxygen atoms perpendicular to the plane of the molecule. As an example, 1,3,6,8 (1,3,2′,4′)-tetra-CDD elutes ahead of 1,3,7,9 (1,3,1′,3′)-tetra-CDD:

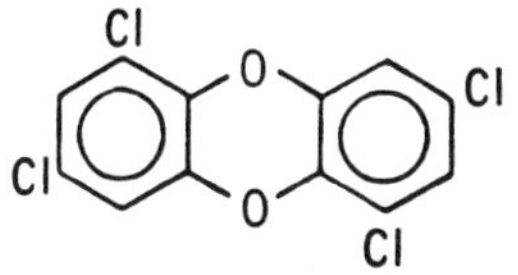

mirror plane

1368 (132'4') - TCDD

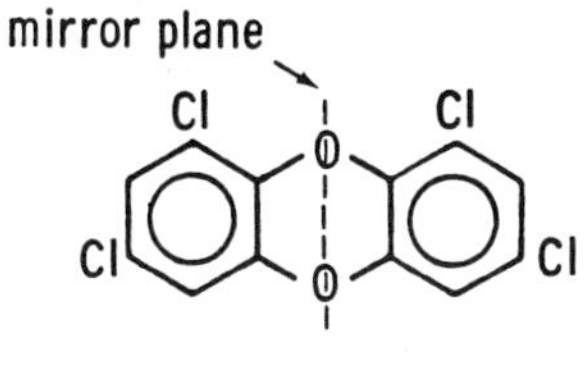

1379 (131'3') - TCDD

In a similar manner, 1,2,6,7 (1,2,3′,4′)-tetra-CDD elutes ahead of 1,2,8,9 (1,2,1′,2′)-tetra-CDD:

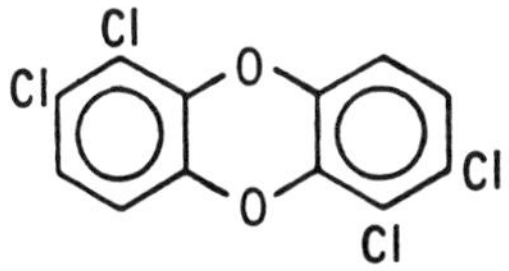

mirror plane

1267 (123'4') - TCDD

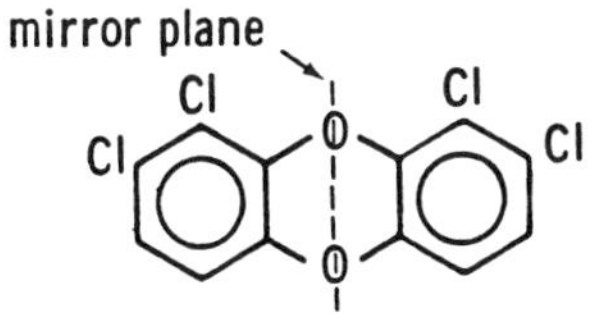

1289 (121'2') - TCDD

It should be noted that observation 1 (above) was stated by Buser [11] when he observed the following elution order for substitutions on ring 2 when ring 1 was completely chlorinated:

13- < 14- < 23- < 12- < 124- < 123- < 1234-

Observation 1 can be extended to include Buser's results for the elution order for three chlorines on a ring. The 1,2,4- substitution with one pair of vicinal chlorines would be expected to elute ahead of the 1,2,3- substitution, which has two pairs of vicinal chlorines. Observation 2 then takes the elution orders derived here and by Buser [11] and adds to them the observation related to molecular symmetry.

It should be emphasized that these observations are for the situations where one establishes the substitution on ring 1 and then determines the elution order for all possible 2-chlorine substitutions on ring 2. The concepts cannot yet be extended with good reliability to determining the elution order of tetra-CDD having different substitutions on both rings 1 and 2.

This whole procedure may be repeated for the elution order obtained by Buser and Rappe [5] for the 2:2-substituted tetra-CDD using a highly polar Silar-10C capillary column. The results from these correlations are shown in Table 3.7 and Figure 3.13. Once again the elution orders in Figure 3.13 maintain internal consistency, with one exception. The only significant difference between

Table 3.7. Elution Temperatures for the Thirteen 2:2-Substituted Tetra-CDD:Silar-10C

Name	*Alternative Name*	*Elution Temp. (°C)*
1,3,6,8	1,3,2′,4′	223.7
1,3,7,9	1,3,1′,3′; 2,4,2′,4′	226.4
1,3,7,8	1,3,2′,3′; 2,3,2′,4′	229.3
1,3,6,9	1,4,1′,3′; 1,4,2′,4′	230.9
1,2,6,8	1,2,2′,4′; 1,3,3′,4′	232.7
1,4,7,8	1,4,2′,3′	234.0
2,3,7,8	2,3,2′,3′	234.5
1,2,7,9	1,2,1′,3′; 2,4,3′,4′	236.0
1,2,7,8	1,2,2′,3′; 2,3,3′,4′	239.0
1,4,6,9	1,4,1′,4′	239.0
1,2,6,9	1,2,1′,4′; 1,4,3′,4′	240.7
1,2,6,7	1,2,3′,4′	242.3
1,2,8,9	1,2,1′,2′; 3,4,3′,4′	247.4

Silar-10C

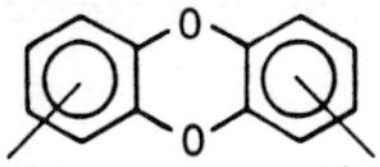

Ring #1 Ring #2

SUBSTITUTION PATTERN RING #2 ASCENDING ORDER OF ELUTION	SUBSTITUTION PATTERN—RING #1			
	13	14	23	12
	24			24
		24 = 13	24 = 13	
	13			13
	23	23	14*	23
	14	14	23*	14
	34			34
		34 = 12	34 = 12	
	12			12

*These two isomers elute closely to each other.

Figure 3.13. Elution Order for the 13 2:2-Substituted Tetra-CDD: Silar-10C.

Figures 3.12 and 3.13 is that on Silar-10C the 2,3- substitution on ring 2 now elutes ahead of the 1,4- substitution. This may give some indication as to the nature of the interaction of the tetra-CDD with the stationary phase.

One of the primary objectives of this work was to apply the concepts learned from the tetra-CDD to understanding the elution order of the 2:2-substituted tetra-CDF. To meet this objective, the complete elution orders for the twenty-one 2:2-substituted tetra-CDF on a SE-54 capillary column (data from Figure 3.10) are given in Figure 3.14. Since SE-54 and OV-101 stationary phases are comparable, it is reasonable to compare the observations made for the elution orders of the tetra-CDD with those for the tetra-CDF. The elution orders in Figure 3.14 are more complex than for the tetra-CDD but some very familiar patterns reappear. All of the substitutions on ring 2 that elute first are either 2,4- or 1,3-. This follows observation 1 for tetra-CDD that substitutions having neither vicinal hydrogens nor vicinal chlorines elute early. Likewise, substitutions on ring 2 eluting last are either 1,2- or 3,4-, in both cases having both vicinal hydrogens and vicinal chlorines.

When considering all the 2:2-substituted tetra-CDF isomers, two basic elution orders emerge from Figure 3.14. The distinguishing characteristic between the two orders appears to be whether or not there is a chlorine in the 1- position of ring 1. We have therefore defined two groups, the first containing those substitutions where the 1- or 1′- position is chlorinated. In a similar manner group 2 contains substitutions that are not chlorinated in the 1- or 1′- position:

GROUP 1

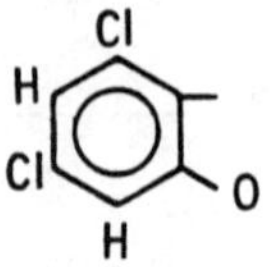

Cl, H, Cl, H, O

13-or 1'3'-

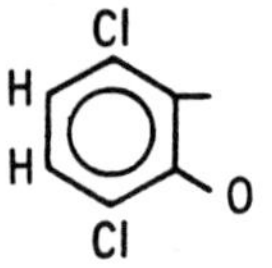

Cl, H, H, Cl, O

14- or 1'4'-

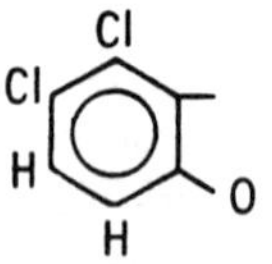

Cl, Cl, H, H, O

12- or 1'2'-

GROUP 2

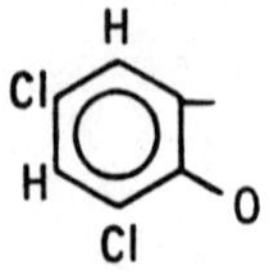

H, Cl, H, Cl, O

24- or 2'4'-

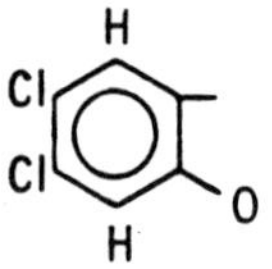

H, Cl, Cl, H, O

23- or 2'3'-

H, H, Cl, Cl, O

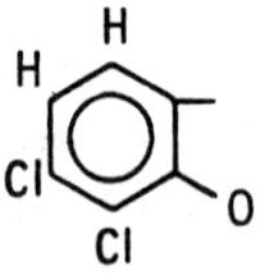

34- or 3'4'-

Note that each group contains:

1. a substitution with no vicinal chlorines or hydrogens (1,3- or 2,4-);
2. a substitution with one pair of either vicinal chlorines or hydrogens (1,4- or 2,3-); and
3. a substitution with both vicinal chlorines and vicinal hydrogens (1,2- or 3.4-).

If the observations from tetra-CDD carry over to tetra-CDF, one would predict the following elution orders within a group:

Group 1: 1,3- < 1,4- < 1,2-

Group 2: 2,4- < 2,3- <3,4-

SE-54

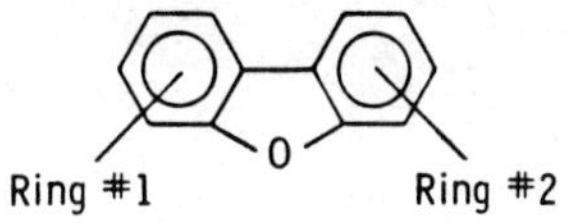

Ring #1 O Ring #2

SUBSTITUTION PATTERN RING #2 ASCENDING ORDER OF ELUTION	SUBSTITUTION PATTERN—RING #1					
	24	23	34	13*	14*	12*
↓	13	13	13	24	24	24
	14	14	14	23	23	23
	24	24	24	34	34	34
	12	12	12	13	13	13
	23	23	23	14	14	14
	34	34	34	12	12	12

*These substitutions have a chlorine in the 1-position and are classified as Group #1. The remaining substitutions (24, 23, 34) are classified as Group #2.

Figure 3.14. Elution Order for the 2:2-Substituted Tetra-CDF: SE-54.

As an example for tetra-CDF, the elution order for substitutions on ring 2 when ring 1 is 1,3- (group 1) is as follows:

2,4- < 2,3- < 3,4-	<	1,3- < 1,4- < 1,2-
members of group 2		members of group 1

Not only are the elution orders within the groups in exactly the predicted order, but the groups are segregated from each other. It should be noted in this example that ring 1 has a substitution from group 1 and that the three latest-eluting substitutions on ring 2 are also from group 1. In fact, in all cases where the substitution on ring 1 are of group 1, the later-eluting substitutions on ring 2 are from group 1 and vice versa. This is consistent with observation 2, in which later-eluting isomers contain a mirror plane through the oxygen(s). As an example, 1,3,7,9 (1,3,1′,3′)-tetra-CDF and 2,4,6,8 (2,4,2′,4′)-tetra-CDF elute after 1,3,6,8 (1,3,2,′,4′)-tetra-CDF in a manner similar to 1,3,7,9 (1,3,1′,3′)-tetra-CDD eluting after 1,3,6,8 (1,3,2′,4′)-tetra-CDD:

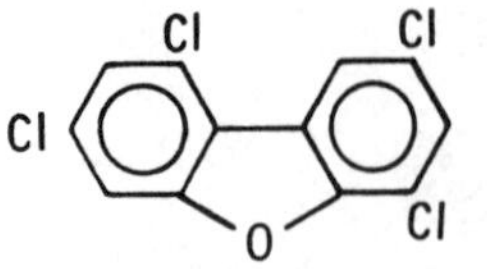

1368(132'4')-TCDF

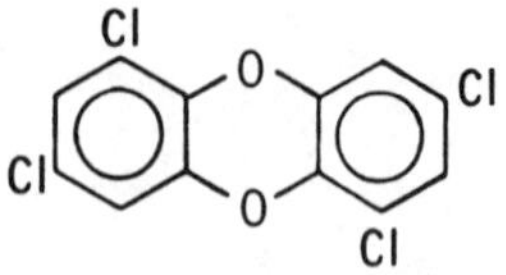

1368 (132'4')-TCDD

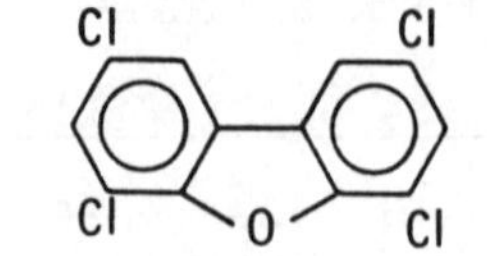

2468 (242'4')-TCDF

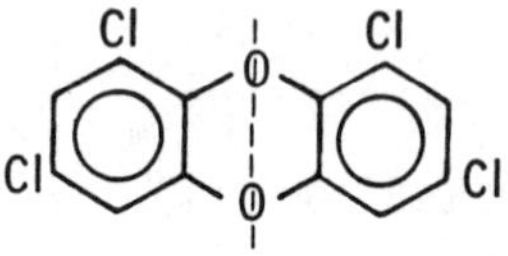

1379 (131'3')-TCDD

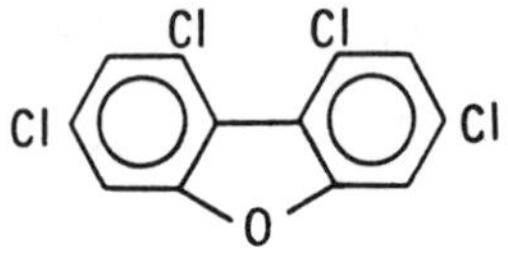

1379 (131'3')-TCDF

and 3,4,6,7 (3,4,3′,4)-tetra-CDF and 1,2,8,9 (1,2,1′,2′)-tetra-CDF elute after 1,2,6,7 (1,2,3′,4′)-tetra-CDF again in a manner similar to 1,2,8,9-tetra-CDD eluting after 1,2,6,7-tetra-CDD as discussed earlier.

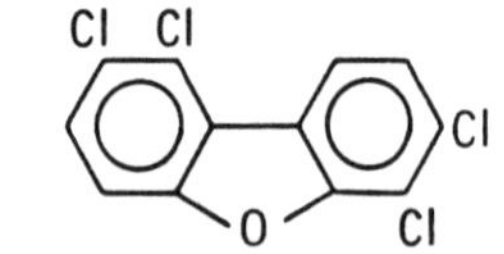

1267 (123'4')-TCDF

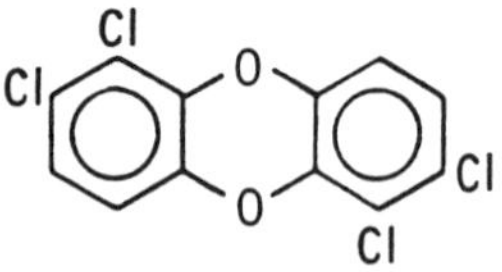

1267 (123'4')-TCDD

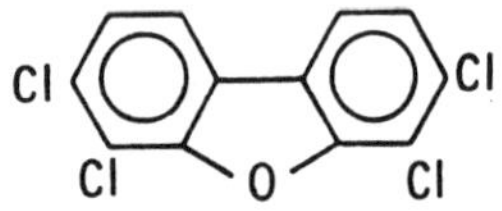

3467 (343'4')-TCDF

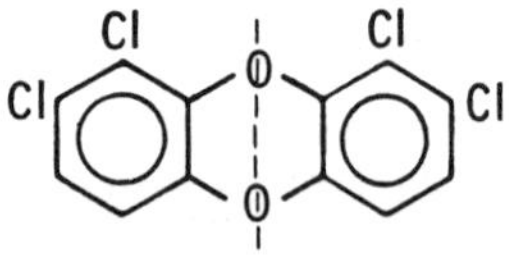

1289 (121'2')-TCDD

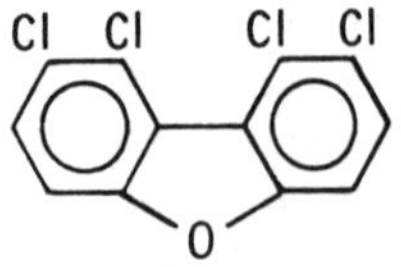

1289 (121'2')-TCDF

Considering the other case where ring 1 is now 2,4- (a member of group 2) the elution order for ring 2 is observed to be:

group 1

1,3- < 1,4- < 2,4- < 1,2- < 2,3- < 3,4-

group 2

Notice that the elution order within a group is as predicted, but now there is some overlap of the two groups. The segregation in the elution order for groups 1 and 2 when ring 1 was of group 1 possibly could be related to a steric effect. If both rings 1 and 2 are of group 1, then chlorines occupy the 1- and 9- (or 1- and 1′-) positions and are near each other:

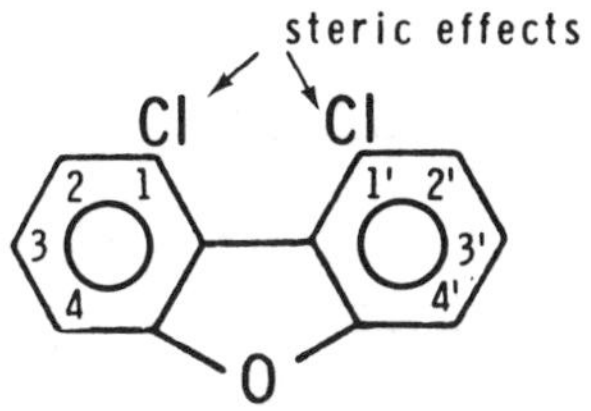

steric effects

Cl Cl

2 1 1' 2'

3 3'

4 4'

O

This effect may be similar to the hindered rotation in PCB having ortho and ortho′ chlorines, which prevents these PCB from becoming planar. This effect could force these tetra-CDF out of planarity and possibly cause them to elute late. This phenomenon will be discussed further when we deal with the 3:1-substituted tetra-CDF.

What is most important here is that the consistency in the data in figure 3.14 gave added confidence in the structural assignment of the 2:2-substituted tetra-CDF. This type of check on the data was one of the primary objectives of this portion of this study.

In a manner similar to the tetra-CDD one can also consider the data for tetra-CDF generated on a SP-2330 capillary column. This is a highly polar column, similar to the Silar-10C column used by Buser and Rappe [5]. The elution orders for the various 2:2-substituted tetra-CDF are given in Figure 3.15. Once again, there was consistency in the elution orders for substitutions on ring 2. However, it should be pointed out that when the substitution on ring 1 was of group 2, the elution order for ring 2 segregated into first group 1 and then group 2 without the overlap observed for the SE-54 column. When ring 1 was of group 1, the elution order of ring 2 showed an overlap of group 2 and group 1, again in contrast to the SE-54 results. This behavior must have been due to the nature of the interaction of the tetra-CDF with the polar SP-2330 column, while the consistency in the data in Figure 3.15 serves as a check on the structural assignments of the tetra-CDF.

SP-2330

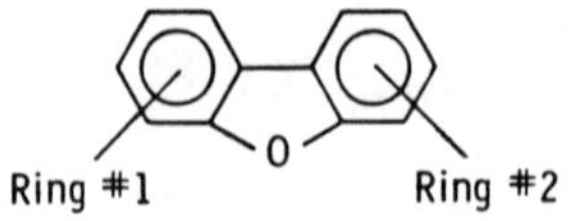

Ring #1 O Ring #2

SUBSTITUTION PATTERN RING #2 ASCENDING ORDER OF ELUTION ↓	SUBSTITUTION PATTERN—RING #1					
	24	23	34	13*	14*	12*
	13	13	13	24	24	24
	14	14	14	23	23	23
	12	12	12	13	13	34**
	24	24	24	34	34	13**
	23	23	23	14	14	14
	34	34	34	12	12	12

*See note on Figure 3.14, starred substitutions are of Group #1 and unstarred are of Group #2.

**These isomers elute closely to each other.

Figure 3.15. Elution Order for the 2:2-Substituted Tetra-CDF: SP-2330.

The elution order for the eight 3:1-substituted tetra-CDD on OV-101 as reported by Buser and Rappe [5] is given in Table 3.8. Figure 3.16 shows the elution order for the substitutions on the two rings. As expected from observation 1 and the findings of Buser [11], the 1,2,4- substitution (one pair of vicinal chlorines) always elutes ahead of the 1,2,3- substitution (two pairs of vicinal chlorines). Figure 3.17 shows the comparable data for the 3:1 substituted tetra-CDF. The elution orders for the substitutions on ring 2 with ring 1 having either a 2- or 3- position chlorine are consistent in light of the results for the 3:1-substituted tetra-CDD. What is most significant is that when ring 1 has a 1-chlorine, the 2,3,4-substitution on ring 2 elutes ahead of both 1,3,4- and 1,2,4-. There is some precedent for this unusual behavior. It was postulated before that the 1- and 9- (or 1- and 1′-) positions of tetra-CDF can possibly experience steric effects if both positions were substituted with chlorines. Thus with the 1- position filled on ring 1, the only substitution on ring 2 that does not have the 1′- position filled is the 2,3,4- substitution. All other substitutions on ring 2 have the 1′-position filled, which could cause these tetra-CDF to be bent out of planarity. This could result in late elution of these isomers as observed in Figure 3.17 and also observed for the 2:2-substituted tetra-CDF.

Table 3.8. Elution Temperatures for the Eight 3:1-Substituted Tetra-CDD:OV-101

Name	*Alternative Name*	*Elution Temp. (°C)*
1,2,4,7	1,2,4,3′	218.3
1,2,4,8	1,2,4,2′	218.3
1,2,4,6	1,2,4,4′	218.6
1,2,4,9	1,2,4,1′	218.6
1,2,3,6	1,2,3,4′	220.8
1,2,3,7	1,2,3,3′	221.2
1,2,3,8	1,2,3,2′	221.2
1,2,3,9	1,2,3,1′	221.8

OV-101

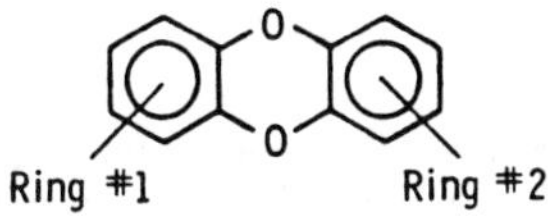

O
O
Ring #1
Ring #2

SUBSTITUTION PATTERN—RING #1

SUBSTITUTION PATTERN RING #2 ASCENDING ORDER OF ELUTION ↓	3	2	1	4
	124	124	124	124
	123	123	123	123

Figure 3.16. Elution Order for the Eight 3:1-Substituted Tetra-CDD: OV-101.

SE-54

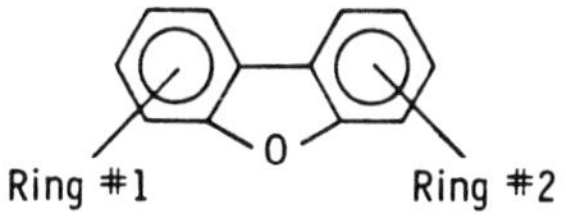

O
Ring #1
Ring #2

SUBSTITUTION PATTERN—RING #1

SUBSTITUTION PATTERN RING #2 ASCENDING ORDER OF ELUTION ↓	3	2	1	4
	134	134	234	134
	124	124	134	124
	123	123	124	123
	234	234	123	234

Figure 3.17. Elution Order for the Sixteen 3:1-Substituted Tetra-CDF: SE-54.

Table 3.9 and Figure 3.18 show the elution characteristics of the 3:1-substituted tetra-CDD as reported by Buser and Rappe [5] on Silar-10C. The elution order as depicted in Figure 3.18 is comparable to that obtained on OV-101 as shown in Figure 3.18. The elution order for the 3:1-substituted tetra-CDF obtained on a SP-2330 column are shown in Figure 3.19. Once again, the early elution was observed for the tetra-CDF isomer in which ring 1 has a 1- chlorine and ring 2 is 2,3,4-substituted.

In summary, the elution data for the 3:1-substituted tetra-CDF are readily explainable in terms of the observations noted for the tetra-CDD. In addition, the steric effect postulated to explain the late elution of 2:2-substituted tetra-CDF having chlorine in the 1- and 9- (1- or 1′-) positions also appears to be a factor for the 3:1-substituted tetra-CDF. This leads to the final observation for tetra-CDF that chlorines in both the 1- and 9- (1- or 1′-) positions result in increased retention of these tetra-CDF.

Table 3.9. Elution Temperatures for the Eight 3:2-Substituted Tetra-CDD:Silar-10C

Name	*Alternative Name*	*Elution Temp. (°C)*
1,2,4,7	1,2,4,3′	230.9
1,2,4,8	1,2,4,2′	230.9
1,2,4,6	1,2,4,4′	235.0
1,2,4,9	1,2,4,1′	235.0
1,2,3,8	1,2,3,2′	235.0 or 235.5
1,2,3,7	1,2,3,3′	235.5 or 235.0
1,2,3,6	1,2,3,4′	236.1
1,2,3,9	1,2,3,1′	239.7

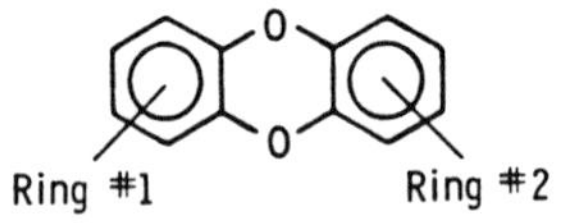

Figure 3.18. Elution Order for the Eight 3:1-Substituted Tetra-CDD: Silar-10C.

SP-2330

Ring #1 O Ring #2

SUBSTITUTION PATTERN RING #2 ASCENDING ORDER OF ELUTION	SUBSTITUTION PATTERN—RING #1			
	3	2	1	4
	134	134	234	134
	124	124	134	124
	123	123	124	123
	234	234	123	234

Figure 3.19. Elution Order for the Sixteen 3:1-Substituted Tetra-CDF: SP-2330.

CONCLUSION

This chapter has dealt with the examination of tetra-CDF using techniques that can be used even when minute amounts of sample are available. The parameters that have been examined in some detail include the photolytic degradation behavior and the GC retention behavior. Both techniques yield explainable results. Photolysis results can be predicted to an extent that the technique can be used to confirm the identity of an unknown tetra-CDF. The elution behavior of tetra-CDF is also predictable, with three basic observations of vicinal atoms, symmetry and steric effects used to explain the data.

REFERENCES

1. Rappe, C. In: *Halogenated Biphenyls, Terphenyls, Naphtalenes, Dibenzofurans and Related Products,* R.D. Kimbrough, Ed. (New York: Elsevier/North Holland Biomedical Press, 1980), pp. 48–68.
2. Buser, H.R., and C. Rappe. *Chemosphere* 3:157 (1979).
3. Buser, H.R., and H. Bosshardt. *Chemosphere* 1:109 (1978).
4. Mazer, T., and F.D. Hileman. *Chemosphere* 11:651 (1982).
5. Buser, H.R., and C. Rappe. *Anal. Chem.* 52:2257 (1980).
6. Hicks, O., Monsanto Company, St. Louis, MO. Personal communication.
7. Gara, A., K. Anderson,C. A.Nilsson and A. Norstrom. *Chemosphere* 4:365 (1981).
8. Rappe, C., Department of Organic Chemistry, University of Umea, Umea, Sweden. Personal communication.

9. Hey. D.H. In: *Vistas in Free Radical Chemistry,* W.A. Waters, Ed. (New York: Pergamon Press, 1959).
10. Rappe, C., A. Gara and H.R. Buser. *Chemosphere* 12:981 (1978).
11. Buser, H.R. *J. Chromatog.* 114:95 (1975).

4

Synthesis of Uniformly Labeled ^{13}C-Polychlorinated Dibenzofurans

Robert A. Bell

Polychlorinated dibenzofurans (PCDF) have been identified as components of the rice bran oil that produced symptoms known as "yusho" in residents of southwestern Japan in 1968, and Taichung, Taiwan, in 1979 [1,2]. As these materials have been shown to arise from high-temperature oxidation of certain polychlorinated biphenyls (PCB) [3,4], and as they have been found at low levels in environmental samples [5], analytical procedures for their detection are needed.

This chapter describes efforts to improve the identification and quantification of PCDF. This research has been directed toward developing a set of isomer standards including some fully labeled ^{13}C-PCDF that must be regarded as essential components of any method based on gas chromatography/mass spectrometry (GC/MS).

BACKGROUND

Dibenzofurans are tricyclic planar molecules containing eight possible positions for chlorination. The basic nucleus and numbering scheme are shown in Figure 4.1. "Isomer" will be used to designate any of the 135 possible chlorinated dibenzofurans [e.g., 2,3,7,8-tetrachlorodibenzofuran (tetra-CDF), or 2-CDF]. "Homolog" will refer to isomers within a series containing equal numbers of chlorine atoms (e.g., 1,2,3,4-, 2,3,4,6- and 2,3,7,8-tetra-CDF). Also shown in Figure 4.1 are the contributions of each homologous series to the total number of isomers. Note, for instance, that tetra-CDF comprise 38 separate isomers, each with distinct chromatographic characteristics. Our work is aimed at the development of analytical procedures able to distinguish individual isomers.

Of late, capillary GC columns have been prepared that allow the separation of a large number of PCDF isomers. Quantification, however, by electron capture detectors (ECD) is fraught with many difficulties. Although ECD are exceedingly sensitive to chlorinated materials, they offer no discrimination between

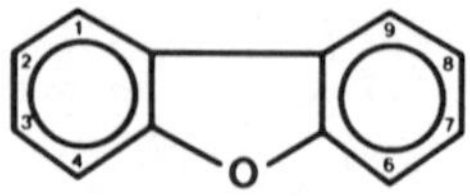

DIBENZOFURAN

DEGREE OF CHLORINATION	NO. OF ISOMERS
MONO -	4
DI -	16
TRI -	28
TETRA -	38
PENTA -	28
HEXA -	16
HEPTA -	4
OCTA -	1
	135 TOTAL

Figure 4.1. Basic Nucleus and Numbering Scheme of Dibenzofurans.

PCDF and the plethora of other chlorinated hydrocarbons that might also be present in a mixed sample. Thus, laborious and complicated extraction/separation methods have been developed [6]. These suffer from nonuniform recoveries over a range of isomers and require a high degree of technical expertise to produce meaningful data.

A second approach is the use of MS as the detection system for capillary GC. At medium resolution, most interferences (other chlorinated hydrocarbons) may be excluded on the basis of precise mass. Further, fragmentation patterns of compounds of interest may be used to verify certain aspects of structure [7]. Once GC retention time has been established by external standard, individual isomers may be measured readily by single-ion monitoring (SIM). This method avoids some of the extensive cleanup/separation steps necessary for a pure GC analysis and thereby offers the potential for better recoveries. However, increases in MS resolution (hence the ability to exclude more interferences based on precise mass) result in diminished signals as fewer ions are reaching the detector. High resolution may therefore produce some decrease in sensitivity. For complicated environmental samples, however, GC/MS methods appear to be most promising.

Both methods of detection are subject to yet another problem: discrimination between internal standard and sample unknown. Ideally, an internal standard should duplicate the chemical and chromatographic behavior of the unknown, but remain identifiable by some other technique as separate material.

Pure isomers of tetra-CDF are optimal standards for identification of GC retention times, but are by definition indistinguishable from unknowns by ECD or MS. ^{13}C-labeled individual isomers are mass-shifted by the amount of ^{13}C they contain, and are thus easily distinguished by MS. This mass-difference from the normal carbon counterparts produces no substantial difference in retention time or in other chromatographic characteristics.

It is necessary to fully label dibenzofurans with ^{13}C to completely separate the envelopes of masses created by ^{35}Cl and ^{37}Cl. As shown in Figure 4.2, the envelope for penta-CDF extends over nine mass units at levels of $\geq 1\%$. Partial ^{13}C labeling (e.g., only one ring—yielding a six-mass-unit shift) may result in overlap, especially in the more highly chlorinated isomers. To avoid any such convolution of standard and unknown, all carbon positions have been enriched $\geq 99\%$ with ^{13}C. The mass spectrum of the unlabeled carbon and ^{13}C-2,3,7,8-tetra-CDF is found in Figure 4.3. Note the lack of any mass contribution at m/e 306 in the ^{13}C-enriched sample. Also shown is a computer-simulated parent peak for tetra-CDF with 90 and 99% ^{13}C enrichment to verify the degree of label in the actual sample.

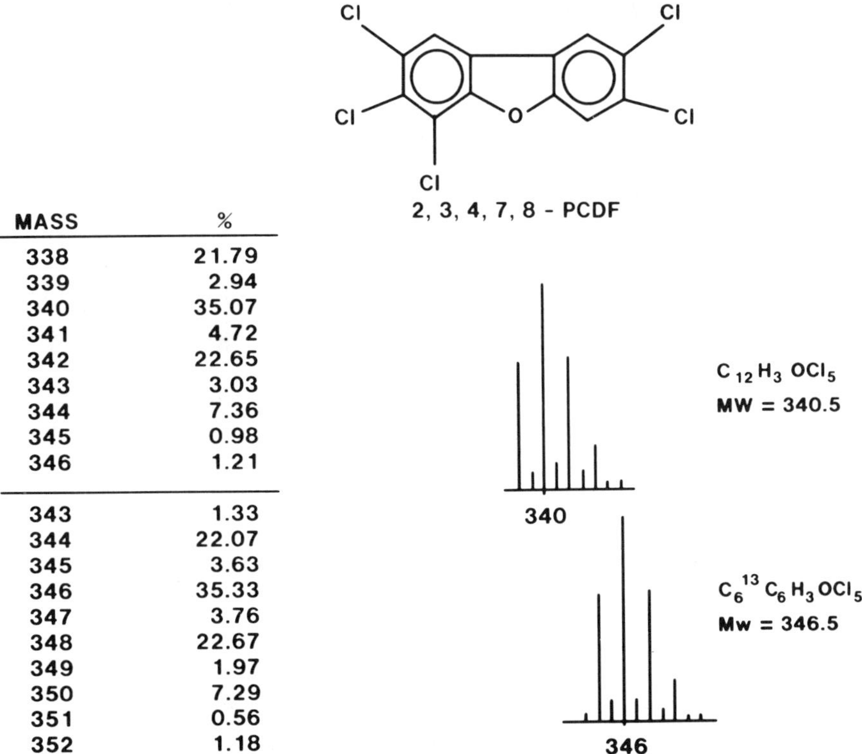

MASS	%
338	21.79
339	2.94
340	35.07
341	4.72
342	22.65
343	3.03
344	7.36
345	0.98
346	1.21
343	1.33
344	22.07
345	3.63
346	35.33
347	3.76
348	22.67
349	1.97
350	7.29
351	0.56
352	1.18

Figure 4.2. Envelope for Penta-CDF.

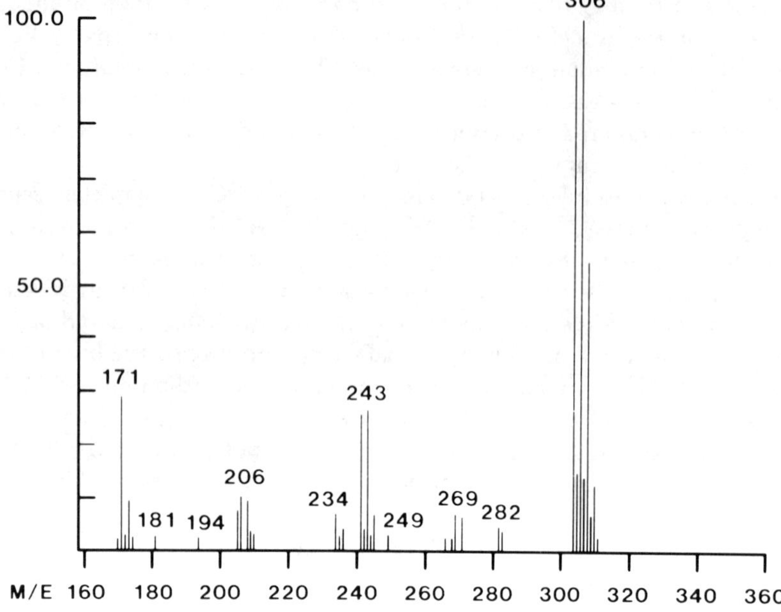

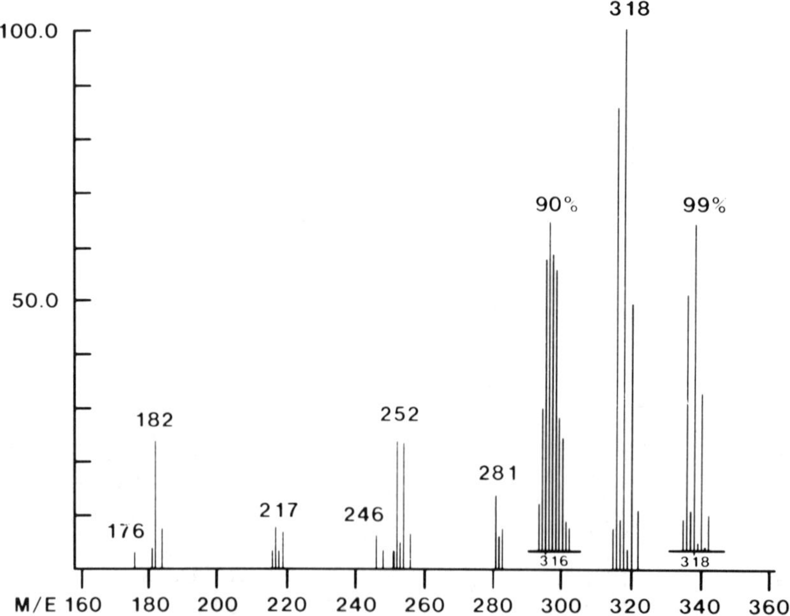

Figure 4.3. Mass Spectra for 2,3,7,8-Tetra-CDF (top) and ^{13}C-2,3,7,8-Tetra-CDF (bottom).

Coelution of individual isomers results in erroneous identification and, hence, faulty quantification. To avoid this possibility, we are synthesizing individual unlabeled isomer standards for all PCDF from tetra- through octa-CDF. These eighty-seven standards are used to quantify mixtures of ^{13}C-labeled PCDF and to establish unambiguous capillary GC retention data.

SYNTHESIS

To house reactions and storage of potentially toxic materials a special synthesis facility was constructed. Access is limited to those trained in proper procedures for its use. A floor plan is shown in Figure 4.4. Synthetic pathways were chosen to minimize both the handling of suspect materials and the appearance of side products.

Diphenyl ethers proved to be excellent penultimate reagents. Cyclization was effected through known photochemical or organometallic methods [9,10]. Ether synthesis was accomplished by phenoxide displacement on an iodonium salt or through the Ullman ether procedure. Yields were quite high with iodonium salts and moderate for Ullman reactions. ^{13}C-labeled starting materials, benzene and phenol, were checked for isotopic purity by MS before use.

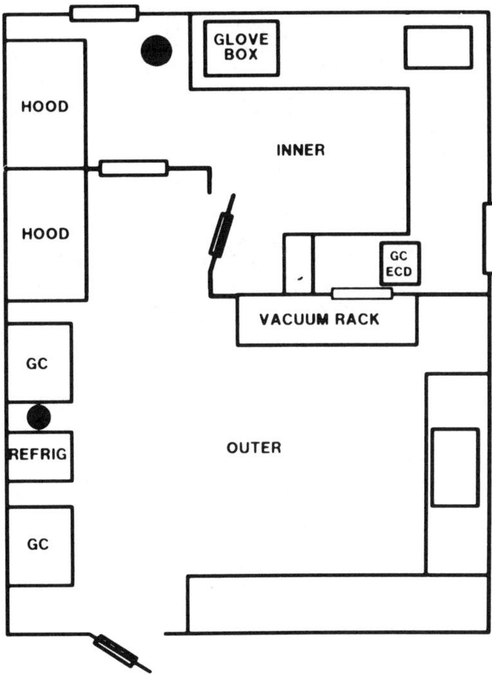

Figure 4.4. Floor Plan of Isomer Synthesis Laboratory.

The first ^{13}C-labeled CDF formed was 1,2,3,4-tetra-CDF. Initial attempts with unlabeled materials indicated that the necessary ether (2,3,4,5-tetrachlorodiphenyl ether) would be difficult to synthesize from phenol and benzene. In particular, the perchlorination of a blocked phenol could not be accomplished without substantial loss of the ortho-blocking group (Br, t-butyl). The blocking group was necessary to reintroduce an ortho hydrogen required for cyclization. A second pathway via pentachlorophenol (penta-CP), from perchlorination, and diphenyl iodonium hexafluorophosphate (Figure 4.5), led to 2,3,4,5,6-pentachlorodiphenyl ether. This was cyclized via photolysis in hexane to yield 1,2,3,4,-tetra-CDF and a small amount of a tri-CDF (presumably 1,2,4-tri-CDF) [8]. The ^{13}C-labeled analog was formed in similar fashion.

Figure 4.6 indicates the synthetic path for ^{13}C-2,3,7,8-tetra-CDF. All reactions proceed in reasonable yields except the Ullman ether synthesis, which for the labeled material was ~30%. After purification by column chromatography over Kieselgel 60 silica gel, the 3,4,3,',4'-tetrachlorodiphenyl ether was cyclized to 2,3,7,8- and 1,2,7,8-tetra-CDF by palladium acetate in acetic acid/methanesulfonic acid. Synthetic details may be found in the Experimental section.

2,3,7,8-tetra-CDF can be chlorinated readily to higher CDF using antimony pentachloride in CCl_4 [8]. As expected, the mixture of isomers obtained includes both penta- and four- hexa-CDF, which exhibit 2,3,7,8,-substitution. The ^{13}C-2,3,7,8-tetra-CDF mixture was chlorinated to form a mixture of about 15–20

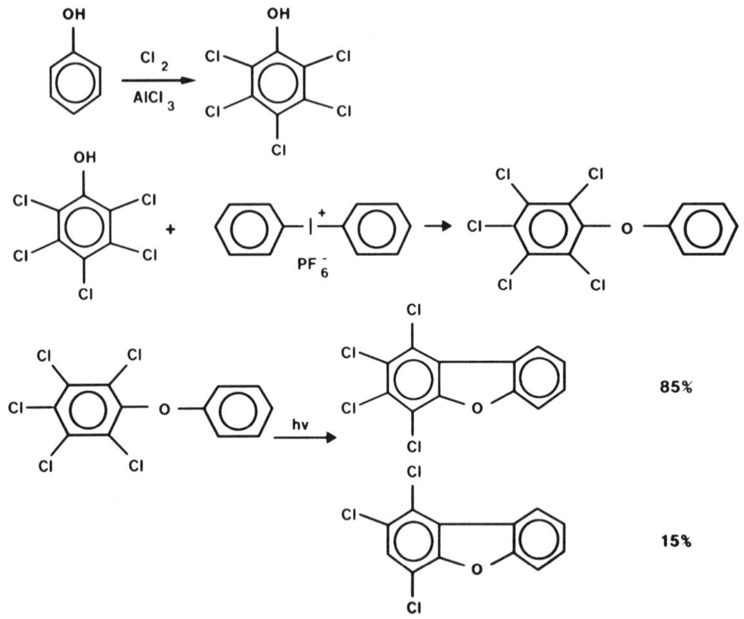

Figure 4.5. Synthetic Route to ^{13}C-1,2,3,4-Tetra-CDF.

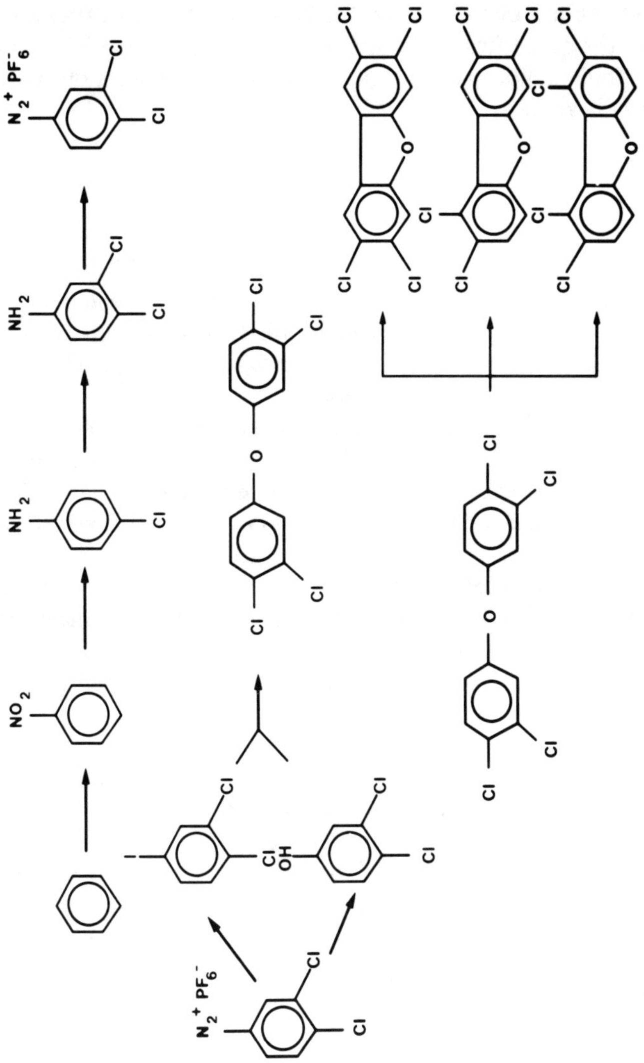

Figure 4.6. Synthetic Route to ^{13}C-2,3,7,8-Tetra-CDF.

higher isomers. These were identified and quantified by comparison to individual unlabeled carbon isomers. Any differences in response factor between ^{13}C and unlabeled carbon are thus canceled when these ^{13}C standards are used to quantify unlabeled carbon PCDF in environmental samples.

We believe the use of ^{13}C-labeled PCDF standards constitutes a significant improvement in the quantification of these compounds. Through the availability of individual isomer standards, the risk of incorrect identification due to coelution is substantially reduced.

EXPERIMENTAL

In consideration of the reported toxicity of 2,3,7,8-tetra-CDF in various animal species [11,12] it is recommended that these compounds be treated as potentially hazardous. The isolation facility constructed for this work contained a carbon-filtered high-velocity fume hood in which synthesis was done, and a glove box in which solid samples were weighed and stored.

Mass spectral data were obtained at 70 eV on a Vg ZAB spectrometer in the positive ion EI mode with a Varian 1400 capillary (DB-1) gas chromatograph feeding directly into the source. GC was obtained on a Shimadzu GC-Mini-2 fitted with a ^{63}Ni ECD and a 60-m glass SP-2330 column. Nitrogen flowrate was 2.2 mL/min through the capillary and a split ratio of 50:1 was used. Column temperature was isothermal at 240°C.

Uniformly labeled (UL) ^{13}C-phenol and -benzene was obtained from KOR Isotopes, Inc. UL ^{13}C-benzene was also obtained from Cambridge Isotope Laboratories, Inc. All labeled starting materials were examined by MS before synthesis to verify isotopic purity.

UL ^{13}C-Diphenyl Iodonium Hexafluorophosphate

A mixture of 1.0 g (11.9 mmol) UL ^{13}C-benzene, 1.27 g (5.9 mmol) potassium iodate, 2.5 mL acetic anhydride and 3 mL dichloromethane was placed in a 50-mL round-bottom flask equipped with a thermometer and stirred magnetically. The flask was cooled to −10°C using a dry ice/acetone bath; the contents were kept under nitrogen to avoid water condensation. Concentrated sulphuric acid 1.27 mL) was added dropwise with stirring, while maintaining the temperature of the mixture at −10 to −5°C. After the addition, the mixture was held at −10°C for 2 hr, then allowed to warm slowly to room temperature. It was then stirred overnight.

With cooling, 10 mL of water was added to the reaction mixture. The dichloromethane layer was separated and discarded. To the aqueous layer was added an excess of hexafluorophosphoric acid (65% in water). The precipitate that formed was collected by filtration, washed with cold water and dried overnight in vacuo in the dark. The yield was 2.5 g of beige crystals, 70%.

UL ^{13}C-Penta-CP

A 0.5 g-sample of UL ^{13}C-phenol (5 mmol) was combined with 50 mL anhydrous aluminum chloride in a 100-mL Fisher-Porter pressure vessel. The apparatus was charged with 50 psi of chlorine gas and heated to 70°C. As chlorination occurred a precipitate formed. Heating was increased to 150°C and the charge was vented and replaced with fresh chlorine. After a total reaction period of 1.5 hr the vessel was cooled to room temperature, yielding a yellowish solid. This was recrystallized from ethanol to yield 0.69 g of UL ^{13}C-penta-CP, 51% yield.

UL ^{13}C-2,3,4,5,6-Pentachlorodiphenyl Ether

A 300-mg sample of UL ^{13}C-penta-CP (1.1 mmol) was dissolved in 15 mL water and 3 mL of 1 *M* aqueous potassium hydroxide. A 470-mg aliquot of UL ^{13}C-diphenyl iodonium hexafluorophosphate was added slowly with magnetic stirring. After the addition was complete the mixture was heated to gentle reflux for 1 hr. The mixture was cooled to room temperature and extracted twice with 25-mL portions of diethyl ether, which were combined and dried over anhydrous magnesium sulfate. After filtration this solution was evaporated to yield 370 mg of yellowish microcrystals, which proved to be a 1:1 complex of pentachlorodiphenyl ether and iodobenzene (80%). A 200-mg sample of the complex was sublimed at 140°C, 0.01 mm Hg, to yield 90 mg of UL ^{13}C-2,3,4,5,6-pentachlorodiphenyl ether as a white crystalline solid (82% based on 200-mg complex).

UL ^{13}C-1,2,3,4-Tetra-CDF

In fashion similar to that of Norstrom et al. [9], 70 mg of pentachlorodiphenyl ether was dissolved in 100 mL of *n*-hexane and purged with N_2 for 15 min. The solution was irradiated for approximately 30 min using a medium-pressure mercury lamp (254 nm radiation) with a quartz condenser. Progress of the cyclization was monitored by GC to maximize tetra-CDF. Evaporation of the solvent and fractional recrystallization of the residue yielded 24 mg of yellowish powder, which by GC analysis was 85% 1,2,3,4-tetra-CDF and 15% of an unidentified tri-CDF. The retention time of the tetra-CDF was identical to 1,2,3,4-tetra-CDF (normal carbon) formed by palladium cyclization from 2,3,4,5-tetrachlorodiphenyl ether.

UL ^{13}C-Nitrobenzene

Following the method of Davis and Cook [13], 8.4 g of UL ^{13}C-benzene was combined with 35 mL dichloromethane and placed in a 100-mL round-bottom flask filled with a reflux condenser. A 15-ml portion of 80% sulfuric acid was added and the contents were stirred at gentle reflux (41 °C). A 7.35-g aliquot of 90% nitric acid (0.105 mol) was added dropwise. After the addition was complete the mixture was stirred for 0.5 hr. The dichloromethane layer was separated, passed through a short silica gel column to remove occluded acid, washed with saturated potassium carbonate solution and dried over magnesium sulfate. On evaporation of solvent, 12.5 g of UL ^{13}C-nitrobenzene was recovered (97%).

UL ^{13}C-Aniline Hydrochloride

A 5-g sample of UL ^{13}C-nitrobenzene (38.8 mmol) was combined with 50 mL of 95% ethanol and 40 mg of 5% palladium on carbon in a 100-mL round-bottom flask. The contents were heated to 70 °*C* on a water bath and stirred magnetically. Hydrazine hydrate, 3.06 g (61.1 mmol), was added dropwise, resulting in gas evolution and frothing. When the addition was complete the mixture was stirred for 0.5 hr at 70 °C. The solution was allowed to cool to room temperature, passed through a silica gel column to remove palladium and treated with anhydrous hydrogen chloride. Partial evaporation of solvent followed by addition of hexane precipitated 5.0 g (36.9 mmol) of UL ^{13}C-aniline hydrochloride (95%).

UL ^{13}C-*p*-Chloroaniline

Following the method of Crocker and Walser [14], 3.0 g of UL ^{13}C-aniline hydrochloride (22.1 mmol) was dissolved in 10 mL concentrated hydrochloric acid and added to a solution of 3.77 g copper(II) chloride dihydrate (22.1 mmol) in 25 mL concentrated hydrochloric acid. A yellowish-gold precipitate formed. The suspension was warmed to 90 °C, and oxygen and anhydrous hydrogen chloride were slowly bubbled through the mixture. After 24 hr the mixture was cooled in ice and a solid was filtered off. The solid was dissolved in water, filtered to remove insolubles and added to an excess of cold dilute ammonia solution. The product formed a white precipitate and was filtered from the deep blue mother liquor, yielding 1.9 g (64%), melting point (m.p.) = 70 °C.

UL ^{13}C-3,4-Dichloroaniline

Following the procedure of Suthers et al. [15], 1.80 g UL ^{13}C-*p*-chloroaniline (13.5 mmol) was mixed with 3.60 g anhydrous aluminum chloride (27 mmol, Baker analyzed reagent) in a 100-mL Fisher-Porter vessel. The apparatus was warmed to 75 °C, whereupon the mixture formed a greenish liquid. Introduction of chlorine gas produced a maroon color. The reaction process was monitored by GC on aliquots drawn periodically. After 1 hr the yield appeared maximized and the reaction was cooled to room temperature. Water was added with cooling (exothermic) and the solution was filtered. The filtrate was added to an excess of cold dilute ammonia solution, which was subsequently extracted with diethyl ether (two 50-mL portions). These were combined, dried over magnesium sulfate, and evaporated to dryness. The solid was dissolved in methanol and fractionally recrystallized to yield 1.25 g of 3,4-dichloroaniline (55%), m.p. = 71°C.

UL ^{13}C-3,4-Dichlorobenzene Diazonium Hexafluorophosphate

A 1.0-g sample of UL ^{13}C-3,4-dichloroaniline (5.95 mmol) was mixed with 20 mL of 1 *M* HCl with stirring and maintaining the temperature at 10°C. A 0.415-g (6.01 mmol) aliquot of sodium nitrite dissolved in 10 mL of water was added dropwise. After 0.5 hr of stirring, the solution was filtered and the filtrate was added to 25 mL of cold water containing 2.7 g of hexafluorophosphoric acid (65% in water, 12 mmol). The white precipitate that formed immediately was filtered, washed with cold water, with diethyl ether, and then dried in vacuo. The yield was 1.65 g (85%).

UL ^{13}C-3,4-Dichloroiodiobenzene

A 625-mg sample of iodine (2.54 mmol) and 330 mg of lithium iodide (2.54 mmol) were dissolved in 25 mL of dry tetrahydrofuran. With stirring, 800 mg of UL ^{13}C-3,4-dichlorobenzene diazonium hexafluorophosphate was slowly added. Nitrogen was evolved and stirring was continued until all gas formation had ceased (0.5 hr). The solution was reduced in volume to 5 mL of hexane, and 25 mL water was added. With stirring, sodium sulfite was added until the color of iodine disappeared. The hexane layer was removed, dried over magnesium sulfate and reduced in volume to 5 mL. This was chromatographed over Kieselgel 60 using hexane as an eluant to obtain 567 mg of 3,4,-dichloroiodobenzene (83%).

UL ^{13}C-3,4-di-CP

Following the procedure of Horning et al. [16], 170 mg of potassium carbonate (1.23 mmol) was added to 10 mL trifluoroacetic acid. An 800-mg aliquot of UL ^{13}C-3,4-dichlorobenzenediazonium hexafluorophosphate (2.46 mmol) was added and the solution was stirred at reflux until it no longer gave a positive test for diazonium salts (alkaline β-naphthol). The solution was added to 50 mL of water and stirred for 2 hr. The solution was extracted using two 25-mL portions of diethyl ether, which were combined and treated with 50 mL of 1 *M* KOH solution. The basic layer was separated, acidified with 1 *M* HCl and extracted with two 25-mL portions of diethyl ether, which were dried over magnesium sulfate. On evaporation, 250 mg of 3,4-dichlorophenol was recovered (60%).

UL ^{13}C-3,4,3′,4′-Tetrachlorodiphenyl Ether

Ullman coupling of 3,4,-dichloroiodobenzene and 3,4-di-CP was accomplished through the copper phenoxide modification of Whitesides et al. [17]. UL ^{13}C-copper(I) 3,4-dichlorophenoxide was prepared at low temperature from methyl copper via literature methods. To the solution (1.48 mmol) under N_2 in pyridine at −30°C was added 500 mg of UL ^{13}C-3,4-dichloroiodobenzene (1.79 mmol) dissolved in 5 mL pyridine. The solution was slowly heated under N_2 and allowed to gently reflux for 16 hr. The solution was cooled and reduced in volume to approximately 2 mL on a rotary evaporator. With cooling, 50 mL of 1 *M* HCl was added. Extraction of this mixture with hexane yielded a fraction that was reduced in volume and chromatographed over Kieselgel 60 (hexane as eluant) to yield 145 mg of UL ^{13}C-3,4,3′,4′-tetrachlorodiphenyl ether (32%).

UL ^{13}C-2,3,7,8-Tetra-CDF

A modification of the procedure of Norstrom et al. [10] was used to cyclize diphenyl ether to a dibenzofuran. A 30-mg aliquot of UL ^{13}C-3,4,3′,4′-tetrachlorodiphenyl ether (0.097 mmol) was dissolved in 5 mL glacial acetic acid. To this solution were added 43.6 mg of palladium acetate (Alfa-Ventron, 0.195 mmol) and 0.35 mL of methane sulfonic acid. The solution was refluxed gently for 2 hr, during which palladium metal appeared as a black precipitate. The warm solution was allowed to settle, was filtered through a short silica gel tube, and was allowed to cool. An off-white flocculent precipitate formed, which was separated by filtering the solution through glass wool. The precipitate dissolved in diethyl ether, which was evaporated to yield 8.3 mg of a mixture of tetra-CDF in the following ratio (by ECD/GC):2,3,7,8-, 80%; 1,2,7,8-, 19%, 1,2,8,9-, 1%. These figures are not corrected for possible variations in response factors. The yield was 28% based on starting diphenyl ether, but clearly substantial quantities of tetra-CDF remain in the acetic acid mother liquor.

Mixture of UL ^{13}C-PCDF from Chlorination of 2,3,7,8-Tetra-CDF

A 0.85-mg sample of the above mixture of tetra-CDF (0.0028 mmol) was dissolved in 1.0 mL carbon tetrachloride. A 10.7-μL aliquot of antimony pentachloride was added (24.9 mg, 0.083 mmol), and the solution was passed dropwise through a 4-in.-long Pyrex® hot tube apparatus stabilized at 270°C. Nitrogen was used as a carrier gas at approximately 10 mL/min. The products were allowed to condense in additional carbon tetrachloride, forming a pale yellow solution. Color was removed by chromatography over silica gel. On evaporation, a small amount of white precipitate formed. Gas chromatography on the capillary column described yielded (in order of elution on SP-2330):

- 1,2,7,8
- 2,3,7,8
- 1,2,3,7,8
- 2,3,4,8,9
- 1,2,8,9
- 1,2,3,4,7,8
- 1,2,3,6,7,8
- 2,3,4,7,8
- 1,2,3,8,9
- 1,2,3,7,8,9
- 1,2,3,4,6,7,8
- 2,3,4,6,7,8
- 1,2,3,4,6,8,9
- 1,2,3,4,7,8,9
- 1,2,3,4,6,7,8,9

Isomers were identified by comparison of retention times with individual natural carbon isomers prepared from appropriate chlorinated diphenyl ethers.

ACKNOWLEDGEMENTS

Thanks are extended to Dr. Woody Ligon and Ralph May for mass-spectral analysis and to Dr. Anders Garå for assistance in the preparation of unlabeled isomers.

REFERENCES

1. Buser, H.R., C. Rappe and A. Gara. *Chemosphere* 7:439–449 (1978).
2. Kashimoto, T., H. Miyata, S. Kuita, T. Tung, S. Hsu, K. Chang, S. Tang, G. Ohi, J. Nakagawa, and S. Yamamoto. *Arch. Environ. Health* 36:321–326 (1981).
3. Buser, H.R., H.P. Bosshardt and C. Rappe. *Chemosphere* 7:109–119 (1978).
4. Morita, M., J. Nakagawa and C. Rappe. *Bull. Environ. Contam. Toxicol.* 19:665–670 (1978).
5. Rappe, C., H.R. Buser, D.L. Stalling, L.M. Smith and R.C. Dougherty. *Nature* 292: 524–526 (1981).
6. Albro, P. W., and C. E. Parker. *J. Chromatogr.* 197:155–169 (1980).
7. Buser, H.R. Thesis, University of Umeå, Umeå, Sweden (1978).
8. Mazer, T., F.D. Hileman, R.W. Noble and J.J. Brooks. *Anal. Chem.* 55:104–110 (1982).
9. Norstrom, A., K. Andersson and C. Rappe. *Chemosphere* 5:21–24 (1976).
10. Norstrom, A., K. Andersson and C. Rappe. *Chemosphere* 5:419–423 (1976).
11. Moore, J.A., E.E. McConnell, D.W. Dalgard and M.W. Harris. *Ann. N.Y. Acad. Sci.* 320:151–163 (1979).

12. McKinney, J.D., K.Chae, B.N. Gupta, J.A. Moore and J. A. Goldstein, *Toxicol. Appl. Pharmacol.* 36:65–80 (1976).
13. Davis, G., and N. Cook. *ChemTech* 7:626–629 (1977).
14. Crocker, H.P., and R. Walser. *J. Chem. Soc.* (C)1982–1986 (1970).
15. Suthers, B.R., P.H. Riggins and D.E. Pearson. *J. Org. Chem* 27:447–451 (1962).
16. Horning, D.E., D.A. Ross and J.M. Muchowski. *Can. J. Chem.* 51:2347–2348 (1973).
17. Whitesides, G.M., J.S. Sadowski and J. Lilburn. *J. Am. Chem. Soc.* 96:2829–2835 (1974).

Environmental

5

Offsite Transport of 2,3,7,8-Tetrachlorodibenzo-*p*-dioxin from a Production Disposal Facility

L.J. Thibodeaux

This chapter summarizes the results of transport calculations performed in an effort to assess the quantities and concentration levels of tetrachlorodibenzo-*p*-dioxin (tetra-CDD) leaving an herbicide production facility that practiced onsite disposal. The plant, located in Jacksonville, Arkansas, had been manufacturing 2,4-D and 2,4,5-T since approximately 1958. Waste material generated by the plant was usually placed in metal drums and buried in several locations on the plant site. Besides the buried quantities, tetra-CDD existed on soil surfaces and in water bodies located within the plant boundary.

Approximately 1000 soil, air, water and sediment samples were taken by state government, federal government and chemical company representatives after the presence of tetra-CDD was detected in spring 1979. Based on the levels of tetra-CDD observed in these various samples, it was necessary to employ transport models in an effort to assess the quantities of the chemical leaving the site. Questions involving general exposure considerations can be addressed if estimates of the rate of transport are available. The objectives of this chapter are (1) to demonstrate the utility of simple transport models in estimating the rates of emission of tetra-CDD from various points of origin on the plant site; (2) to identify the major sources and mechanisms of offsite transport; and (3) to delineate measurements that need to be taken to better quantify offsite transport rates.

ENVIRONMENTAL TRANSPORT REVIEW

Several documents contain reviews of transport of 2,3,7,8-tetra-CDD in or into soil, water and air [1–3]. Selected material from these documents is presented here. Table 5.1 contains tetra-CDD data extracted from these documents.

The evidence presented indicates that tetra-CDD is primarily deposited in the soil compartment. However, inputs to both air and water occur and are significant in localized areas. A number of transport processes may affect the fate of tetra-CDD in soil, including volatilization and leaching. In addition, erosion and soil transport may result in movement from the original site of application or deposition.

Volatilization is suggested as the loss mechanism for 29–46% of tetra-CDD in 350 days when applied to a soil. Other reports attribute losses to volatilization when applied to silica but no loss on soil in 8 hours. Losses from soybean leaves and oat leaves of 6 and 37%, respectively, in 21 days are reported to be due to volatilization. There is evidence to suggest that photodecomposition is a competitive mechanism, and volatilization may occur, but at a slow rate. It is also suggested that water-mediated evaporation takes place. In addition, it is reported that volatilization of tetra-CDD from water may occur, although TCDD has a low vapor pressure. An average of 71% of the administered radioactivity was recovered after 589 days of incubation of tetra-CDD in lakewater with no sediment present. Volatilization from water is suggested by simple model calculations. Volatilization half-life of 10–20 hr from 1-m water depth are indicated from the calculations. The results on volatilization are inconclusive and further work is needed to verify the role of volatilization from water and soil.

Many studies have addressed the question of the mobility of 2,3,7,8-tetra-CDD in soils. In general, it has been found that dioxins are more tightly bound to soils having high organic matter. The partition coefficient of 22,900 reported in Table 5.1 supports high sorption onto soils. Details of several soil studies reveal some general behavior patterns. Dioxins applied to the surface of such soils gen-

Table 5.1. Selected Physical and Chemical Properties of 2,3,7,8-Tetra-CDD

	Dow [1]	*PEDCo [2]*	*ADL [3]*	*Ad Hoc Report [4]*
Molecular Weight	321.98	322	322	321.87
Melting Point (°C)	302–305	305	303–305	306
Vapor Pressure (mm Hg)[a]	Relatively involatile		10^{-6}–10^{-7}	1.7×10^{-6}
Water solubility (μg/L)[a]	0.2–0.6	0.2	0.2	
Log of Partition Coefficients				
Octanol/water			7.14	
Soil/organic matter and water			4.36	
Biota/water			5.04	

[a]Temperature unknown.

erally remain in the upper 6–12 in. They will migrate deeper in more sandy soils, to depths of 3 ft or more. Dioxins may appear in normal water leachate from soils that have had several dioxin applications. Several investigators conclude that dioxins would pose no threat to groundwater supplies because they would not be mobilized by rainwater or irrigation water in areas possessing high organic content soils. However, transport of tetra-CDD with soil particles may occur and result in surface water contamination.

Essentially no information is available on the transport of tetra-CDD in the atmosphere. It is likely to be suspended by winds as particles of tetra-CDD–laden dust and dirt.

There is some evidence that tetra-CDD is transported between water and sediment in aquatic environments. The primary mechanism for tetra-CDD contamination of water is erosion of contaminated soil or plants. As in the case of soil, tetra-CDD reaching the aquatic environment would likely be strongly sorbed to sediment. One study found a half-life of 600 days in sediment and loss was largely attributed to desorption and consequent volatilization.

DESCRIPTION OF THE VERTAC SITE

Approximately 3000 barrels of waste material were stored on the site. The following is a brief outline of problem areas within the site [5] (Figure 5.1).

Reasor-Hill Dump Site

This is probably the oldest dump site, located on the southwest corner of the plant. A flowing creek, Rocky Branch, runs along the western side of this site, and leachate has been observed running out of the ground and into the creek. In May 1979 a clay cap had been placed over the site, which has apparently stopped much, if not all, of the leachate. Also, several sumps were dug between the site and the creek to collect leachate.

Oxidation Pond

This pond was constructed in 1964 after several massive fish kills on Bayou Meto, into which Rocky Branch and the Jacksonville sewage treatment plant (STP) effluent ultimately flow. It was constructed to pretreat the wastewater before it was released. Water from the pond runs to the Jacksonville STP, and then to Bayou Meto. The water is high in phenols, and the sludge in the bottom of the pond has shown dioxin content. There is evidence that there is some leaching of material from the bottom of the pond through the soils toward Rocky Branch.

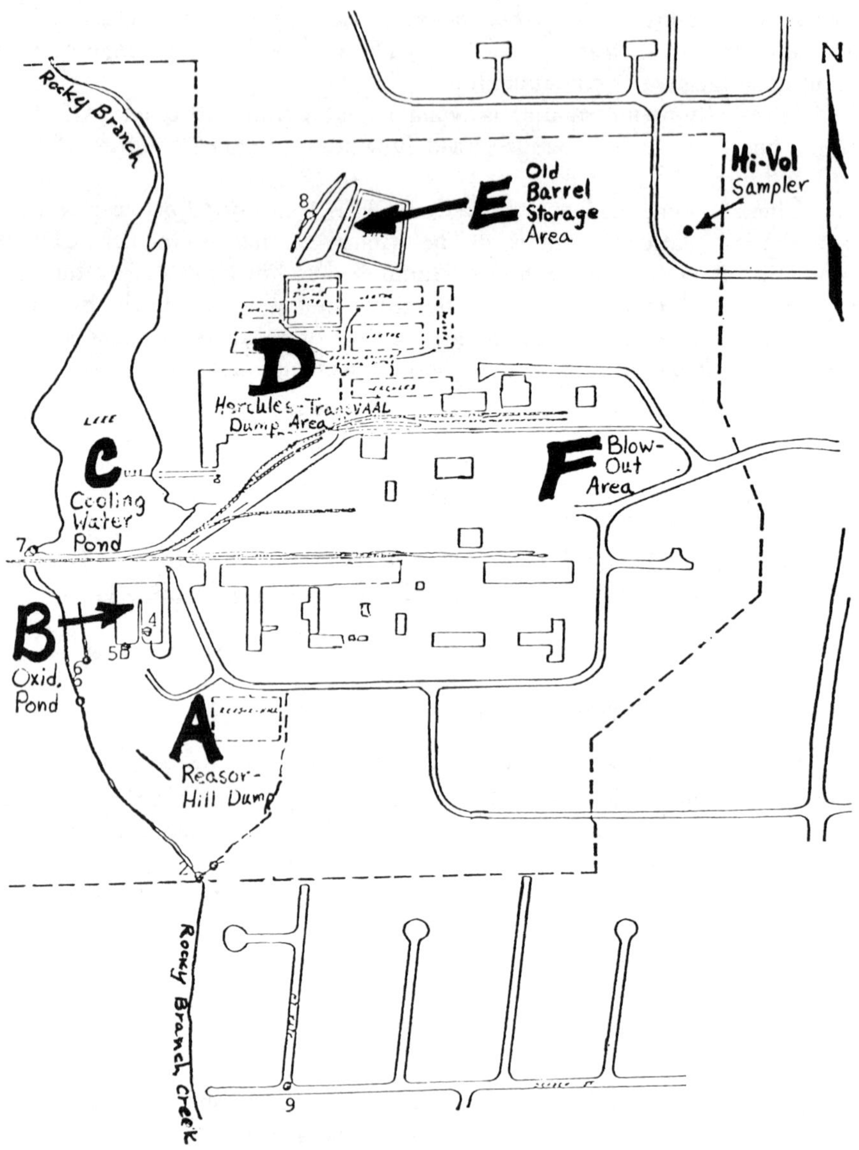

Figure 5.1. VERTAC Site Map, May 1979.

Cooling-Water Pond

This pond was created by damming Rocky Branch to provide cooling water to the plant. Evidence indicates the presence of dioxin in sediment in the pond, probably as a result of the washing of contaminated dirt into the pond and creek upstream from the pond by rain.

Hercules-Transvaal Dump Area

This is a large dirt area immediately north of the processing area into which were dumped metal drums of waste material, similar to the Reasor-Hill dump. This site was not capped at the time of the federal court suit in April 1980.

Old Barrel Storage Area:

This was an uncovered dirt area immediately north of the Hercules-Transvaal dump area in which barrels of waste material were stored in the open. Photographs of the area taken in May 1979, indicate that a large pond of brownish fluid formed between the barrels and the retaining dike constructed west of the barrels. This was probably waste material that leaked from the barrels. The barrels have now been removed to covered buildings north of the plant site.

"Blow-Out" Area

This is a dirt area near the so called "Sputnik" portion of the plant, into which accidental "blow-outs" apparently sprayed dioxin-laden material. There is apparently surface contamination, and some very high concentrations were observed there. Dirt from this area is washed by rain into a drainage ditch that runs around the east side of the plant and ultimately into Rocky Branch.

TRANSPORT CALCULATIONS

There appears to be much evidence in the literature to suggest that tetra-CDD is somewhat mobile across all the environmental interfaces and that it moves with the various phases in convective manner either in solution or attached to particles. This information is derived primarily from laboratory simulations and field surveys. The science of chemodynamics has advanced to a state that it is now possible to make some reasonable estimates to tetra-CDD flux rates, half-life and residence times in selected compartments of the environment. The following is a set of transport calculations, based on measured concentrations and simple models, that quantifies mass rate of movement of tetra-CDD from various locations on the VERTAC site to the greater environment. Most calculations are based

Table 5.2. Concentration of Tetra-CDD in Various Media

Media	*Sample Size (number)*	*Average (ng/kg)*	*Range (ng/kg)*
Air Particulate	2	1,100	±350
Water (Includes Offsite)	23	14	ND[a] to 47
Soil	5	1,300	ND to 2,900
Creek Sediment	5	770	ND to 1,800
Pond Sediment	3	22,100	±2,100

[a]ND = not detected.

Table 5.3. Surface Source Areas and Emission Rate Summary

Source	*Area (m^2)*	*Emission Rate (g/yr)*
Blow-out Area, Volatilization	753	120.–1,200.
Blow-Out Area, Entrainment	753	28.–37.
Rocky Branch Creek, Dissolved		0.89–2.1
Reasor-Hill Dump	1,129	0.1–1.0
Rocky Branch Creek Sediment		0.094–0.22
Cooling Water Pond	15,050	0.015–0.016
Total		150.–1,240.

on the steady-state assumption and represent offsite transport as conditions existed in 1978 and 1979. Concentrations of tetra-CDD in air, water, soil and sediment samples from on-site locations appear in Table 5.2. Table 5.3 contains surface areas of various contaminated sites. These areas were estimated from aerial photos [5].

Rocky Branch Creek

The creek that flows along the western edge of the site can transport tetra-CDD offsite. As the water flows into the cooling pond, over the dam and back into the stream bed (see Figure 5.1), quantities of the chemical in both dissolved state and attached to suspended sediment can be entrained. The convective flux of the suspended fraction is obtained by:

$$W_A = \omega_A \rho_{32} Q \qquad (1)$$

where

W_A = rate leaving the southern boundary (g/hr)

ω_A = mass fraction of TCDD on the sediment (g/g)

ρ_{32} = sediment concentration in the water (g/m^3)
Q = stream flowrate (m^3/hr)

From a similar lake in Arkansas a ρ_{32} = 7.0 mg/L was chosen, and Q on April 25, 1980, was 12.6 × 10^6 gallons per day. Employing the range of concentrations (sediment, creek ω_A) in Table 5.2, Equation 1 yields an average rate of 0.094 g/yr and a maximum rate of 0.22 g/yr.

No tetra-CDD was detected in solution in the creek water. Based on the solubility range in Table 5.1, the correlation of Chiou et al. [6], with sediment organic fraction of 2.9%, yields sediment partition coefficients of 19,400–10,500. These values are slightly lower than the value of 22,900 represented in Table 5.1. The appropriate dissolved tetra-CDD convective flux equation is:

$$W_A = \omega_A Q / K_{A32}^* \qquad (2)$$

where K_{A32}^* = sediment/water partition coefficient, ω_A / ρ_{A2} (cm^3/g).

A value of 15,000 results in an average rate of 0.89 g/yr and maximum of 2.1 g/yr. Creek sediment concentrations in Table 5.1 were also used in this calculation.

Vaporization from Soil Surface

Some degree of success has been achieved in modeling the vaporization rate of surface applied chemicals by using:

$$W_A = 0.268 A V^{0.78} P_A^* M_A / L^{0.11} T \qquad (3)$$

where

W_A = flux (g/hr)
A = area on the contaminated surface (m^2)
V = wind speed (m/hr)
P_A^* = pure-component vapor pressure of the chemical (mm Hg) at the soil surface temperature
M_A = molecular weight
L = fetch of the contaminated surface (m)
T = soil surface temperature (°K)

This equation assumes that the air boundary layer controls the evaporation rate and that the chemical applied to the soil surface exerts its pure-component vapor pressure. This equation is one form of several such equations that were developed from numerous studies of evaporation from chemical pools, lakes and the sea surface.

Calculations were made assuming that the blow-out area met the conditions of the model. Vaporization of some pesticide loadings similar to that observed for tetra-CDD (see Table 5.2) can be modeled fairly well with Equation 3. Using a wind speed of 5 knots, 25 °C and the range of vapor pressures in Table 5.1 yields 120–1200 g/yr vaporized from the blow-out area.

Entrainment of Soil Particles

An air particulate sampler was placed at the northeast corner of the site (see Figure 5.1). Between August 22 and October 26, 1979, the sampler was operated for thirty-six days, resulting in two dust samples. The tetra-CDD concentration in the dust collected was very near the average concentration observed on the soil (see Table 5.2).

Deposition of entrainment of dust particles by wind is a well known micrometeorological process. Flux of a chemical species or particles to and from a surface is quantified by a rate equation of the form:

$$W_A = AV \omega_A \rho_{31} \tag{4}$$

where

V = deposition (velocity) or entrainment coefficient (cm/hr)

ω_A = weight ratio of tetra-CDD in the dust (g/g)

ρ_{31} = concentration of dust particles in the air very close to the soil surface (g/m^3)

A value of 2 cm/sec was chosen for V based on ranges of reported values for these coefficients. The field measured value of 54 $\mu g/m^3$ was used for ρ_{31}. Based on the average tetra-CDD loading and the standard deviation value, a range of 28–37 g/yr is suggested for this route of transport. The calculation is based on the contaminated soil in the blow-out area.

Vaporization from Landfill Cells

It has been suggested by model calculations and by limited experimental and field data that volatile chemical species placed in landfill cells and retained in a free state (i.e., not in container) can vaporize readily and be transported to the soil surface by a combination of diffusive and convective processes [7]. There were indications at the site, particularly at the Reasor-Hill dump, that liquids escaped from the burial containers (i.e., drums corroded) and existed free inside the landfill cells. The following equation was used to estimate vapor emission rates:

$$W_A = 12{,}180AD_{A1}\epsilon^{4/3}p_a{}^*M_A/hT \qquad (5)$$

where

D_{A1} = molecular diffusivity of tetra-CDD in the air-filled soil pore spaces (m/hr)

ϵ = porosity of the soil cover (m^3/m^3)

h = soil cover depth (m)

This equation is primarily a result of theoretical models for diffusion through porous media; however, limited experimental results for volatile chemicals from landfill cells and chemical piles support its validity. Assuming a soil cover thickness of one meter, a cell temperature of 15°C, soil porosity of 0.25 and tetra-CDD diffusivity 0.0506 cm^2/sec results in 0.1 and 1.0 g/yr when vapor pressures of 10^{-7} and 10^{-6} mm Hg are used. This model also assumes that the chemical exerts its pure-component vapor pressure within the cell. Biogas production within the cells can possibly increase the rate by a factor of eight.

Vaporization from Pond Surface

The pond sediment appears to be the most contaminated source of tetra-CDD on the site (see Table 5.2). Surface draining for the northern half of the site is into the cooling water pond. Contaminated dirt from the Hercules-Transvaal dump area, the old barrel storage area, the process area and the blow-out area is likely deposited in the pond and settles to the bottom, resulting in the highly contaminated sediment.

Several mechanisms must be operative for tetra-CDD to vaporize from the pond surface. If the chemical is readily available in the sediment it must: (1) desorb into the pore-water (an equilibrium process in this model); (2) diffuse through the bottom-water boundary layer; (3) be transported to the surface; and (4) desorb from the lake surface into the air. Assuming no background concentration of tetra-CDD in the air, the following transport model applies:

$$W_A = A\,{}^3K_{A2}\,\omega_A/K_{A32}{}^* \qquad (6)$$

where the overall transport coefficient ${}^3K_{A2}'$ is:

$$1/{}^3K_{A2} = 1/{}^3k_{A2}' + 1/{}^1k_{A2}' + 1/H_\rho{}^2k_{A1}' \qquad (7)$$

the respective individual transport coefficients are:

- ${}^3k_{A2}'$ for the bottom water (m/hr);
- ${}^1k_{A2}'$ for the surface water (m/hr);
- ${}^2k_{A1}'$ for the air side of the surface (m/hr); and
- H_ρ = Henry's constant (i.e., $H_\rho = \rho_{A1}/\rho_{A2}$).

Employing published expressions for estimating Henry's constant [8], $H_\rho = 0.1471\ H_2O/L$ air, and surface water and air-side coefficients [8] $^1k_{A2}' = 2$ cm/hr and $^2k_{A1}' = 1050$ cm/hr as reasonable for a 5-knot wind. Thibodeaux and Becker [9] give the following equation for estimating the bottom-water transport coefficient (cm/min):

$$^3k_{A2}' = (2.2 \times 10^{-4})C_D(\rho_1/\rho_2)V_x^2H^{5/4}/LM_A{}^{1/2} \tag{8}$$

where

C_D = drag coefficient, 0.00166

ρ_1/ρ_2 = density ratio of air and water

V_x = wind speed (cm/min)

H = average water depth (cm)

L = average wind fetch (cm)

This relationship for the bottom-water coefficient is based on dissolution experiments performed in a laboratory wind-water tank and an air-water momentum transfer model. Using a 5-knot wind, fetch of 138 m and an average depth of 1 m yields a coefficient of 0.0079 cm/hr. Clearly, the bottom-water coefficient controls. This coefficient was not a consideration in calculations performed by others [3]. Completing the calculation with an overall coefficient of 0.78 m/hr, the concentration range for pond sediment in Table 5.2, the surface area in Table 5.3 and $K_{A32}^* = 15{,}000$ results in 0.015–0.016 g/yr.

Other On-site Sources

Other on-site sources likely contribute to offsite transport. Vapor emissions are likely to be associated with the Hercules-Transvaal dump area, the old barrel storage area, the oxidation pond and the process area in general. The rates of emission from these sites are less easily quantified due to lack of specific information. Other reports [1,3] suggest limited tetra-CDD mobility in leachate or groundwater transport; however, rate determinations are difficult to make. Tetra-CDD transport offsite by the wastewater route was not considered, nor was drag-out by people, vehicles, etc., during movement from the site.

CONCLUSIONS

Results of the offsite transport calculations are summarized in Table 5.3. The calculations show that 150–1250 g/yr of tetra-CDD left the site in 1978–1979 before implementation of remedial measures by the operator. Vaporization of tetra-CDD from the soil surface in the blow-out area is the major contributing source.

Attempting chemodynamic calculations, such as outlined above for tetra-CDD at a specific site, reveals areas where more information is needed before such calculations can be made with high degrees of certainty. In the case of tetra-CDD, the lack of reliable data on water solubility and vapor pressure contributed to a sizable range of uncertainty in the results. Chemical property uncertainties aside, major conceptual approaches preclude reliable prediction of emission rates. If emission rate predictions are poor, risk assessment is little more than speculation. A study by Schaum and Falco on this particular site for tetra-CDD required emission data to anlyze human exposure [10].

Samples for chemical analysis of soil, water, sediments and air are insufficient. This information must be complemented by pertinent physical measurements to include:

- stream flow,
- flux measurement techniques,
- sediment contaminant depth,
- soil contaminant depth,
- soil and sediment properties,
- soil surface temperatures,
- wind speed, and
- gradient measures (wind, temperature and species concentration).

The body of knowledge in environmental transport processes is small. Interface transport coefficient observations under field conditions are almost nonexistent. Basic research and development programs in this area plus the thermodynamics of equilibria of soil-waste systems, multicomponent (nonhomologous) mixture behavior, and equilibria, need to be undertaken and continued. The area of mass balance and system models is rich and likely requires less developmental effort.

REFERENCES

1. "Criterion Document, 2,3,7,8-Tetrachlorobenzo-*p*-dioxin," Dow Chemical Company, Midland, MI.
2. "Dioxins: Sources, Transport, Exposure, and Control," Preliminary Draft, PEDCo Environmental, Inc., Cinncinnati, OH (1979).
3. Perkaw, J.A., A. Eschenroeder, M. Goyer, J. Stevens and A. Wechsler. "An Exposure and Risk Assessment for 2,3,7,8-Tetrachlorodibenzo-*p*-dioxin," Office of Water Regulations and Standards, U.S. EPA, Washington, D.C. (1980).
4. "Report of the Ad Hoc Study Group on Pentachlorophenol Contaminants," EPA/SAB/78/001, Environmental Health Advisory Committee, Science Advisory Board, U.S. EPA, Washington, D.C. (1978).
5. "Document File, Arkansas Department of Pollution Control and Ecology vs. Vertac Chemical Corporation and Hercules, Inc.," U.S. District Court No. LR-C-8-110, Little Rock, Arkansas (1980).
6. Chiou, G.T., L.J. Peters and V.H. Freed. *Science* 206:16 (1979).

7. Thibodeaux, L.J. *J. Haz. Mat.* 4:3 (1981).
8. Thibodeaux, L.J. *Chemodynamics—Environmental Movement of Chemicals in Air, Water and Soil* (New York: John Wiley & Sons, Inc., 1979).
9. Thibodeaux, L.J. and B. Becker. *Environ. Prog.* (Fall 1982).
10. Schaum, J., and J. Falco. "Exposure Analysis of Vertac Facility," Exposure Assessment Group, Office Health and Environmental Assessment, ORD, U.S. EPA, Washington, D.C. (1981).

6

2,3,7,8-Tetrachlorodibenzo-*p*-dioxin and 2,3,7,8-Tetrachlorodibenzofuran Residues in Great Lakes Commercial and Sport Fish

J.J. Ryan, P.-Y. Lau, J.C. Pilon and D. Lewis

2,3,7,8,-Tetrachlorodibenzo-*p*-dioxin (2,3,7,8-tetra-CDD) has been the subject of a great deal of scientific and public interest, mainly due to its potent toxicological properties. Early attempts [1] to detect this chlorinated aromatic hydrocarbon in aquatic species were hampered by the lack of sensitive, specific and reliable techniques. However, with the advent of superior extraction, purification and chromatographic procedures coupled with highly sensitive and specific mass spectrometers, the measurement of contaminants such as dioxins and furans at low parts per trillion has become feasible in the last few years.

As a result, in late 1978 the U.S. Environmental Protection Agency (EPA) reported preliminary observations [2] on the presence of 2,3,7,8-tetra-CDD in several species of fish originating from rivers that flow into Lake Huron in the vicinity of Saginaw Bay. Details of this survey were published [3] and revealed that 26 of 36 samples comprising six different species of fish were positive for 2,3,7,8-tetra-CDD. Values of the positive samples ranged from 4 to 695 ppt, with 10 fish samples containing more than 40 ppt 2,3,7,8-tetra-CDD. Channel catfish and carp (bottom feeders) had the highest mean values (157 and 55 ppt, respectively). The New York state Department of Health reported their preliminary findings (news release, April 24, 1979) that sampling seven fish from Lake Ontario demonstrated measurable levels of 2,3,7,8,-tetra-CDD in two fish. At the same time the authors' analysis of Great Lakes fish indicated that tetra-CDD residues were present, but the reliability of the method was uncertain, because the precision, accuracy and detection limits had not been established. As a result, the methodology was examined further, modified and subsequently included in an international interlaboratory comparison with twelve other laboratories. This exchange of fish samples demonstrated that, provided an internal standard was used, many laboratories experienced in low-level organic residue analysis could measure tetra-CDD at parts-per-trillion levels with a relative standard deviation of 15–25% [4].

Further reports of dioxin in the environment came from the Canadian Wildlife Service [5], which found levels of up to 800 ppt tetra-CDD in herring gull (*Larus argentatus*) eggs from the Great Lakes with average values around 50 ppt. Highest levels were found in samples from Lake Ontario and the Saginaw Bay area of Lake Huron. This information has been summarized [6].

Because all of the data on tetra-CDD in fish and other biological samples were incomplete and still fragmentary, we decided to investigate the extent of 2,3,7,8-tetra-CDD contamination in commercial fish from Lake Ontario. Fish were chosen from Lake Ontario, as this lake is generally believed to contain the highest amount of chemical contamination of all the Great Lakes [7]. We were also aware that two other groups, the New York state Department of Health and Ontario Ministry of Environment (MOE), were conducting surveys on fish that contained substantial sampling of sport fish. Because we are mainly interested in fish used as food, our sampling was mostly of the commercial variety. This chapter gives the results of 2,3,7,8-tetra-CDD levels in commercial fish taken from Lake Ontario in 1980 and compares the levels with a smaller sampling of commercial fish taken in 1979 and sport fish taken in 1980. Preliminary data on the presence of 2,3,7,8-tetrachlorodibenzofuran (tetra-CDF) in some of these fish are also given.

MATERIALS AND METHODS

Sampling

Commercial fish were caught with seine nets throughout 1980. A total of 62 samples of Great Lakes commercial fish were collected, mostly (56) from the northeastern part of Lake Ontario near Kingston, Bay of Quinte and the Thousand Islands. This area has the only substantial commercial fishing operation on Lake Ontario. Four samples originated from Lake Erie and two from the Welland Canal. The kinds and numbers of fish samples are outlined in Table 6.1. Ten commercial species of fish were collected and analyzed. A pooled sample concept was used whereby enough fish were taken to prepare a homogenized composite of 2–3 kg of muscle fillet. This is our standard procedure for surveying contaminants in fish and was also necessary due to the small size and corresponding fillet from seven of the species. Catfish samples were chosen from previous analyses based on comparison of polychlorinated biphenyl (PCB) values and length such that the PCB level would be below 5 ppm. No carp are presently taken commercially because of restrictions on the sale of this fish due to PCB levels. As far as possible, at least 20% of the fish samples were collected from each of the spring (ice breakup to June 1), summer (June 1 to September 1) and fall (September to

Table 6.1. 2,3,7,8-Tetra-CDD Levels (ppt) in 62 Samples of Great Lakes Fish (56 from Lake Ontario) Along with Average Values (± standard deviation) of PCB, Mirex and Fat Content

			TCDD Levels				
Species	*Number Analyzed*	*Number Positive for Tetra-CDD*	*Mean of Positives (ppt)*	*Range Positives (ppt)*	*PCB (ppm)*	*Mirex[a] (ppb)*	*Fat (%)*
Rock Bass	6	0			0.21±0.10	7.9±7.0	0.47±0.19
Sunfish	9	0			0.11±0.06	2.5±0.0	0.67±0.16
Black Crappie	5	0			0.18±0.08	4.0±3.4	0.66±0.46
White Perch	3	1		6.3	0.56±10.31	13.3±5.8	3.88±1.22
Yellow Perch	10	2	3.8	3.2–4.3	0.12±0.05	5.5±3.9	0.82±0.56
Brown Bullhead	7	2	6.0	3.4–8.6	0.13±0.03	4.6±3.7	1.45±0.29
White Sucker	5	2	3.0	2.0–4.0	0.44±0.15	10.0±0.0	1.33±0.31
Catfish	3	3	15.5	12.8–17.7	3.63±0.52	103.0±21.0	12.50±1.84
Eel	6	5	19.8	6.4–38.5	4.90±2.78	148.0±88.0	36.60±4.54
Smelt	8	6	20.0	11.3–32.9	1.28±0.50	24.0±23.0	3.41±0.96

[a]Samples not detected were taken to be 2.5 ppb (one-half the detection limit of 5 ppb).

freeze) seasons. Fillet tissue without skin was ground in a food mixer, placed in plastic sample containers and frozen.

Smelt samples from 1979 were obtained, prepared and analyzed similarly to 1980 commercial fish. Sport fish caught in 1980 were available for analysis from two related studies and comprised either muscle fillet or whole fish samples.

Analysis

Aliquots (10 g) of the homogenized composite were analyzed for 2,3,7,8-tetra-CDD according to the method outlined by Ryan and Pilon [8] and Ryan et al. [4]. In summary, tissues were extracted with chloroform-methanol, the solvent was exchanged for hexane, and lipid degradation and removal were accomplished by partitioning against concentrated sulfuric acid. The extract was then applied to a mini-Florisil column, PCB were removed with the hexane-dichloromethane (98%/2%), and all the dioxins and furans were then eluted with dichloromethane. The tetra-CDD fraction was separated from other dioxin congeners and the extract was purified further using reverse-phase high-performance liquid chromatography (HPLC) by eluting with methanol. The tetra-CDD fraction was then injected onto a fused-silica capillary gas chromatographic (GC) column (DB-5, chemically bonded SE-54). The latter was coupled directly to a Varian-MAT 311A mass spectrometer (MS) operating in the electron impact and single-ion monitoring (m/z 320, molecular ion of tetra-CDD) mode at a resolution (10%) valley) of 1000. Multiple ion monitoring at m/z 257, 320 and 322 at high resolution (8000–10,000) on a VG-Micromass ZAB-2F instrument was used to confirm positive samples. An internal standard of ^{13}C-2,3,7,8-tetra-CDD at a level of 50 ppt (500 pg) was added before extraction to correct for losses in the workup. Recoveries of 58.9% (standard deviation = 18.6; n = 75) were obtained. Detection limits varied between 2 and 10 ppt, depending on the background from individual fish samples and the percent recovery. Our method distinguishes 2,3,7,8-tetra-CDD from other compounds as well as the 21 other tetra-CDD isomers, except for a possible two or three similar GC-eluting isomers. However, comparison of our validated method with other validated methods more specific for 2,3,7,8-tetra-CDD has demonstrated [4] no difference in levels of 2,3,7,8-tetra-CDD in several fish samples. The fish were analyzed in sets of six with each set containing one or two quality control samples such as a blank or fortified reagent blank or fish.

Fish were analyzed for 2,3,7,8-tetra-CDF in an analogous fashion to that for 2,3,7,8-tetra-CDD. The HPLC peak for 2,3,7,8-tetra CDF elutes slightly earlier than that for 2,3,7,8-tetra-CDD, so a wider HPLC fraction was taken. The GC conditions were also similar to those for tetra-CDD, again with a slightly earlier elution for tetra-CDF. The MS conditions for tetra-CDF were single-ion monitoring at m/z 304 and 306 at a resolution of 1000. Confirmation of the tetra-CDF levels by high-resolution MS has not yet been carried out. The values have been corrected for mechanical and absorptive losses by using ^{13}C-labeled 2,3,7,8-tetra-

CDD as internal standard. The detection limits for tetra-CDF are similar to those for tetra-CDD, but the specificity of the method for the 2,3,7,8- isomer is uncertain, since only a few of the 38 tetra-CDF isomers are available for comparison.

PCB and Mirex were determined in tissue, briefly as follows. Fish were extracted with ethyl acetate and the extract defatted on a gel permeation column. After exchanging the solvent for hexane, the extract was purified further on a 2% deactivated Florisil (previously activated at 130°C for 24 hr) column and the PCB and Mirex were eluted with hexane—a step that separates them from DDT. Measurement is effected by GC on a 2% OV-1 plus 4% OV-210 column with electron capture detection—a column that separates the earlier-eluting PCB peaks from Mirex. Quantification of PCB is based on Aroclor 1254 as standard using the three major peaks eluting subsequent to DDE. Detection limit is about 0.01 ppm for PCB and 5 ppb for Mirex.

Lipid content of fish tissue was determined according to the method of Schmitt et al. [9]. A 10-g aliquot of tissue was first blended with anhydrous sodium sulfate and the powdery mixture was extracted with 20% acetone in isooctane. An aliquot of the centrifuged supernatant was evaporated to dryness under a stream of nitrogen and weighed. Values are quoted in percent.

RESULTS AND DISCUSSION

The dioxin, PCB, Mirex and fat content values for the 62 commercial fish samples from 1980 are summarized in Table 6.1. 2,3,7,8-Tetra-CDD was found in 21 of the 62 samples (34% positive), with values ranging between 2.0 and 38.5 ppt. Of these 62 samples, 12 (19%) had 2,3,7,8-tetra-CDD levels above 10 ppt and 5 (8%) were above 20 ppt. Of the ten species of commercial fish, seven had detectable levels of 2,3,7,8-tetra-CDD. The only isomer of tetra-CDD (22 are possible) present in all the positive samples was one that had all the characteristics of 2,3,7,8-tetra-CDD. Figure 6.1 shows a GC/MS tracing of a post-cleanup extract of smelt containing 25.7 ppt 2,3,7,8-tetra-CDD. Ions monitored were 320 and 322 for native tetra-CDD and 332 for the isotopically labeled internal standard, each offset one min from each other on the chart. Only one signal for tetra-CDD is present in this extract; it occurs at the same time as that for a standard of the 2,3,7,8-tetra-CDD isomer, i.e., no other tetra-CDD isomers are present. The tetra-CDD values for the different fish species appear to fall into three groups. Rock bass, sunfish and black crappie had no detectable levels; white perch, yellow perch, brown bullhead and white sucker were found to contain levels below 10 ppt; and catfish, eel and smelt had the highest levels.

All samples of fish contained readily measurable levels of PCB. The first seven species in Table 6.1 all showed values less than 1.0 ppm and the last three species (catfish, eel and smelt) had values almost always greater than 1.0 ppm. The PCB values appeared to parallel the fat content in most cases, as those with low or high PCB also had low or high lipid. The levels of Mirex were much more variable and, in many cases, the value was at or near the detection limit of 5 ppb.

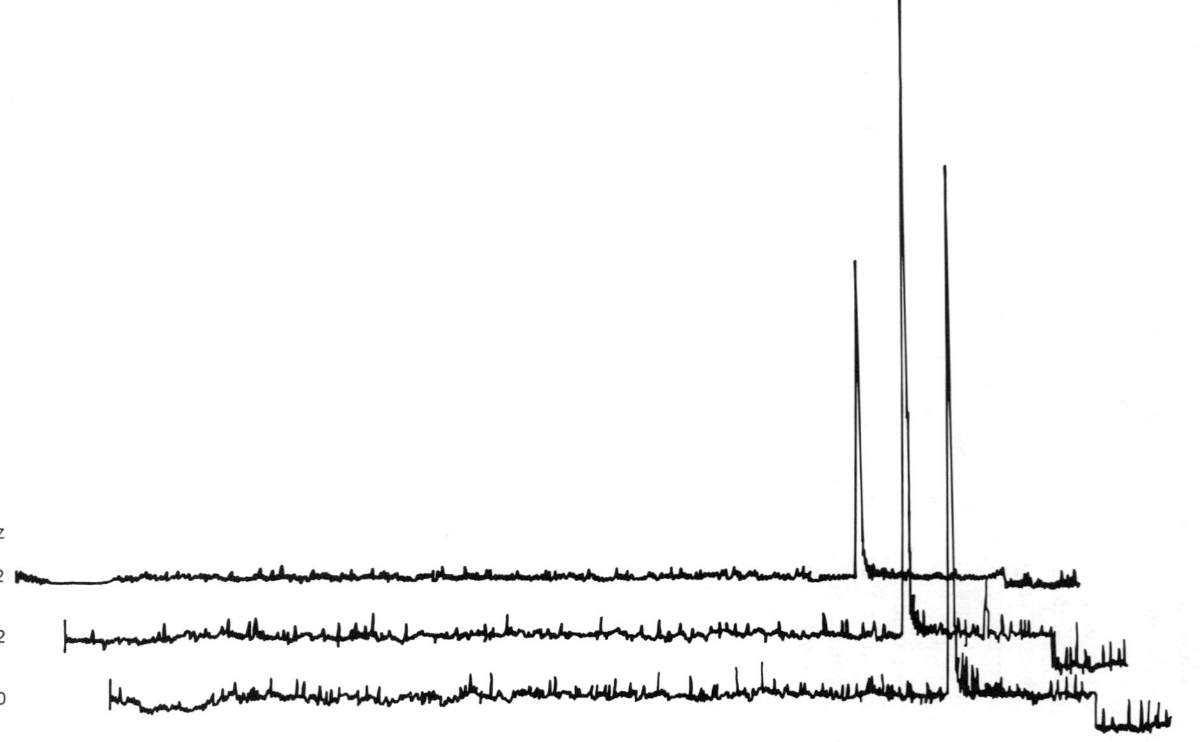

Figure 6.1. GC/MS Schematic Chromatogram of a Cleaned Extract of Smelt Monitored at Three Ions; m/z 332 for the Internal Standard and m/z 320 and 322 for Native Tetra-CDD—Each Ion Offset One Min on Chart for Clarity. Calculated Value of 2,3,7,8-Tetra-CDD Was 25.7 ppt.

Table 6.2. Individual Values of PCB, Mirex, and Fat Content for Three Fish Species Showing a High Level of 2,3,7,8-Tetra-CDD Contamination

Fish	*TCDD (ppt)*	*PCB (ppm)*	*Mirex (ppb)*	*Fat Content (%)*
Catfish	12.8	3.03	80	13.5
	16.1	3.88	110	13.7
	17.1	3.98	120	10.4
Eel	6.4	7.29	290	31.9
	9.2	8.78	170	37.9
	14.7	0.95	40	32.2
	30.4	3.61	160	42.7
	38.5	4.42	160	38.9
Smelt	11.3	1.86	ND[a]	2.6
	12.5	1.69	ND[a]	3.1
	14.2	1.37	40	4.1
	23.3	1.16	50	4.5
	25.7	1.79	50	3.1
	32.9	1.20	40	4.3

[a]ND = not detected at limit of detection (5 ppb).

The average values for this contaminant in Table 6.1 are based on values of 2.5 ppb for those samples in which no Mirex was detected. Only in catfish, eel and, to a lesser extent (four of eight), smelt was Mirex readily detected and quantified.

Three fish species from the commercial fish of 1980 showed high levels of chlorinated hydocarbon contamination. The individual values for 14 of these fish are tabulated (Table 6.2) for samples that were positive for 2,3,7,8-tetra-CDD. For catfish, all three samples collected in November 1980 had high dioxin, lipid, Mirex and PCB (>3 ppm). The six smelt samples that were positive for 2,3,7,8-tetra-CDD had a fat level greater than 2%, a PCB content greater than 1 ppm and some of the highest dioxin values. Four of the six were caught from the eastern part of Lake Ontario and two from the Welland Canal. Interestingly, the two of eight smelt samples in which no dioxin was found and in which the PCB content was low (<1.0 ppm) originated from Lake Erie. Five of the six eel samples were positive for tetra-CDD. This species had the highest PCB, Mirex and, by far, the highest lipid content.

For comparative purposes, a smaller sampling of salmonid sport fish from 1980 and commercial smelt from 1979 were analyzed for 2,3,7,8-tetra-CDD and PCB and these results are listed in Table 6.3. All salmonids collected from the Great Lakes in 1980 had readily measurable levels of 2,3,7,8-tetra-CDD, with average values greater than those of commercial fish. The three trout samples represented whole fish and this type of sampling is believed to give tetra-CDD levels 30–50% higher than those obtained by fillet sampling. The only negative sampling from the salmonid sport fish originated from the Pacific Coast. All tetra-CDD positive salmonids also had high levels of PCB. The smelt samples

Table 6.3. 2,3,7,8-Tetra-CDD and PCB Levels in Sport Fish (1980) and Smelt (1979)

Species	*Origin*	*TCDD (ppt)*	*PCB (ppm)*
Lake Trout[a]	Lake Ontario	58	7.28
	Lake Huron	37	5.03
Rainbow Trout[a]	Lake Ontario	33	1.77
Coho Salmon	Lake Ontario	28[b]	7.39
	Pacific Coast	ND[c] (4)	0.03
Smelt	Lake Ontario	11	
		16	
		11	
	Lake Erie	ND[c] (2)	

[a]Whole fish.

[b]Also contained 36 ppt of hexa-CDD (three isomers) and 93 ppt of octa-CDD.

[c]ND = not detected at bracketed detection limit.

from 1979 also contained tetra-CDD, but the average values of the positives were somewhat lower than those smelt collected in 1980. Again, as in 1980, it is noteworthy that the Lake Erie sample contained no tetra-CDD.

A class of contaminants with similar chemical and toxicological properties to the chlorinated dibenzo-*p*-dioxins is the chlorinated dibenzofurans. Using similar methodology for estimating 2,3,7,8-tetra-CDD in fish, the analogous furan, 2,3,7,8-tetrachlorodibenzofuran (2,3,7,8-tetra-CDF), was determined in some of the fish in which 2,3,7,8-tetra-CDD was known to occur. These results are given in Table 6.4 for commercial fish from 1980 and for the salmonids. These preliminary data indicate that 2,3,7,8-tetra-CDF is also present in many of the fish samples from the Great Lakes at the same order of magnitude as 2,3,7,8-tetra-CDD and with a wide individual variation. Two eel samples not listed in Table 6.4 that contained 30 and 39 ppt 2,3,7,8-tetra-CDD were found to contain no detectable 2,3,7,8-tetra-CDF. The sampling of fish for tetra-CDF is somewhat biased, as most analysis has been done on samples known to contain 2,3,7,8-tetra-CDD above 10 ppt and probably does not reflect the average situation. In several cases, more than one isomer of tetra-CDF was present, but the exact configuration was uncertain due to lack of suitable standards and more definitive methodology.

Five of the commercial fish and all of the Great Lakes salmonid samples from 1980 were above the Canadian Health Protection Branch (HPB) regulation of 20 ppt for dioxin. Two of five commercial fish and three of four salmonids also surpassed a regulation of 2.0 ppm for PCB. Individual fish in these samples and other pooled samples below 20 ppt could be significantly higher, since the composite sampling procedure could result in lower-level samples diluting higher-level samples. Hence the values for single fish could show a wider variation. A guideline of 20 ppt 2,3,7,8-tetra-CDD in fish has been set by the New York state

Table 6.4. Preliminary Data Comparing 2,3,7,8-Tetra-CDF Levels (ppt) in Fish to 2,3,7,8-Tetra-CDD

	Tetra-CDF			
Fish	*2,3,7,8*	*Other Isomers*[a]	*Total*	*2,3,7,8-Tetra-CDD*
Smelt	34	78 (2)	112	23
	19		19	26
	3.2		3.2	14
	16		16	11
Catfish	54	45 (1)	99	16
White Perch	14.7		14.7	6.3
Lake Trout	8.5		8.5	58
	24		24	37
Rainbow Trout	12	188 (2)	200	33
Ontario Salmon	79	74 (1)	153	28
Pacific Salmon	ND[b] (10)			ND (4)

[a]Number of tetra isomers present (not including 2,3,7,8-tetra-CDF) are in parentheses.

[b]ND = not detected at bracketed detection limit.

Department of Health. Their regulation is based on a weekly individual consumption of fish of 6–8 oz (150–200 g) whereas the Canadian regulation is based on a 4-oz (100-g) consumption.

The data in Table 6.2 for the three commercial fish species having high tetra-CDD levels were treated to a statistical analysis involving simple linear regression of the tetra-CDD content (dependent variable) on either the fat or PCB level (independent variables). For both the smelt and the eel sampling, there was a positive correlation between tetra-CDD and fat level and an inverse negative one between tetra-CDD and PCB (correlation coefficients of 0.54 and −0.52 for smelt and 0.65 and −0.45 for eels). Catfish, however, had opposite correlation coefficients: tetra-CDD on fat was negative ($r = -0.78$) and on PCB was positive ($r = 0.95$). No clear statistical relationship is evident for these fish species; this probably is due to the small sample size. The average PCB and Mirex levels and the positive tetra-CDD levels in Table 6.1 expand a range of 45,59 and 6.7, respectively. If these same values are adjusted on an equal-fat basis, the ranges for PCB and Mirex diminish to 4.2 and 5.3, respectively, while that for TCDD remains relatively unchanged at 10.9. A more empirical approach is to state that samples with a high lipid content had high PCB and Mirex contamination and a high probability for the presence of dioxin. Not enough sampling was available to make any attempt to correlate dioxin content with season of year or the area of Lake Ontario—a consequence of the expense of doing this resource-intensive analysis.

The data reported in Tables 6.1 to 6.3 indicate that the levels and incidence of 2,3,7,8-tetra-CDD can be categorized by fish species and possibly by locale. These observations for fish also are similar to the situation in herring gulls. Eggs of this fish-eating bird have been found to contain the highest concentrations of 2,3,7,8-tetra-CDD in samples from Lake Ontario, Saginaw Bay and Lake Huron, and lower amounts in the other Great Lakes areas [6]. A further observation in finding tetra-CDD levels in fish appears to be one of size. Zabik et al. [10] found a positive correlation between PCB level and size (length and/or weight) for carp from Lake Huron. Some species of adult fish with high levels of 2,3,7,8-tetra-CDD also have a large size (e.g., catfish and trout) and many with a small adult size (less than 250–300 mm) have little or no dioxin contamination. This trend is particularly evident for the related *Ictalurus* species, where bullhead is low but the larger catfish is high. Moreover, related information from New York state Department of Health and MOE from such sport species as brown, rainbow and lake trout, and chinook, atlantic and coho salmon tend to support this classification of tetra-CDD level by fish size.

There are at least two possible sources of 2,3,7,8-tetra-CDD and -CDF contamination in Great Lakes fish. The simplest explanation is that they are being emitted into the environment from a point source, entering the water and being accumulated in certain species. Certainly, the production of chlorinated phenols, related pesticides and PCB at several locations near the Great Lakes over many years and the subsequent emission of these contaminants directly in effluents or indirectly through slow leaching at waste sites would support this explanation. Another possibility, although less likely, is through fly ash generated by incineration of municipal solid waste. Fly ash contains a wide spectrum of dioxins and furans, with 2,3,7,8-tetra-CDD at least a minor component ($<5\%$) of the tetrachlorinated portion. Bioaccumulation of 2,3,7,8-tetra-CDD and -CDF isomers in the fish over the other isomers would lead to the single tetra-CDD or a few TCDF peaks as are found in fish. The fact that only one isomer of tetra-CDD and few of tetra-CDF are found in certain locales and not all Great Lakes makes the second explanation less feasible.

This work has shown that 2,3,7,8-tetra-CDD is a relatively common contaminant of Lake Ontario commercial fish collected in 1980 (about 25% of 62 samples contained levels >10 ppt). Certain species, such as catfish, eel and smelt, had the highest levels and these were associated with a high PCB and lipid content of the fish. Comparison of these data with a more limited sampling of smelt from 1979 and salmonid fish from 1980 indicated slightly lower levels of tetra-CDD in the former and higher levels in the latter categories of fish. Preliminary data are also presented on the presence of residues of 2,3,7,8-tetra-CDF in these fish samples at the same order of magnitude as 2,3,7,8-tetra-CDD.

ACKNOWLEDGMENTS

The authors would like to thank H. McLeod of Health and Welfare Canada and A. Gervais of Fisheries and Ocean Canada for sample collection and delivery. P.

Calway of the latter department kindly carried out the PCB and Mirex determinations. The skillfull technical help of L. Panopio and B. Kennedy effected some of the sample extraction and cleanup.

REFERENCES

1. Zitko, V. *Bull. Environ. Contam. Toxicol.* 7:105 (1972).
2. Anon. *Pest. Toxic Chem. News* 6:23 (November 8, 1978).
3. Harless, R.L., E.O. Oswald, R.G. Lewis, A.E. Dupuy, D.D. McDaniel and H. Tai. *Chemosphere* 11:193 (1982).
4. Ryan, J.J., J.C. Pilon, H.B.S. Conacher and D. Firestone. *J. Assoc. Off. Anal. Chem.* 66:(1983).
5. *Pest. Toxic Chem. News* 9:23 (December 10, 1980).
6. Norstrom, R.J., D.J. Hallett, M. Simon and M.J. Mulvihill. In: *Chlorinated Dioxins and Related Compounds: Impact on the Environment,* O. Hutzinger, R.W. Frei, E. Merrian and F. Pocchiari, Eds. (Oxford: Pergamon Press, 1982), p. 173.
7. Gilman, A.P., G.A. Fox, D.B. Peakall, S.M. Teeple, T.R. Carroll and G.T. Haymes. *J. Wildlife Manage.* 41:458 (1977).
8. Ryan, J.J., and J.C. Pilon. *J. Chromatog.* 197:171 (1980).
9. Schmitt, C.J., J.L. Ludke and D.F. Walsh. *Pest. Monit. J.* 14:136 (1981).
10. Zabik, M., C. Merrill and M.J. Zabik. *Bull. Environ. Contam. Toxicol.* 28:592 (1982).

7

Polychlorinated Dibenzo-*p*-dioxins, Dibenzofurans and Other Polynuclear Aromatics Formed during Incineration and Polychlorinated Biphenyl Fires

C. Rappe, S. Marklund, P.-A. Bergqvist and M. Hansson

In 1977 it was first reported that polychlorinated dibenzo-*p*-dioxins (PCDD) and dibenzofurans (PCDF) could be found in fly ash samples from municipal incinerators and industrial heating facilities [1,2]. During 1978–1982 papers, reports and reviews were published, confirming the original findings [3–6]. Up to now, fewer data have been reported on the levels in incineration products like particulates and flue gas condensates: the true emissions [7].

Due to the extreme toxicity of some of the PCDD and PCDF isomers and the large variation in toxicity and biological potency between closely related isomers, risk evaluation should be based on the levels of the highly toxic isomers found in isomer-specific analyses. However, in most studies the results are given in terms of total levels of tetra-, penta-, hexa-, hepta- and octa-CDD and CDF. The value of such studies is limited, especially in this case, where the number of isomers is quite large; more than 30 PCDD and 60 PCDF have been found in fly ash samples [2,8,9]. Other authors make the assumption of an equal distribution among the isomers [10]. This seems to be an erroneous approach. Using a narrow-bore Silar-10C glass capillary column, Buser and Rappe proved that most of the 22 tetra-CDD isomers could be separated and quantified in one injection [9]. Most of the isomers could be found in a fly ash sample. Figure 7.1 shows the separation of the tetra-CDD in a standard mixture and a fly ash sample. Of special interest is the observation that the highly toxic 2,3,7,8-tetra-CDD is approximately 3–4% of the total level of the tetra-CDD. Moreover, the ratio between the highest and smallest peak is approximately 30:1.

We have previously studied the pyrolysis of polychlorinated biphenyls (PCB). When commercial mixtures or well-defined isomers are heated in air in sealed ampules at 600–800° C, the yields of PCDF are normally in the 1–5%

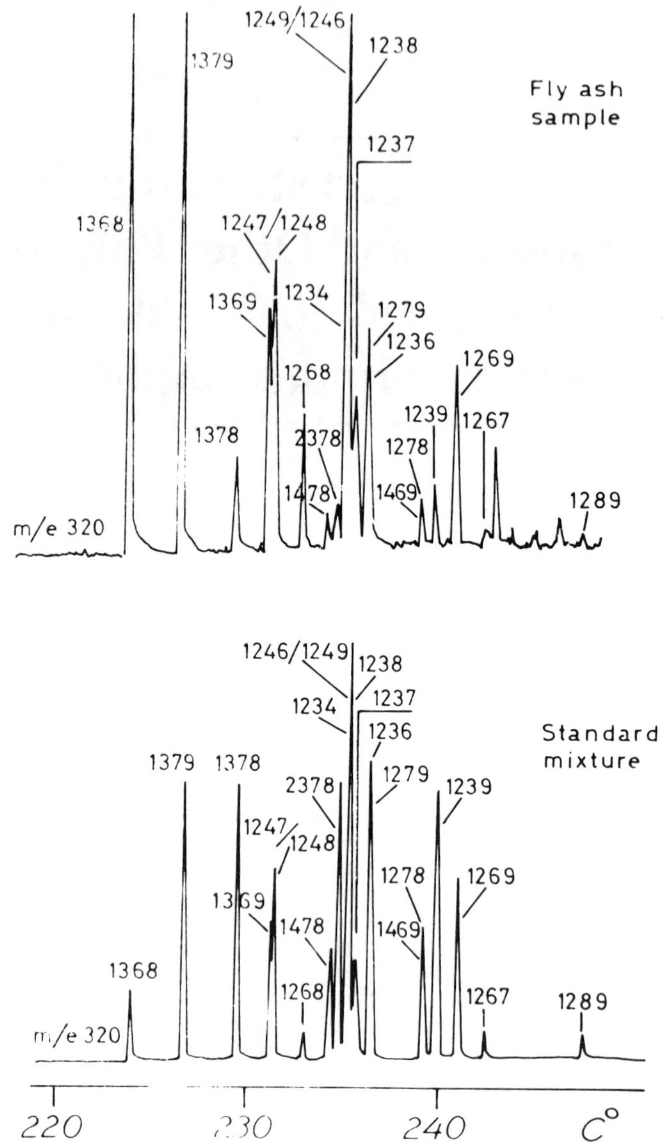

Figure 7.1. Separation of Tetra-CDD in Fly Ash Sample and Standard Mixture (Silar-10C Column).

range, and this method has successfully been used for the preparation of PCDF standards [8,11,12]. These observations indicated that any burning of PCB can be an important environmental source of hazardous PCDF. In 1978 we suggested that all degradation of PCB should be controlled carefully, to avoid accidental formation of PCDF [11]. This statement was later confirmed by two accidental PCB fires, one in a transformer station in downtown Toronto, Canada [13], and the other in a capacitor battery situated 70 km north of Stockholm, Sweden [14]. In the Swedish study, a 50- to 100-fold increase in the levels of PCDF was observed, but the analytical technique did not allow the identification of individual PCDF isomers.

In the present study we report on the analyses of soot samples and wipe tests from three accidental PCB fires and from a test study of thermal destruction of PCB. We have also included analyses of samples from a municipal waste incinerator operating with the fluidized-bed technique and from a test burn of pentachlorophenol (penta-CP) waste in an industrial boiler.

EXPERIMENTAL

Sample Types

Incineration and accidental fires can give rise to a variety of sample types, such as soot, fly ash, bottom ash, particulates, flue gas and flue gas condensates. Some of these sample types have been used in the present investigation.

Extraction and Cleanup

The sample (soot sample or Kleenex® tissue from wipe tests) was spiked with 1–5 ng of $^{13}C_{12}$-2,3,7,8-tetra-CDD, $^{37}Cl_4$-2,3,7,8,-tetra-CDF and $^{37}Cl_8$-octa-CDD. Thereafter, the sample was treated with 10 mL of 1 *M* hydrochloric acid for one hour (shaking). The slurry was filtered by suction in a Buchner funnel, washed with water and dried in the air.

The dried material was transferred to a Soxhlet extractor and extracted with toluene (about 100 mL) for 8–48 hr at a speed of 1–2 mL/min. The toluene was evaporated to 100 μL in a rotary evaporator with some external heating; the rest of the solvent was eliminated by a purified N_2 flush without external heating. The residues were dissolved in 2 mL of *n*-hexane and added to a silica gel column prepared from 0.5 g of Kieselgel 60, 70–230 mesh (Merck) stored at 130°C and eluted with 5 mL of *n*-hexane. The hexane was evaporated to 100 μL using a stream of nitrogen; 2 μL of this solution was used for the analysis of polychlorinated biphenylenes (PCBP). PCDD and PCDF were segregated from other polychlorinated impurities using an Alox column (1.0 g basic Alox activ.1, Woelm, ICN, West Germany, stored at 190°C). The first fraction, 10 mL *n*-hexane: methylene chloride (98:2) was discharged. The second fraction, 10 mL *n*-hexane: methylene chloride (1:1) was collected and evaporated to dryness in a stream of nitrogen. The residue, dissolved in 100 μL solvent, was used for the gas chromatography/mass spectrometry (GC/MS) analyses.

Isomer-Specific High-Resolution GC/MS Analyses

We used two glass capillary columns, one 25-m SE-54 (0.34 mm i.d.) and one 55-m Silar-10C (0.26 mm i.d.) and a 60-m SP-2330 fused-silica column (0.26 mm i.d.). The GC conditions are summarized in Table 7.1. The glass columns were attached via an open split and 0.5 m of a fused-silica column directly into the ion source of the MS. A Finnigan Model 4021 mass spectrometer was used operating in the EI or NCI mode (CH_4) (Table 7.2).

Table 7.1. GC Conditions

	SE-54	*Silar-10C*	*SP-2330*
Column Length (m)	25	55	60
Column i.d. (mm)	0.34	0.26	0.26
Interface, Length (mm)	0.5	0.5	
Interface, i.d. (mm)	0.16	0.11	
Pressure (Helium) (psi)	21	55	40
Injector Temp. (°C)	230	230	230
Temp. Program	80°C for 2 min	80°C for 2 min	100°C for 2 min
	80–160°C at 20°C/min	80–160°C at 20°C/min	100–200°C at 20°C/min
	160–260°C at 2°C/min	160–240°C at 2°C/min	200–240°C at 3°C/min
	260°C for 10 min	240°C for 10 min	240°C for 10 min

Table 7.2. MS Conditions

	EI	*NCI*
Electron Energy (eV)	70	120
Emission Current (mA)	34	34
Electron Multiplier (kV)	1.5	1.5
Sensitivity (A/V)	10^{-8}	10^{-8}
High Vacuum (torr)	5×10^{-7}	5×10^{-5}
Forepressure Ionizer (torr)	0.01	0.30
Ionizer Temp. (°C)	270	180
Manifold Temp. (°C)	130	130
Transmitter Line Temp. (°C)	270	270
Interface Line Temp. (°C)	250	250
Reagent Gas	—	CH_4

Standards and Reference Compounds

The present investigation has concentrated on the following groups of compounds:

- PCDD,
- PCDF,
- PCBP,
- polychlorinated pyrenes (PCPY), and
- polychlorinated chrysenes (PCCY).

The structures for these compounds are shown in Figure 7.2.

The PCDD and PCDF standards used (tetra- to octa-) have been described elsewhere [9,15,16]. No standards were available for mono-, di- and tri-CDF. A solution of 2,3,6,7,-tetra-CBP was obtained from D. Stalling, Columbia, Missouri. No other qualitative or quantitative PCBP, PCPY or PCCY standards are available.

Individual MS response factors have been determined for 28 different PCDF [17], and were used for the quantification of these isomers. For unidentified isomers, the response factors for 2,3,7,8-tetra and 2,3,4,7,8-penta-CDF were used.

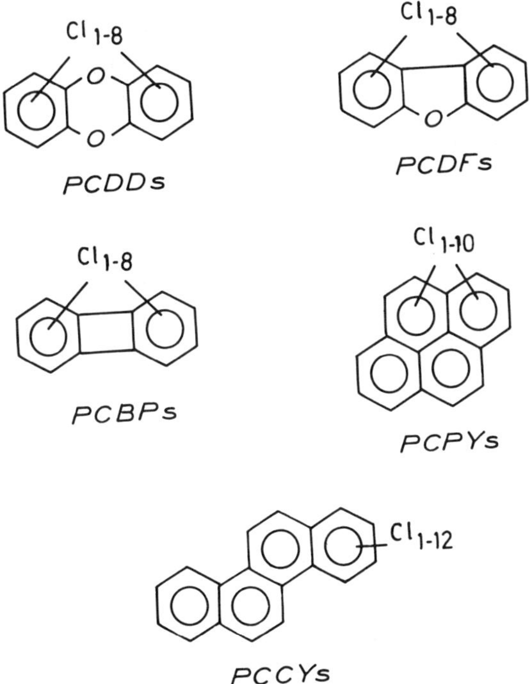

Figure 7.2. Structures for the Compounds Investigated.

RESULTS

Fluidized-Bed Municipal Waste Incinerator

We have analyzed four samples from an incinerator burning municipal waste using a fluidized bed. The incinerator is situated in Eksjö, Sweden, and operates at approximately 700°C, with a retention time of about 2 sec. The samples we analyzed were extracts of particulates and condensates taken during two different sampling periods; each sample corresponds to an emission of about 3 m^3.

From Table 7.3 it can be seen that there is a large variation in the PCDD and PCDF levels between the two sampling periods. It is also clear that the levels in the condensates are higher than those found in the particulates. Within each class of compounds, there are a large number of peaks (20 peaks for tetra-CDF and 16 peaks for tetra-CDD), but tetra-CDD (and 2,3,7,8-tetra-CDD) could only be found in one of the four samples, a condensate. It is interesting to notice the difference in tetra-CDD isomers in this sample (Figure 7.3, top) and the "normal" fly ash sample in Figure 7.1. The toxic 2,3,7,8- and 1,3,7,8- isomers are much more abundant in this sample, while 1,3,6,8-tetra-CDD is a smaller peak. The lower curve in Figure 7.3 are the tetra-CDF analyzed on a Supelco SP-2330 fused-silica column. On this column, 2,3,7,8-tetra-CDF can be separated from the 2,3,4,8-isomer [17].

Table 7.3. Analyses of Samples from a Fluidized-Bed Municipal Incinerator (ng/sample), Sample Volume = 3 m^3

	Condensates				*Particulates*			
	1		*2*		*1*		*2*	
Tetra-CDD								
All	<0.5		68	(12)[a]	<0.5		<0.5	
2,3,7,8-	<0.1		3		<0.1		<0.1	
Tetra-CDF								
All	205	(20)	1574	(20)	110		75	
2,3,7,8-	17		35		9		4	
Penta-CDD	<0.5		<1		<0.5		<0.5	
Penta-CDF	17	(4)	260	(5)	30	(5)	13	(2)
Hexa-CDD	<0.1		<0.1		<0.1		<0.1	
Hexa-CDF	30	(3)	240	(6)	75	(5)	13	(3)
Hepta-CDD	20	(2)	74	(2)	60	(2)	35	(2)
Hepta-CDF	22	(3)	45	(3)	74	(3)	25	(3)
Octa-CDD	11		25		40		30	
Octa-CDF	4		17		27		25	

[a]Number of isomers is given in parentheses.

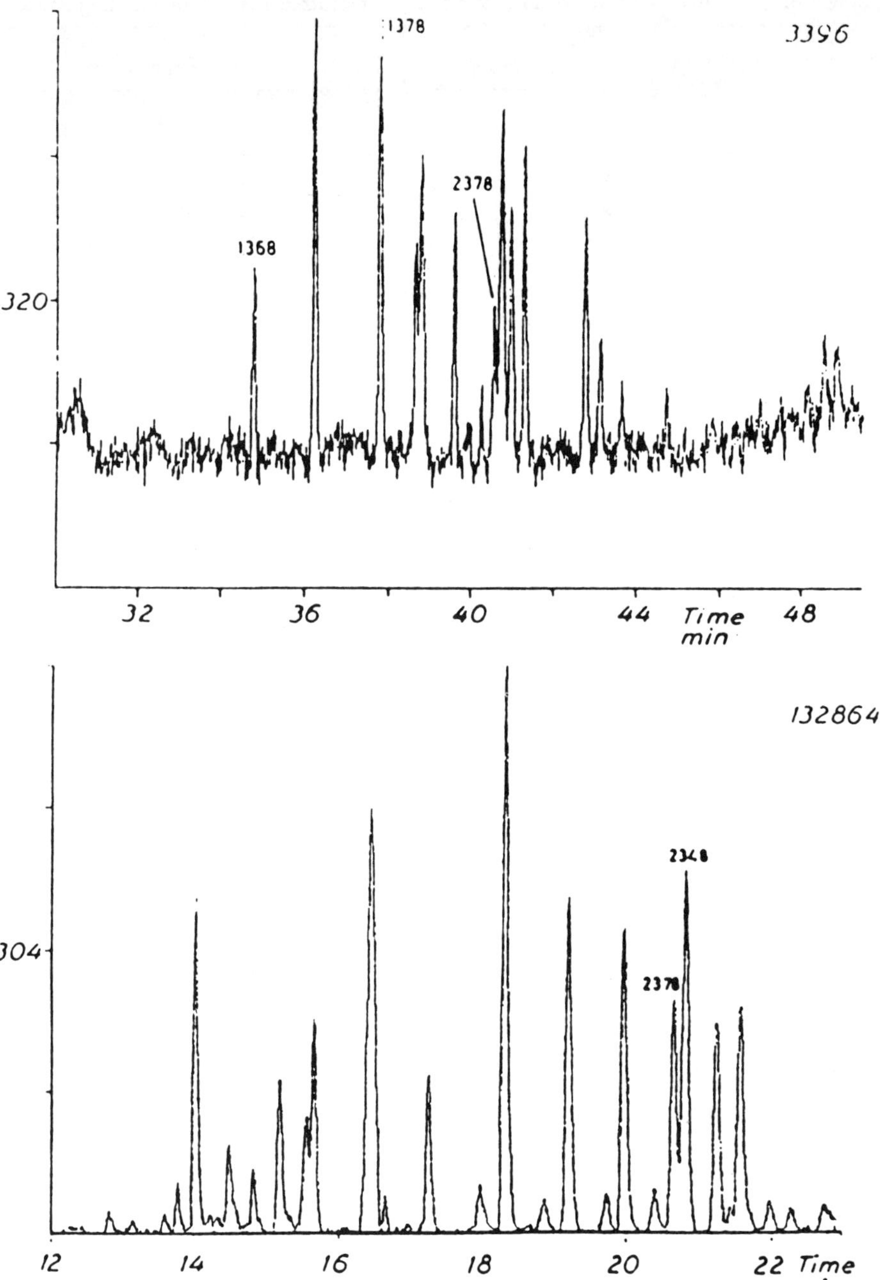

Figure 7.3. Tetra-CDD (Silar-10C Column, top) and -CDF (SP-2330 Column, bottom) in Condensate from Fluidized-Bed Incinerator.

Table 7.4. PCDF/PCDD Ratios in Samples from Fluidized-Bed Municipal Incinerator

	Condensates		*Particulates*	
	1	*2*	*1*	*2*
Cl_4	>500	>25	>100	>100
Cl_5	>30	>250	>60	>25
Cl_6	>300	>1000	>500	>200
Cl_7	1	2	1.2	0.8
Cl_8	0.36	0.68	0.60	0.83

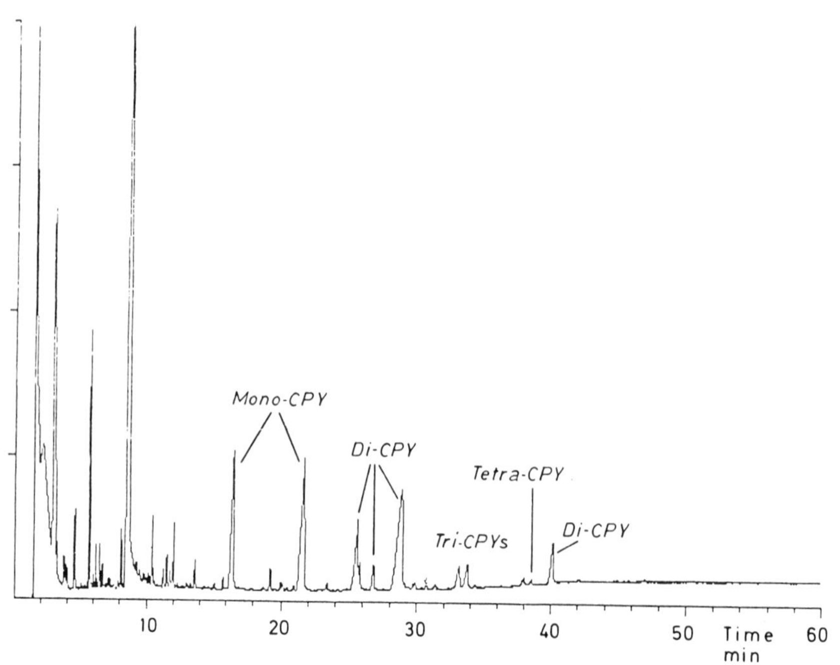

Figure 7.4. Compounds in Condensate from Fluidized-Bed Incinerator.

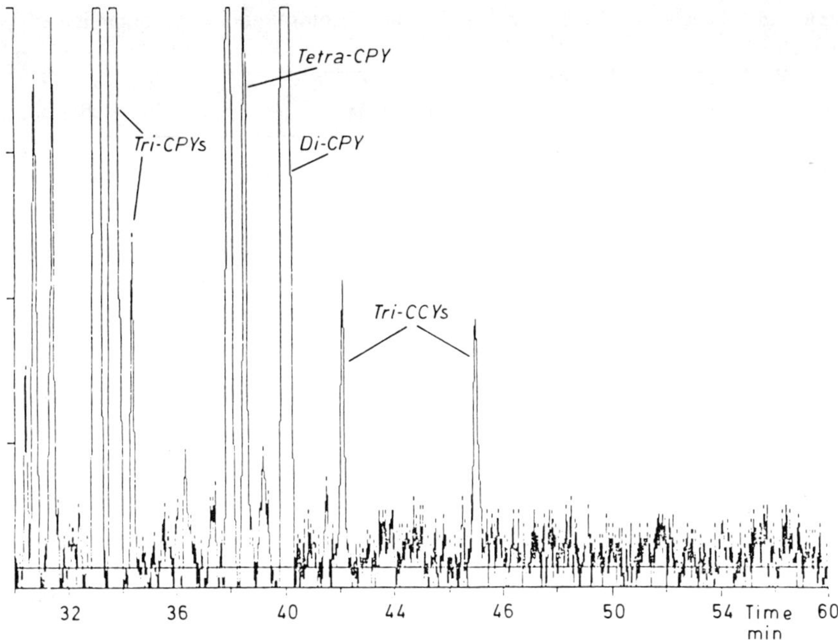

Figure 7.4. *(continued)* Compounds in Condensate from Fluidized-Bed Incinerator.

We have also investigated the PCDF/PCDD ratio in this incinerator (Table 7.4). Among the tetra-, penta- and hexa- compounds, we found a marked predominance for the PCDF over the PCDD, but for the octa- compounds the ratio was reversed. We also found tetra-CDF to be the dominating chlorination level among the PCDF, while hepta-CDD was the dominating chlorination level among the PCDD. These observations indicated differences in precursors or formation pathway between PCDD and PCDF.

Examination of the total ion current (TIC) curves from the analysis of the second fraction from the Alox column reveals that PCDF and PCDD are only minor components; the major products are PCPY and PCCY (Figure 7.4). Such compounds have never found any industrial use; consequently, it can be assumed that they are formed [together with other polynuclear aromatic hydrocarbons (PAH)] during combustion in this municipal incinerator. Polyvinyl chloride (PVC) is a very common source of organochlorine compounds in municipal waste.

Table 7.5. Levels of PCDD and PCDF from Burning Penta-CP–Contaminated Waste (= μg/g)

	Baghouse Ash	*Bottom Ash*
Tetra-CDD		
All	0.96	0.01
2,3,7,8-	<0.005	<0.001
Penta-CDDs	1.4	0.02
Hexa-CDDs	2.0	0.04
Hepta-CDDs	0.7	0.10
Octa-CDD	0.2	0.14
Tetra-CDF		
All	0.90	ND
2,3,7,8-	0.10	ND
Penta-CDF		
All	1.5	ND
1,2,3,7,8-	1.5	ND
2,3,4,7,8-	0.10	ND
Hexa-CDF		
All	0.15	ND
1,2,3,4,7,8-	0.02	ND
Hepta-CDF	0.06	ND
Octa-CDF	0.006	ND

Incineration of Penta-CP Waste in an Industrial Boiler

A test burn of penta-CP waste was performed in an industrial boiler in the United States. Penta-CP is a well known precursor to octa-CDD [18]. Samples of baghouse ash and bottom ash were analyzed and the results are collected in Table 7.5. (see also Figures 7.5 and 7.6, lower curves).

In a baghouse ash, total level of PCDD was 5.3 μg/g, but the amount of octa-CDD was only 0.2 μg/g. The major constituents were lower-chlorinated PCDD like hepta-, hexa- and penta-CDD. These compounds are probably formed by dechlorination reaction from octa-CDD (the penta-CP dimerization product) as previously described [19].

Figure 7.5 compares the dioxin content in the baghouse ash and a "normal" fly ash. The similarity is striking. In both cases 2,3,7,8-tetra-CDD is a very minor constituent.

In the bottom ash we found the dioxin profile to be different (see Table 7.5). The highest values were found for the hepta- and octa-CDD; the lower-chlorinated dioxins had much lower values. The dechlorination reaction discussed above apparently takes place in the flames.

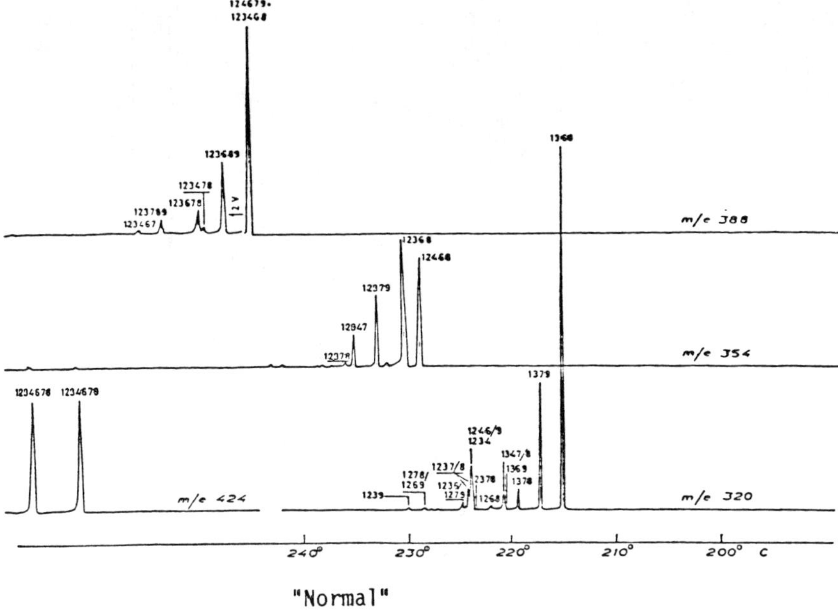

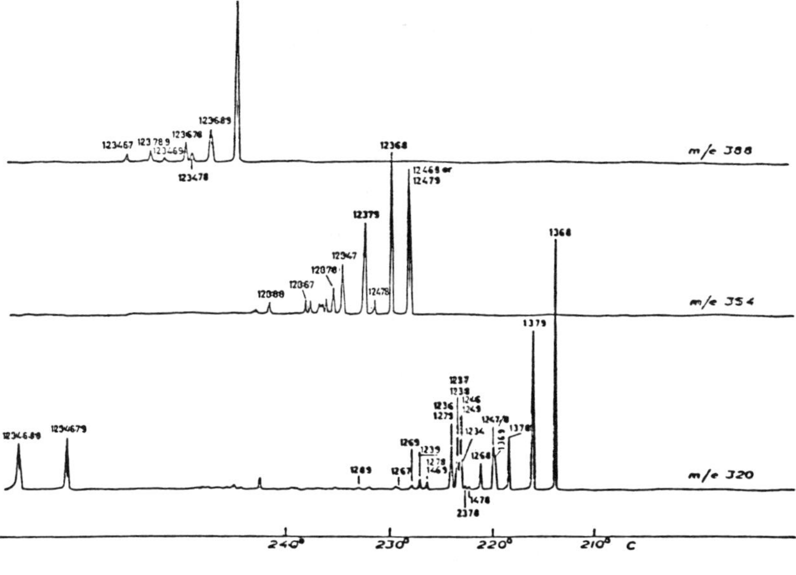

Figure 7.5. PCDD Found in Baghouse Ash from Penta-CP Incineration and "Normal" Fly Ash.

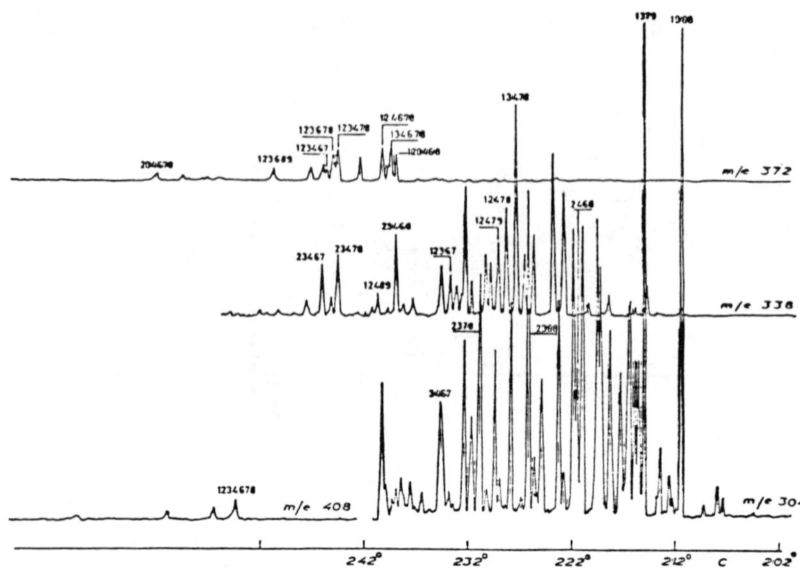

"Normal"

Figure 7.6. PCDF Found in Baghouse Ash from Penta-CP Incineration and "Normal" Fly Ash.

Table 7.6. PCDF/PCDD Ratio in Samples from Burning Penta-CP-Contaminated Waste

	Baghouse Ash	*Bottom Ash*
Cl_4	0.96	0
Cl_5	1.05	0
Cl_6	0.14	0
Cl_7	0.08	0
Cl_8	0.03	0

We also found the baghouse ash to be contaminated by PCDF at a total level of 2.5 $\mu g/g$ (see Table 7.5 and Figure 7.6). For the tetra- and penta- compounds we found approximately equal amounts of PCDF and PCDD (a PCDF/PCDD ratio of ~ 1.0, see Table 7.6). This is completely unexpected, since CP can form PCDD by

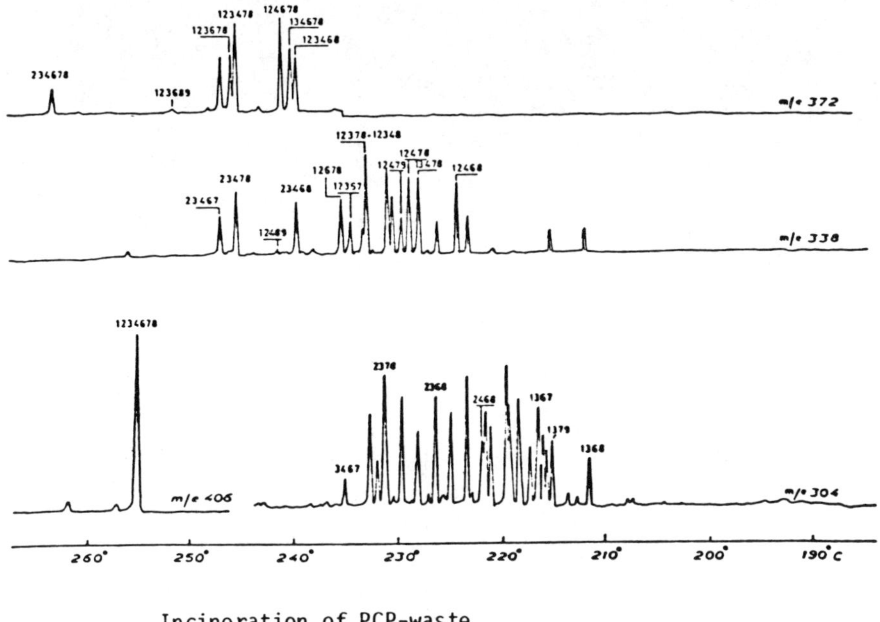

Figure 7.6. *(continued)* PCDF Found in Baghouse Ash from Penta-CP Incineration and "Normal" Fly Ash.

a normal dimerization reaction, but they are not known to be precursors to PCDF. However, the level of octa-CDD is much higher than the level of octa-CDF.

In Figure 7.6 we can see that the PCDF pattern in this sample is different from the pattern found in a "normal" fly ash sample.

No PCDF could be found in the bottom ash, where the total level of PCDD was 0.31 $\mu g/g$. This observation is another indication for a difference in the formation pathway between PCDD and PCDF; the PCDF might be formed preferentially in the flames.

Examination of the traces m/z 270 (tri-CDF), 304 (tetra-CDF) and 338 (penta-CDF) shows that other products in the baghouse ash are a group of compounds eluting much later than PCDF (Figure 7.7). PCPY have the same molecular weight as the PCDF (di-CPY = 270; tri-CPY = 304; tetra-CPY = 338). Although we are lacking standard compounds, we assume that PCPY are the major products observed. A multitude of unchlorinated PAH have been reported in the emissions from this test burn [20]. We assume a similar pathway for the thermal formation of chlorinated and nonchlorinated aromatics.

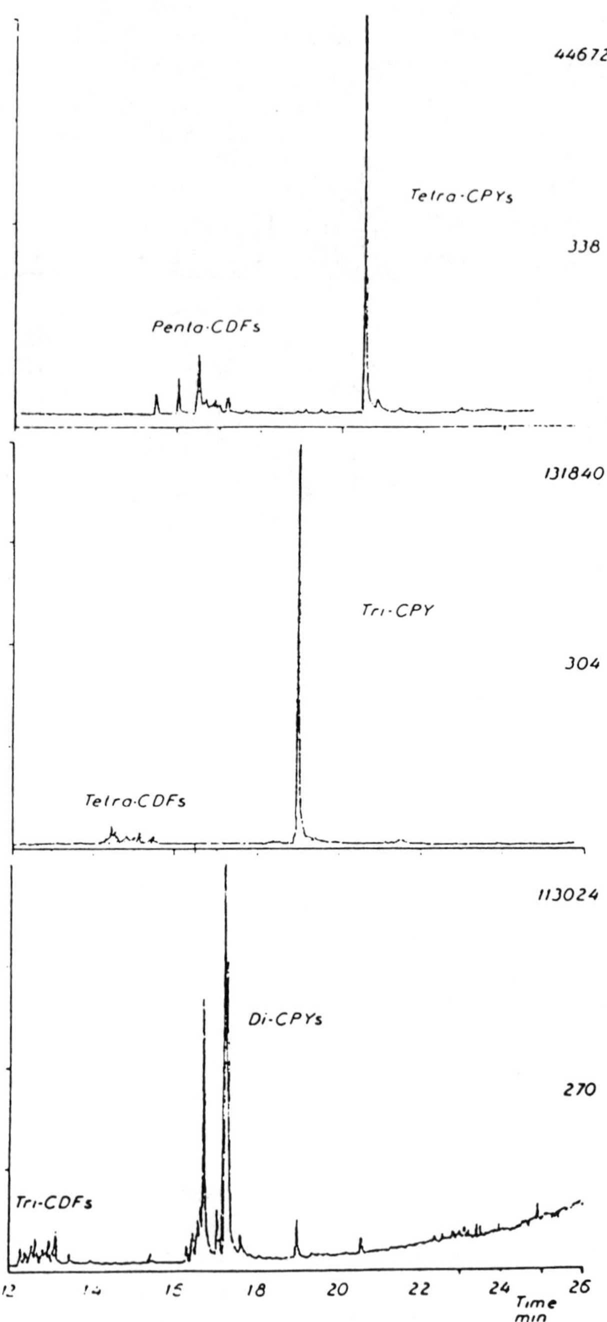

Figure 7.7. PCDF and PCPY in Baghouse Ash Sample from Incineration of Penta-CP Waste.

Table 7.7. Amounts of PCDF and PCDD Found in PCB Destruction (Rotary Kiln)

	Condensate (ng/l)	XAD-2 Filter 1 (ng/g)	XAD-2 Filter 2 ng/g	Filter 1 (ng)	Filter 2 (ng)	Electro-Filter (ng/g)	Sintered Cement (ng/g)
Tri-CDF	<0.04	<0.2	<0.05	<0.6	<0.2	<25	<500
Tri-CDD	<0.1	<0.2	<0.05	<1.9	<0.2	<12	<500
Tetra-CDF	<0.2	<0.1	<0.2	<6.3	<0.1	<1.0	< 50
Tetra-CDD	<0.2	<0.1	<0.05	<3.0	<0.1	<1.0	<50
Penta-CDF	<0.04	<0.05	<0.05	<6.3	<0.2	<2.5	<50
Penta-CDD	<0.1	<0.05	<0.05	<6.3	<0.2	<1.5	<50
Hexa-CDF	<2.4	<0.7	<0.3	<6.3	<0.9	<2.5	<250
Hexa-CDD	<0.6	<0.7	<0.2	<6.3	<0.9	<2.5	<250
Hepta-CDF	<4.6	<1.4	<3	<6.3	<0.2	<5.0	<50
Hepta-CDD	<0.2	<0.1	<0.05	<0.3	<0.1	<0.3	<5
Octa-CDF	<1.0	<0.3	<0.3	<1.3	<0.4	<0.5	<250
Octa-CDD	<0.4	<0.2	<0.1	<0.6	<0.4	<0.5	<250

Thermal PCB Destruction in Rotary Cement Kiln

We have analyzed a series of samples taken during a test study of the thermal destruction of PCB in a rotary cement kiln in Slemmestad, Norway [21,22]. The test kiln is 155 m long with a diameter of 4 m. It is a "wet" oven operating at 1200–1300°C in the solid material and up to 2000°C in the flames. Normal fuel consumption is 92 metric ton/day for a production of 850 ton/day of clinker. During the test period, 4 m^3 of PCB was burned at a feed rate of 50 kg/hr. The destruction efficiency of PCB was found to be better than 99.9997%. The results from the PCDF and PCDD analyses are collected in Table 7.7. No PCDF or PCDD could be found in any of the samples, but in a few samples we found traces of tetra-CBP. This is in agreement with other reports [21,22].

Accidental PCB Fires

Capacitor Fire, Skovde, Sweden

In March 1982 a violent fire broke out in a 400-V capacitor battery serving a high-frequency oven in a casting line of a metal treatment factory in Skövde, Sweden. The dielectric in the capacitors consisted of mineral oil or PCB; the two types of capacitors were mixed at random. The fire started in a capacitor containing mineral oil, and about 2 hours elapsed before the fire was completely extinguished. The capacitor room, situated in the basement of the foundry, was burned out; the copper bars for electricity in the ceiling were partly melted (m.p. = 1080°C). The smoke spread in the building in an area of 60 × 30 m. The capacitor battery contained 21 capacitors filled with PCB (5 kg each). An inspection after the fire showed 12 of these capacitors had been opened and 9 of them were still sealed.

The following samples were analyzed (wipe tests, Kleenex tissue, 1 dm^2).

1. on the floor, in the capacitor battery;
2. on the floor, close to the capacitor battery;
3. on the wall in the capacitor room, 2 m from the capacitor battery, 3 m above the floor; and
4. on a bench 10 m from the oven, in the ground floor.

The results are collected in Table 7.8. The results show that high levels of PCDF (>0.1 $\mu g/m^2$) could only be found in an area close to the fire. At a distance of 10 m the levels were about 100 times lower. The low levels in sample 3 can be explained by the high temperature at this sampling location (the copper bar was partly melted here).

The fragmentogram of the PCDF (tri- to octa-CDF) is shown in Figure 7.8. Making the assumption of an equal response for tri- and tetra-CDF, the levels of tri-CDF are about the same as those of tetra-CDF. The highly toxic 2,3,7,8-tetra-CDF is one of the major tetra-CDF isomers. In addition to the PCDF, Figure 7.8 shows the presence of another series of chlorinated compounds. The molecular weight and number of chlorine atoms indicate that this series is PCPY (see Figure 7.2). Di-CPY are the dominating compounds at m/z 270. However, final proof cannot be obtained due to lack of synthetic standards. It can be noted that no PCDD and no PCBP could be identified in the samples from this fire.

Table 7.8. Levels of PCDF (ng/m^2) for Samples from the Skövde Fire.

	Sample Number			
	1	*2*	*3*	*4*
2,3,7,8-Tetra-CDF	20	100	1	ND[a]
Total Tetra-CDF	100	600	1	10[b]
Total Penta-CDF	40	100	ND	ND
Total Hexa-CDF	40	60	ND	ND
Total Hepta-CDF	8	8	ND	ND
Octa-CDF	5	5	ND	ND

[a]ND = not detected.
[b] Different isomers from those found in samples 1 and 2.

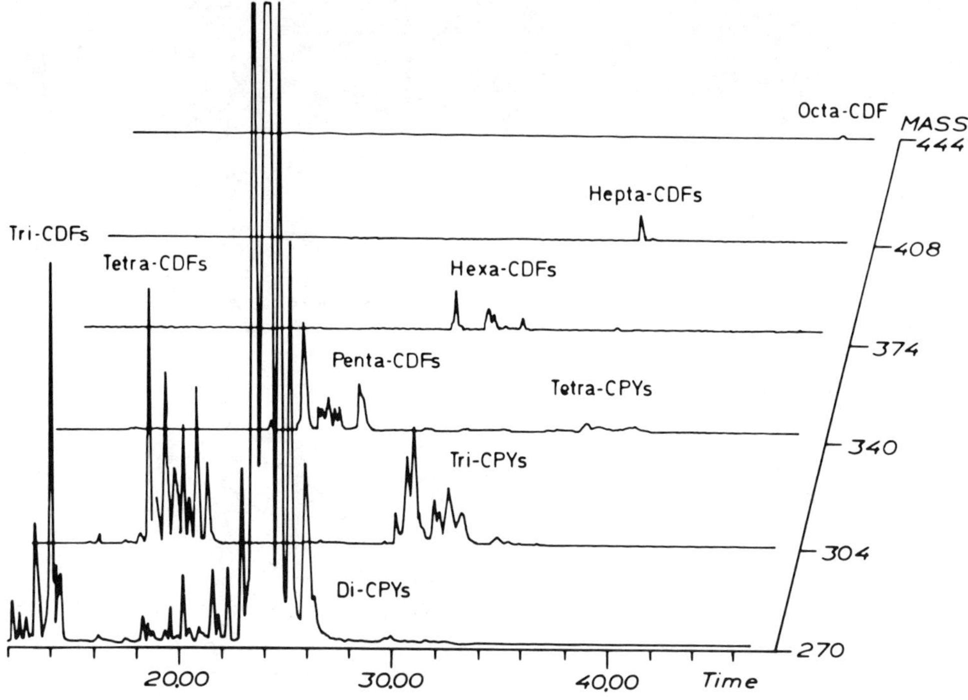

Figure 7.8. Mass Fragmentogram of Extract from Accidental PCB Fire, Skövde, Sweden.

Capacitor Fire, Stockholm, Sweden

In August 1981 an explosion (external arc) took place in a 11-kV capacitor battery in an electrical power station in Stockholm, Sweden. The explosion was caused by malfunction in the capacitor battery. We have analyzed wipe tests from this fire taken about 1 m from the capacitors. First we analyzed the extract after silica gel cleanup (Figure 7.9). In addition to PCB, which were not totally removed from the sample, the chromatogram contained several other peaks. Inspection of the complete mass spectra revealed these peaks to be a series of compounds with a molecular weight corresponding to di-, tri-, tetra-, and penta-CBP, probably formed by a direct cyclization of PCB.

Figure 7.10 shows the mass fragmentograms of the ions corresponding to di- to penta-CBP. Several isomers could be found within each group. The unavailability of synthetic standards creates a problem for identification and quantification, but the presence of PCBP (or the isomeric polychlorinated acenaphthylenes) could be checked by following the M and M+2 traces. For most of the peaks in Figure 7.10 we found the ratio between the M and M+2 ions to be in accord with the expected value for the chlorine cluster.

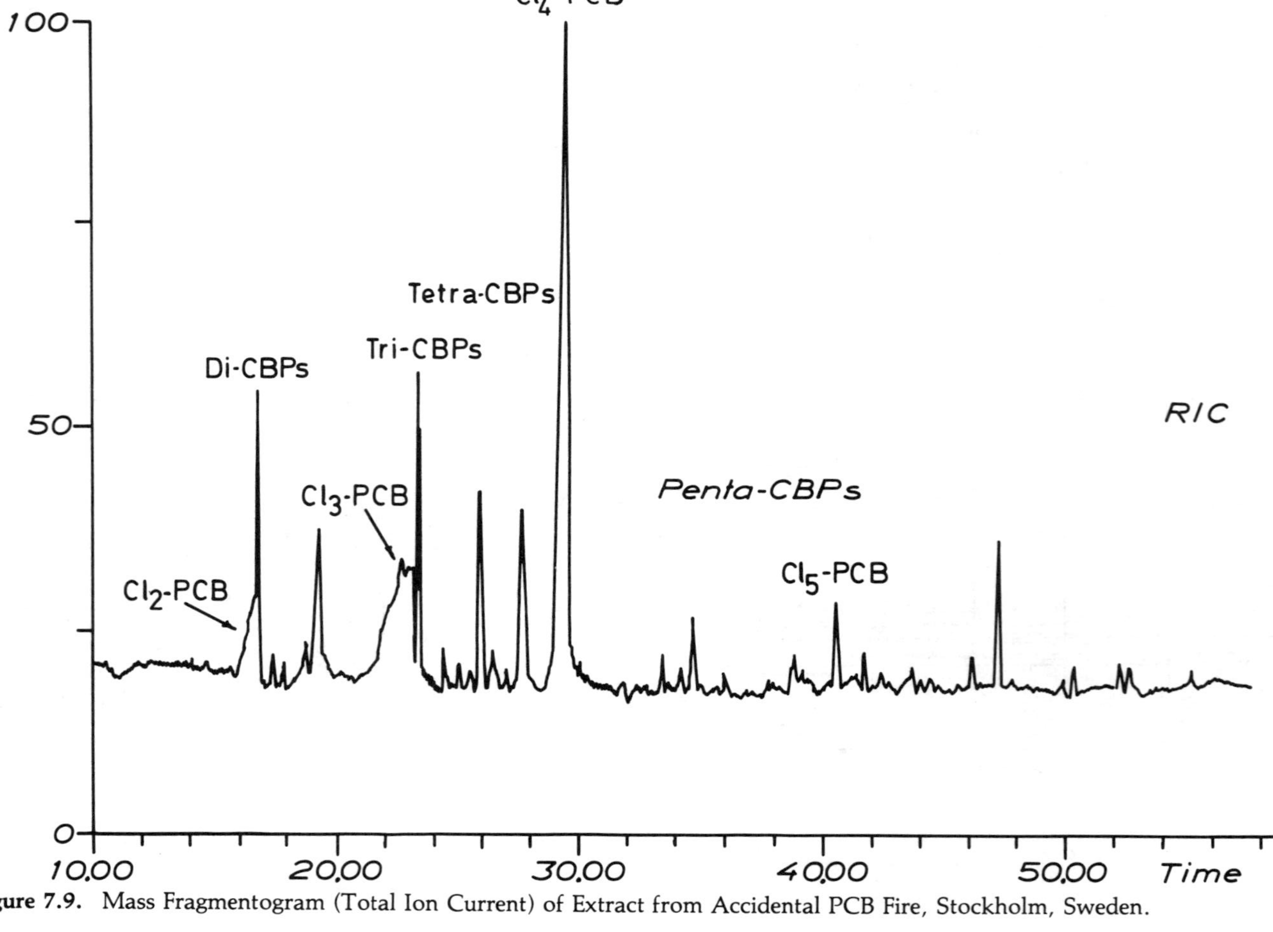

Figure 7.9. Mass Fragmentogram (Total Ion Current) of Extract from Accidental PCB Fire, Stockholm, Sweden.

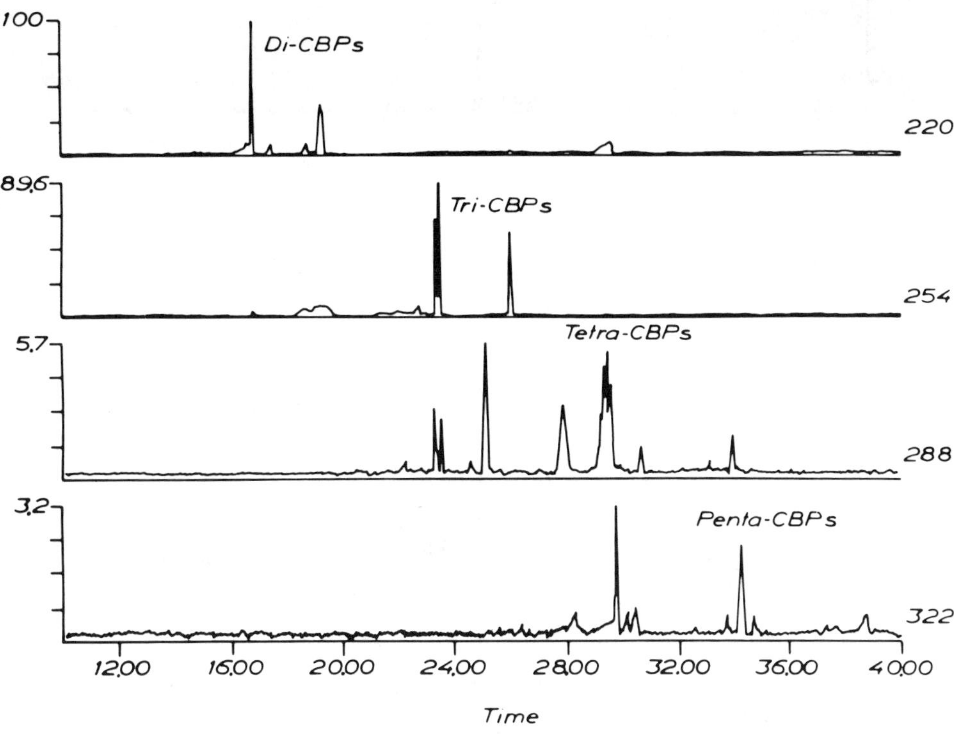

Figure 7.10. Mass Fragmentogram (PCBP) of Extract from Accidental PCB Fire, Stockholm, Sweden.

The new MS/MS technique provides another possibility to check for the presence of PCBP. We have used this method to analyze for tetra-CBP (288) where high levels of tetra-CBP (290) could interfere. We selected 288^+ as the parent ion, scanning for daughter ions. From the synthetic standard 2,3,6,7-tetra-CBP, we found 253^+ and 218^+ to be the major daughter ions. The MS/MS analysis is shown in Figure 7.11. All the major peaks in the extract had the expected daughter ions; the number of tetra-CBP isomers found in this sample was about ten.

The extract from the Alox cleanup step was analyzed for PCDF (and PCDD) in the usual way; the quantitative data are given in Table 7.9. The level of PCDF was found to be 1.4 $\mu g/m^2$, which is approximately two times that of the Skövde fire. However, it should be kept in mind that in the Stockholm fire we should also consider the PCBP present. A round estimation (making the assumption of equal response factors for PCBP and PCDF) gave the levels of PCBP to be about 25–30 $\mu g/m^2$.

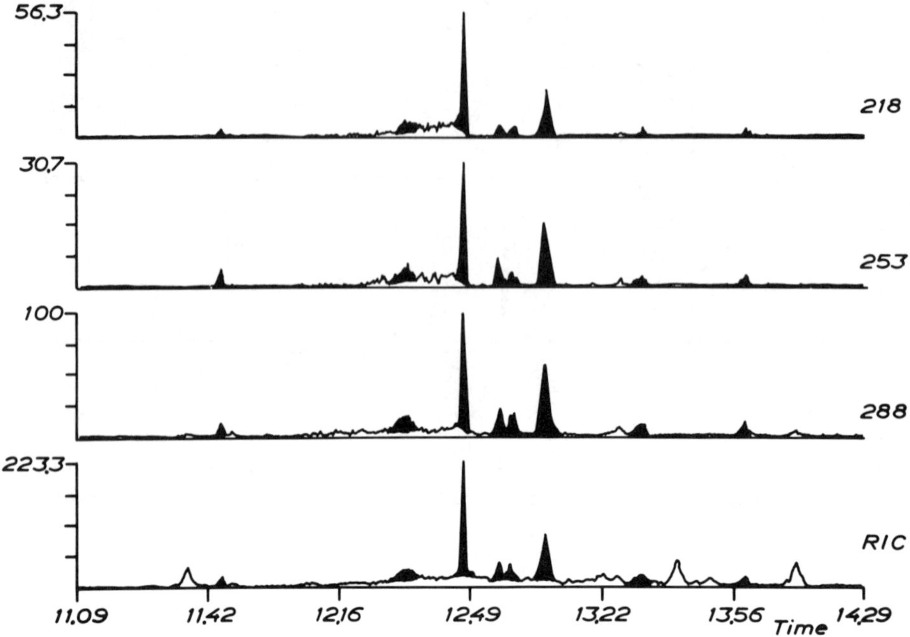

Figure 7.11. Mass Fragmentogram (MS/MS) of Tetra-CBP from Accidental PCB Fire, Stockholm, Sweden.

Figure 7.12 shows the fragmentograms for the ions m/z 202 (mono-CDF and pyrene), 236 (di-CDF and mono-CPY), 270 (tri-CDF and di-CPY), 304 (tetra-CDF and tri-CPY) and 340 (penta-CDF and tetra-CPY). The highest peaks were found for pyrene, mono- and di-CPY, and tetra- and penta-CDF. The observation of PCPY is worth special attention, as PCB are the only known chlorinated constituent in this fire. No PCDD could be found in these samples.

Among the PCDF we can identify a number of isomers. The highly toxic 2,3,7,8-tetra-CDF and 2,3,4,7,8-penta-CDF could both be identified as major constituents. It should be noted that individual response factors [17] have been used to calculate the levels given in Table 7.9.

Transformer Fire, Binghamton, New York

In February 1981 a fire in the State Office Building in Binghamton, New York, caused a transformer to rupture, releasing soot in the whole building. The transformer dielectric consisted of a mixture of PCB (65%) and chlorinated benzenes (35%).

The soot was found to be highly contaminated by PCDF; the total level of PCDF was 2160 μg/g (Table 7.10). No wipe tests have been analyzed. The most toxic isomers (2,3,7,8-tetra-CDF, 1,2,3,7,8- and 2,3,4,7,8-penta-CDF, and

Table 7.9. Amount of PCDF Found in Samples from Accidental PCB Fire, Stockholm, Sweden.

PCDF	*Amount (ng/m^2)*	*PCDF*	*Amount (ng/m^2)*
Tetra-CDF		Penta-CDF	
2,3,7,8-	150	2,3,4,7,8-	45
1,2,7,8-	150	1,2,4,7,8-	38
2,3,6,8-	125	1,2,3,6,7-	15
1,4,6,9-	75	1,3,4,7,8-	11
2,4,6,7-	37	2,3,4,6,7-	7.5
3,4,6,7-	7.5	1,2,4,6,8-	3.8
1,3,6,7-/1,3,6,9-	300	1,2,4,7,8-	3.8
Other	750	1,2,3,6,7-	3.8
Total	1600	1,2,4,8,9-	3.8
		2,3,4,6,8-	2
		1,2,3,7,8-/1,2,3,4,8-	15
		Other	19
		Total	175
		Total Hexa-CDF	<0.5

1,2,3,4,7,8- and 1,2,3,6,7,8-hexa-CDF) were found to be the major constituent within each chlorination level.

In addition to PCDF, we could also identify a series of PCDD (see Table 7.10). Unlike the samples discussed above, we found the highly toxic 2,3,7,8-tetra- and 1,2,3,7,8-penta-CDD to dominate. We assume the chlorinated benzenes to be the precursors to the PCDD in this case. It has been shown that both PCDD and PCDF are formed during pyrolysis of polychlorinated benzenes [23]. In the soot we could also identify a series of PCPY (Figure 7.13) and PCBP.

General Remarks on PCB Fires

Accidental PCB fires or explosions in electrical equipment filled with PCB seem to be quite common. During the period August 1981–November 1982 we have analyzed samples from the following capacitor accidents:

- August 1981, Stockholm, Sweden,
- March 1982, Skövde, Sweden,
- August 1982, Imatra, Finland,
- August 1982, Arvika, Sweden,
- September 1982, Surahammar, Sweden, and
- November 1982, Hallstahammar, Sweden.

All of these fires resulted in formation of PCDF.

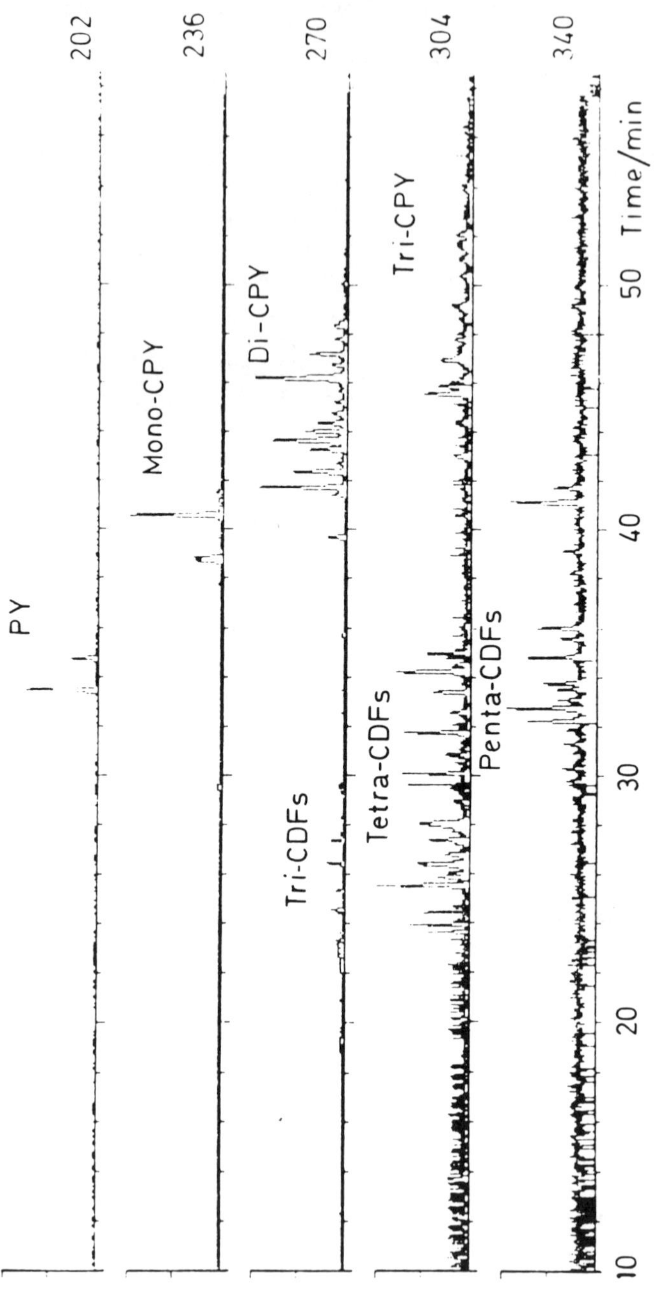

Figure 7.12. PCDF and PCPY from PCB Fire, Stockholm, Sweden.

Table 7.10. Amounts of PCDF and PCDD Found in a Soot Sample from State Office Building in Binghamton, New York.

Compound	*Amount (ng/g)*	*Compound*	*Amount (ng/g)*
Tetra-CDF		Tetra-CDD	
8 isomers	28	5 isomers	1.2
2,3,7,8-	12	2,3,7,8-	0.6
Penta-CDF		Penta-CDD	
20 isomers	670	8 isomers	5.0
1,2,3,7,8-	310	1,2,3,7,8-	2.5
2,3,4,7,8-	48	Hexa-CDD	4.7
Hexa-CDF		Hepta-CDD	7
15 isomers	965	Octa-CDD	2
1,2,3,4,7,8-	310		
1,2,3,6,7,8-	150		
2,3,4,6,7,8-	10		
Hepta-CDF			
4 isomers	460		
Octa-CDF	40		

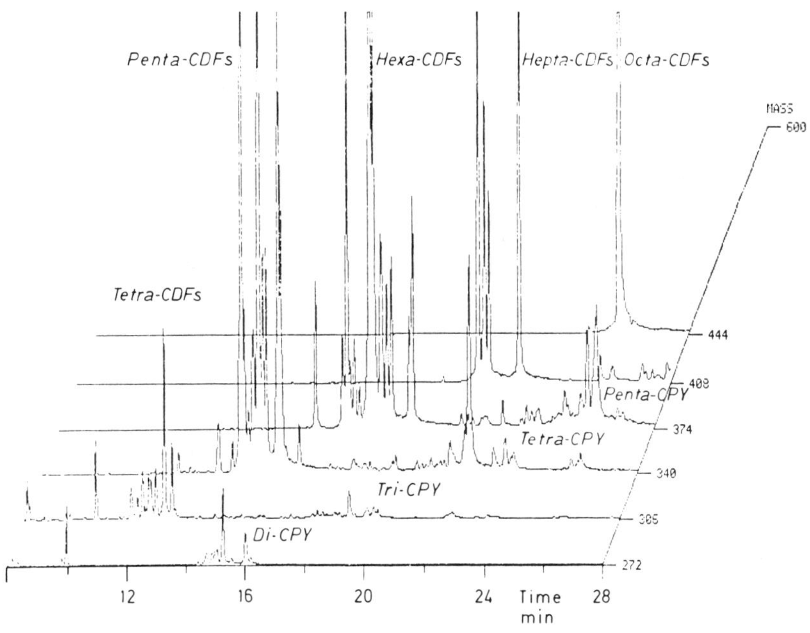

Figure 7.13. PCDF and PCPY in Soot Sample, Binghamton State Office Building.

A comparison between the Skovde fire with the Stockholm fire shows that these two fires produced different products. The Skövde fire is a "normal" fire, where burning of mineral oil generated the heat. The major products formed in this fire were PCDF and PCPY, while no PCBP could be identified.

The Stockholm fire was caused by electrical malfunction. The temperature in this fire (electrical arc) was much higher than in Skovde, and the conditions might have been been partly anaerobic. In this fire the PCBP were the major products while PCDF and PCPY were minor constituents.

CONCLUSIONS

PCDD are only a part of the problems associated with toxic polychlorinated PAH formed during incineration and other thermal reactions. The same or perhaps more attention should be given to PCDF. Attention should also be given to PCBP, PCPY, PCCY and other polychlorinated PAH.

In several cases we found high levels of PCDF where no PCDD could be found. In some cases we found a difference in the PCDD and/or PCDF isomer pattern, indicating different route of formation or different precursors.

The highest levels of PCDD and PCDF are found when special technical products are thermally converted into PCDD and PCDF. Examples of such products are:

chlorinated phenols ⟶ PCDD

PCB ⟶ PCDF and PCBP

chlorinated benzenes ⟶ PCDD and PCDF

Dechlorination of octa-CDD to hepta-, hexa- and other lower-chlorinated dioxins (including tetra-CDD) is another major pathway.

High levels of lower-chlorinated compounds, like mono-, di-, tri-, and tetrachlorinated derivatives of dibenzofurans, pyrenes, chrysenes and other PAH are probably formed in *de novo* reactions similar to the formation of pyrene, benzo[a]pyrene and other PAH in most incineration reactions.

PVC is a very common chlorine source in municipal waste. Chlorinated benzenes have been identified from incomplete combustion of PVC, and they are also frequently found in emission from various incinerators, including municipal waste incinerators. Chlorinated benzenes have been identified as precursors to PCDF and PCDD.

De novo formation of dioxins might also take place according to the pathway outlined above. The levels are in general lower and the actual isomers are the two hepta-CDD and octa-CDD.

Under special conditions flue gas and particulates are the major sources of PCDF, PCDD and other polychlorinated PAH. More attention should be given to the analyses of these emissions.

Accidental PCB fires can result in the formation of PCDF, PCBP and PCPY. Careful inspections of electrical equipment filled with PCB are recommended to prevent these accidents.

Analyses of environmental samples indicate a low general background of PCDF, mainly tetra- and penta-CDF [24,25]. It cannot be excluded that combustion and incineration are the ultimate sources of this background.

REFERENCES

1. Olie, K., P.L. Vermeulen and O. Hutzinger. "Chlorodibenzo-*p*-dioxins and Dibenzofurans as Trace Components of Fly Ash and Flue Gas of Some Municipal Incinerators in the Netherlands," *Chemosphere* 6:455–459 (1977).
2. Buser, H.R., H.-P. Bosshardt and C. Rappe. "Identification of Polychlorinated Dibenzo-*p*-dioxin Isomers Found in Fly Ash," *Chemosphere* 7:165–172 (1978).
3. Lustenhouwer, J.W.A., K. Olie and O. Hutzinger. "Chlorinated Dibenzo-*p*-dioxins and Related Compounds in Incinerator Effluents," *Chemosphere* 9:501–522 (1980).
4. Eiceman, G.A., R.E. Clement and F.W. Karasek. "Variations in Concentrations of Organic Compounds Including Polychlorinated Dibenzo-*p*-dioxins and Polynuclear Aromatic Hydrocarbons in Fly Ash from a Municipal Incinerator," *Anal. Chem.* 53:955–959 (1981).
5. Liberti, A., and D. Brocco. "Formation of Polychlorodibenzodioxins and Polychlorodibenzofurans in Urban Incinerator Emissions," in *Chlorinated Dioxins and Related Compounds*, O.Hutzinger, R.W. Frei, E. Merian and F. Pocchiari, Eds. (Oxford: Pergamon Press, 1982), pp. 245–251.
6. Rappe, C., and H.R. Buser. "Polychlorinated Dioxins and Dibenzofurans in Incinerator Effluents," in *Proceedings of the Second World Congress of Chemical Engineering, Vol. VI*, (1982), pp. 8–11.
7. Cavallaro, A., G. Bandi, G. Invermizzi, L. Luciani, E. Mongini and A. Gorni. "Sampling, Occurrence and Evaluation of PCDDs and PCDFs from Incinerated Solid Urban Waste," *Chemosphere* 9:611–621 (1980).
8. Buser, H.R., H.-P. Bosshardt, C. Rappe and R. Lindahl. "Identification of Polychlorinated Dibenzofuran Isomers in Fly Ash and PCB Pyrolyses," *Chemoshpere* 7: 419–429 (1978).
9. Buser, H.R., and C. Rappe. "High-Resolution Gas Chromatography of the 22 Tetrachlorodibenzo-p-dioxin Isomers," *Anal. Chem* 52:2257–2262 (1980).
10. Olie, K., J.W.A. Lustenhouwer and O. Hutzinger. "Polychlorinated Dibenzo-p-dioxins and Related Compounds in Incinerator Effluents," in *Chlorinated Dioxins and Related Compounds*, O. Hutzinger, R.W. Frei, E. Merian and F. Pocchiari, Eds. (Oxford: Pergamon Press, 1982), pp. 227–244.
11. Buser, H.R., H.-P. Bosshardt and C. Rappe. "Formation of Polychlorinated Dibenzofurans (PCDFs) from the Pyrolysis of PCBs," *Chemosphere* 7:109–119 (1978).
12. Buser, H.R., and C. Rappe. "Formation of Polychlorinated Dibenzofurans (PCDFs) from the Pyrolysis of Individual PCB Isomers," *Chemosphere* 8:157–174 (1979).
13. Ministry of the Environment, Ontario, Canada. News Release (February 13, 1978).
14. Jansson, B., and G. Sundstrom. "Formation of Polychlorinated Dibenzofurans (PCDFs) During a Fire Accident in Capacitors Containing Polychlorinated Biphenyls

(PCB)," in *Chlorinated Dioxins and Related Compounds*, O. Hutzinger, R.W. Frei, E. Merian and F. Pocchiari, Eds. (Oxford: Pergamon Press, 1982), pp. 201–207.
15. Buser, H.R. "Polychlorinated Dibenzo-*p*-dioxins and Dibenzofurans: Formation, Occurrence and Analysis of Environmentally Hazardous Compounds," Thesis, University of Umeå (1978).
16. Garå, A. "Studies of Polychlorinated Dibenzofurans," Thesis, University of Umeå (1982).
17. C. Rappe, S. Marklund, M. Nygren and A. Garå. "Parameters for Identification and Confirmation in Trace Analyses of Polychlorinated Dioxins and Dibenzofurans," Chapter 15, this volume.
18. Sandermann, W., H. Stockmann and R. Carstens. "Über die Pyrolyse des Pentachlorophenols," *Chem. Ber.* 90:690–692 (1957).
19. Rappe, C., S. Marklund, H.R. Buser and H.-P. Bosshardt. "Formation of Polychlorinated Dibenzo-*p*-dioxins (PCDDs) and Dibenzofurans (PCDFs) by Burning or Heating Chlorophenates," *Chemosphere* 7:269–281 (1978).
20. DaRos, B., R. Merrill, H.K. Willard, C.D. Wolbach. "Emissions and Residue Values from Waste Disposal During Wood Preserving," U.S. EPA IERL, Cincinnati, OH (1981).
21. "Forsøk med Forbrenning av Special Avfall i Sementovn ved Norcem Cement, avd. Slemmestad," Slemmestad, Norway (1982).
22. K. Trovaag, "Hazardous Waste Incineration in a Cement Kiln" (London: Plenum Publishing Co., in press).
23. Buser, H.R. "Formation of Polychlorinated Dibenzofurans (PCDFs) and Dibenzo-*p*-dioxins (PCDDs) from the Pyrolysis of Chlorobenzenes," *Chemosphere* 8:415–424 (1979).
24. Rappe, C., H.R. Buser, D.L. Stalling, L.M. Smith and R.C. Dougherty. "Identification of Polychlorinated Dibenzofurans in Environmental Samples," *Nature* 292: 524–526 (1981).
25. Stalling, D.L., L.M. Smith, J.D. Petty, J.W. Hogan, J.L. Johnson, C. Rappe and H.R. Buser. "Residues of Polychlorinated Dibenzo-*p*-dioxins and Dibenzofurans in Laurentian Great Lakes Fish," in *Human and Environmental Risks of Chlorinated Dioxins and Related Compounds*. R.E. Tucker, A.L. Young and A.P. Gray, Eds. (Plenum Press, New York, 1983), pp. 221–240.

8

Assessments of Incineration Processes as Sources of Supertoxic Chlorinated Hydrocarbons: Concentrations of Polychlorinated Dibenzo-*p*-dioxins/Dibenzofurans and Possible Precursor Compounds in Incinerator Effluents

Michael L. Taylor, Thomas O. Tiernan, John H. Garrett, Garrett F. VanNess and Joseph G. Solch

The chemistry and toxicology of polychlorinated dibenzo-*p*-dioxins (PCDD) and dibenzofurans (PCDF) have been summarized in several reviews and texts [1–8]. These reports describe a wide range of toxic effects that have been attributed to PCDD and PCDF on the basis of results of in vivo and in vitro laboratory investigations; these include: lethality, acnegenicity, embryotoxicity, enzyme induction effects, central nervous system (CNS) toxicity, porphyria, hepatotoxicity, and carcinogenicity.

Many of the toxic effects enumerated above are not observed uniformly in all species; in addition, some published whole-animal toxicological data, as well as data derived from in vitro assays, indicate that the various PCDD/PCDF isomers are not equally toxic. It must also be noted that not all of the toxic effects cited above have been documented as occurring in humans. However, several accidents have occurred in recent times in which relatively large numbers of persons were exposed to chemical mixtures containing PCDD or PCDF. Examples of such exposures include the Missouri horse arena exposures [9], the Yusho incident in Japan [10] and a similar incident in Taiwan [11], the industrial exposure that occurred in Nitro, West Virginia [12], and the Seveso, Italy, accident [13]. A cursory examination of the clinical and toxicologial data that have resulted from the investigations of these accidents leads to the conclusion that acute exposures to chemical mixtures containing PCDD or PCDF are apparently not as lethal to man as to certain animal species. However, chloracne and other clinically perceptible symptoms of toxicity (hepatotoxicity, CNS toxicity) were frequently observed among the exposed persons; thus, these clinical findings clearly indicate that chemical mixtures containing PCDD/PCDF are definitely toxic to humans. The long-term effects of exposures to chemical mixtures containing PCDD/PCDF are ill defined at this writing.

All of the instances of human exposure cited above involved chemical mixtures that contained PCDD or PCDF. In fact, not only did the mixtures contain several chemical compounds, including PCDD/PCDF, but in many cases, multiple PCDD/PCDF isomers were present. This is important because, when one attempts to assess the threat that PCDD/PCDF pose to human health, the apparent differences in the toxicities of each of the 75 PCDD isomers and 135 PCDF isomers must be considered. Based on the results of whole animal studies and in vitro assays, the symmetrically substituted tetrachlorinated isomers (2,3,7,8-tetra-CDD and 2,3,7,8-tetra-CDF) appear to be the most toxic. A model for the toxic effects [14] suggests that one can predict the toxicity of a particular PCDD/PCDF isomer based on the number and position of chlorine substituents around the aromatic ring. This model appears to explain the animal toxicities rather well, but the extension of this model to the human is not warranted at this time because of the paucity of clinical data.

It is likely that the toxic hazard posed by a chemical mixture containing PCDD or PCDF actually depends on several factors, including the nature of the major components of the mixture, the concentrations of specific PCDD or PCDF isomers in the mixture, the route of exposure (oral, dermal or inhalation), age, sex, and health of the exposed persons, and various other factors. In any case, the distribution of PCDD/PCDF in environments where human exposure is likely to occur is a matter of considerable concern. While these "chemical" routes of introducing PCDD/PCDF into the environment are still considered to be important, various combustion processes have more recently been identified as important sources of PCDD/PCDF in the environment [15–19]. The relative importance of the latter processes as sources of PCDD/PCDF is currently being assessed. In addition, attempts are being made to define the mechanisms whereby PCDD/PCDF are formed in combustion processes. Several possible mechanisms for formation of PCDD/PCDF during combustion have been proposed [18], including:

1. PCDD/PCDF may exist as components of the refuse that, because of their thermal stabilities, are not destroyed during incineration (especially at the lower combustion temperatures that prevail in many refuse incinerators); thus, these compounds are present in the combustion effluents.
2. PCDD/PCDF may be formed as the result of thermally initiated reactions of molecular species that are present or are produced in the combustion

plasma, and in-situ synthesis of PCDD/PCDF therefore occurs via rearrangement, free-radical condensation, dechlorination, dehydrohalogenations and/or other molecular reactions.

3. PCDD/PCDF are produced as a result of elementary reactions of atoms of appropriate elements present in the active combustion plasma, or directly thereafter in the plume resulting from combustion. The latter process has been termed a *de novo synthesis* of PCDD/PCDF.

Lustenhouwer et al. [18] reviewed the published reports that appear to support one or more of the possible mechanisms cited above. However, many of these previous studies involved mainly the analysis of bulk samples of residues (soot, fly ash), which were obtained from home fireplaces or stoves, or municipal incinerators. Clearly, more definitive sampling techniques, coupled with more extensive analyses to characterize not only the PCDD/PCDF, but also the various possible precursor compounds present in combustion effluents, are required to delineate the mechanisms for formation of PCDD/PCDF in combustion environments. Ultimately, these definitive data will be useful in making accurate risk assessments.

This chapter describes the results of a series of more comprehensive analyses accomplished to determine PCDD/PCDF and possible chlorocarbon precursors [chlorophenols (CP), chlorobenzenes and polychlorinated biphenyls (PCB)] in gaseous and particulate components of the flue gases from a municipal waste-fueled boiler.

DESCRIPTION OF THE MUNICIPAL WASTE-FUELED BOILER AND SAMPLING PROCEDURES

Steam Boiler

The incinerator sampled in this study is a steam boiler fueled solely by refuse. The refuse, after being crudely mixed in a pit, is delivered to the burner by a crane. A schematic diagram of the incinerator is shown in Figure 8.1. Typical incinerator operating temperatures are: furnace (front), 1250–1350°F (677–732°C); furnace (rear), 750°F (400°C); and electrostatic precipitator outlet, 540°F (282°C). Typically, the incinerator achieves a 58% reduction in the weight of the refuse received. During a typical one-month period, the two steam boilers at this plant produced 22,470,400 lb of steam while consuming 5325 tons of refuse.

Sampling Procedures and Samples Obtained

The stack emissions from the incinerator characterized in this study were sampled by Scott Environmental Technology using a Modified Method V train. This train comprised a six-ft-long quartz-lined probe, a four-in. glass particulate filter

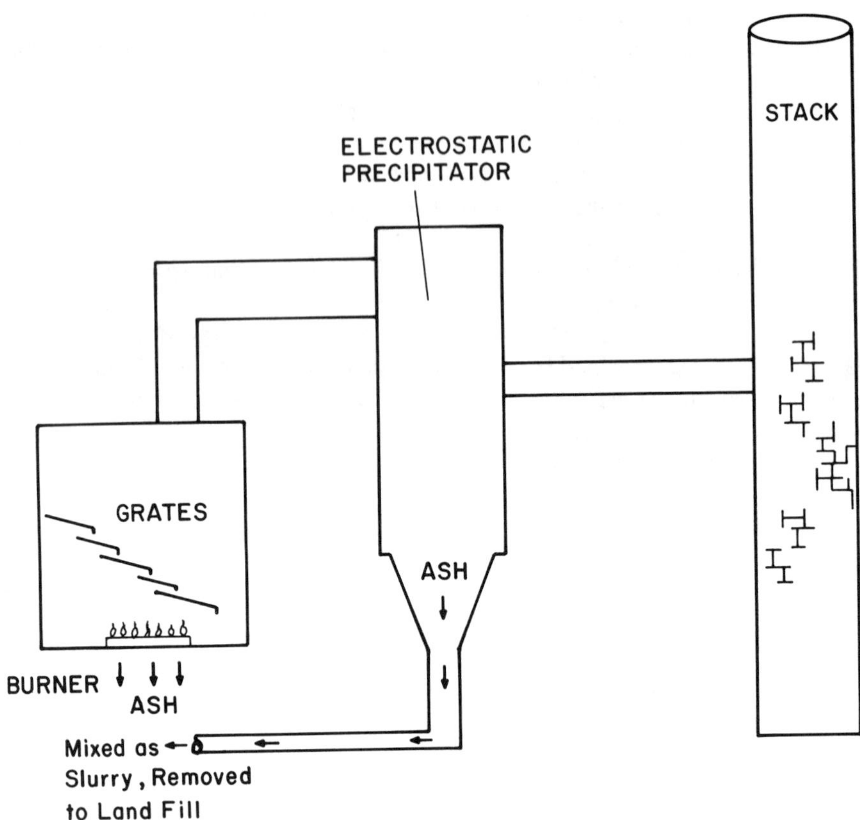

Figure 8.1. Schematic Diagram of Refuse-Fueled Steam Boiler Sampled for Toxic Effluents in this Study.

holder enclosed in a hot box and 3 glass impingers (impinger 1 contained 100 mL water, impinger 2 was empty, and impinger 3 contained silica gel). During sampling, the impingers were chilled in an ice-bath. A glass trap containing approximately 20 g of XAD-2 resin was positioned between impingers 2 and 3. The sample types that resulted from each sampling event are:

1. methanol wash of the front portion of the sampling train, including the quartz-lined probe and the front portion (nearest the probe) of the particulate filter holder;
2. particulate filter and particulates;
3. toluene wash of the front half of the sampling train;
4. toluene wash of the back portion of the sampling train, including the back half of the filter holder, the impingers and connecting glassware;
5. methanol wash of the back portion of the sampling train;

6. XAD-2 resin from the trap; and
7. impinger contents.

Except for sample types 2 and 3, aliquots of each of the above samples were prepared and subsequently analyzed as separate samples. In the case of sample types 2 and 3, the toluene wash from the front half of the train was evaporated to dryness during heating in a sand bath at 55 °C and passing a stream of prepurified nitrogen over the sample. The residue, consisting primarily of particulates removed from the quartz-lined probe, was weighed and combined with sample type 2, the particulate filter and associated particulates. Thus, five discrete samples and one composited sample were obtained for analysis from each of the sampling events.

ANALYTICAL PROCEDURES DEVELOPED AND APPLIED TO ASSAY STACK EFFLUENT SAMPLES FOR CP, CHLOROBENZENES, PCDD/PCDF AND PCB

The analytical procedures developed and applied by our laboratory for determination of CP, chlorobenzenes, PCDD/PCDF and PCB in combustion effluents entail initial extraction of the various samples with appropriate organic solvents, preliminary separation of the compounds of interest using solvent partitioning and liquid chromatography, and analysis of the various processed extract fractions using coupled gas chromatography/mass spectrometry (GC/MS). All of the compounds of interest here, except CP, are injected directly into the GC/MS/. CP are derivatized before GC/MS analysis, as described below.

Procedures for Addition of Internal Standards and Extraction of Samples

Liquid Samples (Impinger Condensate, Probe Rinse, Line Rinses, Solvent Blanks)

Measure the total volume of the liquid sample. Transfer approximately one-fourth of the total sample to a precleaned 125- or 250-mL glass bottle equipped with a Teflon-lined cap. Add appropriate quantities of internal standards [$^{37}Cl_4$-2,3,7,8-tetra-CDD, $^{37}Cl_4$-1,2,3,4,6,7,8-hepta-CDD, $^{37}Cl_8$-octa-CDD, D_6-3,3′,4,4′-tetra-chlorobiphenyl (-CBP), $^{13}C_6$-penta-CP, $^{13}C_6$-hexachlorobenzene) to each sample. Depending on the type of liquid, proceed as described below.

Aqueous Impinger Liquids. Add 50 mL of hexane to the sample, seal the vessel and agitate (a Burell wrist-action shaker was employed) for one hr. Allow the vessel to stand; after the aqueous and organic layers separate completely, transfer the hexane layer to a clean 125-mL glass bottle fitted with a Teflon-lined screw-cap. Repeat the hexane extraction two additional times using a 10-mL aliquot of hexane for each extraction. Combine all hexane extracts. Proceed with cleanup.

Toluene Wash Samples. Proceed directly to cleanup.

Methanol Wash Samples. Add 20 mL of benzene and 30 mL of 1 *N* H_2SO_4 to the sample, cap the vessel and shake for one hr. Allow the vessel to stand; after the layers separate completely, transfer the upper layer to a clean 125-mL glass bottle equipped with a Teflon-lined cap. Repeat the solvent extraction two additional times using a 10-mL aliquot of benzene for each extraction. Combine all benzene extracts. Proceed with cleanup.

Solid Samples (XAD-2, Particulates, Filter) Preclean the Soxhlet extraction apparatus plus a glass extraction thimble containing 1 g silica gel and a plug of glass wool by charging the extraction apparatus with benzene and refluxing for two hr. Remove the benzene and retain for assessment of background contamination. Remove the glass wool from the thimble and retain the glass wool for subsequent use. For extraction of the XAD-2 resin, accurately weigh an aliquot corresponding to approximately one-half of the total sample (typically 5–15 g), and place this into a precleaned thimble. Insert the cleaned glass wool into the glass thimble to hold the resin in place. For extraction of combustion particulates, place an accurately weighed aliquot of particulate into a precleaned extraction thimble. Before extraction of these solids, add to each sample the appropriate quantities of the internal standards: $^{37}Cl_4$-2,3,7,8-tetra-CDD, $^{37}Cl_4$-1,2,3,4,6,7,8-hepta-CDD, $^{37}Cl_8$-octa-CDD, D_6-3,3′,4,4′-tetra-CBP, $^{13}C_6$-hexachlorobenzene and $^{13}C_6$-penta-CP. Subsequent to addition of the internal standards, insert the thimble containing the spiked sample into a precleaned Soxhlet extractor, charge the Soxhlet flask with benzene, and extract the sample for 16 hr. After extraction, remove the flask from the extractor and quantitatively transfer the extraction solution to a 250-mL glass bottle equipped with a Teflon-lined cap. Concentrate the extract to approximately 40 mL by passing a stream of dry nitrogen over the sample while heating in a 55°C sand bath. Proceed with cleanup.

Procedures for Wet Chemical and Elution Chromatographic Cleanup of Sample Extracts

Add 20 mL of concentrated H_2SO_4 to the bottle containing the sample extract, cap the vessel and shake for 2 min. Allow the mixture to stand; after the aqueous and organic layers separate completely, remove and discard the acid layer.

Add 20 mL of doubly distilled water to the bottle containing the sample extract, cap the vessel and shake for 2 min. Allow the bottle to stand until the aqueous and organic layers separate; transfer the aqueous layer to a clean 60-mL bottle equipped with a Teflon-lined cap. Retain this aqueous layer for subsequent analysis for CP.

Add 5 mL of 0.1 *N* KOH to the organic layer and extract the chlorinated phenols for the sample by shaking the mixture for 10 min. Transfer the base layer to the sample bottle containing the water wash from above.

Repeat the base extractions two more times, combining all base extracts in the one bottle. Reserve the base extracts for acetylation of phenols. Continue with cleanup of organic phase for remaining organic constituents.

Using the same procedures as applied above, wash the extract successively with 30-mL portions of concentrated H_2SO_4 and doubly distilled water, in each case discarding the water or acid layer. The acid washing procedure is repeated until the acid layer is virtually colorless. Concentrate to ~1 mL under a stream of dry nitrogen in a water bath (55°C).

Prepare a glass macro-column, 20 mm o.d. × 230 mm in length, tapered to 6 mm o.d. on one end. Pack the column with a plug of silanized glass wool, followed successively by 1.0 g silica, 2.0 g silica containing 33% (w/w) 1 *M* NaOH, and 2.0 g silica. Quantitatively transfer the concentrated extract from above to the column and elute with 90 mL hexane. Collect the entire eluate and concentrate to a volume of ~1 mL in a centrifuge tube, as before.

Construct a disposable liquid chromatography mini-column by cutting off a 5-mL Pyrex® disposable pipette at the 2.0 mL mark and packing the lower portion of the tube with a small plug of silanized glass wool, followed by one g of Woelm basic alumina, which has been previously activated for at least 16 hr at 600°C in a muffle furnace, and cooled in a desiccator for 30 min just before use. Quantitatively transfer the concentrate from above onto the liquid chromatography column, rinse the centrifuge tube consecutively with two 0.3-mL portions of 3% CH_2Cl_2:hexane, and transfer the rinses to the chromatography column.

Elute the column with 7 mL of 3% (v/v) CH_2Cl_2:hexane and retain the eluate for PCB and chlorobenzene analyses.

Elute the column with 20 mL of 50% (v/v) CH_2Cl_2:hexane and retain the eluate for PCDD and PCDF analyses.

Concentrate each of the retained fractions to a volume of approximately 1 mL by heating the tubes in a water bath while passing a stream of prepurified N_2 over the solutions, as described above. Quantitatively transfer the concentrated fractions into separate 2-mL micro-reaction vessels. Evaporate the solutions in each of the micro-reaction vessels almost to dryness, using the procedures just mentioned; rinse the walls of each vessel with 0.5 mL CH_2Cl_2; and reconcentrate just to dryness.

Approximately 1 hr before GC/MS analysis, dilute the residue in each micro-reaction vessel with an appropriate quantity of solvent (depending on the anticipated quantities of analytes in each vessel) and gently swirl the solvent in the vessel to ensure dissolution of PCDD, PCDF, PCB and chlorobenzenes.

If preliminary GC/MS screening analysis of the sample indicates the presence of potential interfering compounds (that is, those that obscure the PCDD, PCDF, PCB or other signals of interest), or other sample matrix constituents that elute from the GC at very long times, then additional sample cleanup or fractionation is required using high-performance liquid chromatography

(HPLC). A Varian Model 5021 Microprocessor-controlled High-performance Liquid Chromatograph equipped with a CDS-111L Data System is employed. Dual Du Pont Zorbax ODS columns (25 cm × 0.6 cm i.d.) are employed. Methanol is employed as the mobile phase in the isocratic mode. The sample is injected into the HPLC, and appropriate fractions, determined in advance by measuring the retention times of pure CDD and CDF standards, are collected.

Acetylation of Sample Extracts for CP Analysis

Check the pH of the solution from the base extractions with pH paper and add HCl if necessary to adjust pH to <3.

Add 5 mL of benzene to the acidified mixture and agitate the sample for 10 min using a wrist-action shaker. Allow the mixture to stand until complete layer separation occurs (centrifuge if necessary).

Transfer the benzene layer to a 60-mL flint glass bottle (fitted with a Teflon-lined cap) that contains approximately 2 g of anhydrous Na_2SO_4.

Repeat the benzene extraction step two more times, combining all extracts and then shake the bottle to dry the benzene solution.

Transfer approximately 10 mL of the benzene solution into a 16- × 100-mm screw-capped test tube fitted with a Teflon-lined cap.

Concentrate the solution in the tube to a volume of approximately 1 mL by passing a stream of purified N_2 over the surface of the liquid, while applying gentle heat (50°C) to the test tube. Add more of the solution and continue to concentrate until all of the solution has been transferred and concentrated.

Add 5 mL of benzene to a 60-mL bottle containing Na_2SO_4, shake to rinse the bottle, and transfer the benzene into the test tube containing the previously concentrated solution. Concentrate as above to approximately 1 mL.

Add 2.0 mL isooctane, 2.0 mL acetonitrile, 25 μL pyridine and 10 μL acetic anhydride to the test tube containing the extract and agitate for 5 min on a wrist-action shaker.

Add 6 mL of 10 m*M* H_3PO_4 to the test tube and agitate for 2 min on a wrist-action shaker.

Remove the organic layer and concentrate it in a Reacti-Vial at room temperature (using prepurified N_2) to dryness just before analysis.

Procedures for High-Resolution GC/MS Analyses of Extracts of Stack Effluent Samples for PCDD/PCDF, Chlorobenzenes, CP and PCB

The instrumentation used to perform quantitative analyses for the chlorocarbons of interest comprises a GC/MS/data system (GC/MS/DS), which includes a Perkin Elmer Sigma III Gas Chromatograph coupled through a specially designed interface to a Kratos MS-25 Double Focusing Mass Spectrometer equipped with a Kratos DS55SM Data System. The gas chromatograph is equipped with a split/splitless injector and a wall-coated open-tubular (WCOT) fused-silica capillary column. The interface mentioned above includes provision for both direct admission of the column effluent into the mass spectrometer source as well as admission of the column effluent via single-stage, all-glass jet separator. The interface has been modified to minimize peak broadening due to excessive dead volume and to optimize the temperature throughout the interface. The mass spectrometer was operated in the multiple ion monitoring mode under the control of the computer-based Kratos DS55SM Data System. Elaborate programs for the DS55SM Data System were developed that permitted rapid, automated retuning of the mass spectrometer during analysis of a sample extract; thus, 20–30 separate ion masses were sequentially monitored during an analysis. Each of the groups of chlorocarbons of interest (PCDD/PCDF, chlorobenzenes, CP and PCB) were analyzed separately. The GC/MS/DS parameters specifically developed for each of these groups are discussed in detail below. A very detailed discussion of the procedures for quantifying PCDD/PCDF will be presented, followed by the descriptions of analogous procedures for each of the other groups of compounds.

PCDD/PCDF

Tetra- through octa-CDD and -CDF include 136 separate isomers (49 PCDD and 87 PDCF). Since pure standards corresponding to all of these compounds are not available in any laboratory, the approach for calibration of the GC/MS/DS system that is considered state-of-the-art is to use the available standards to obtain data presumed to be representative of all of the isomers in the group. Calibration of the GC/MS/DS system, therefore, initially entails characterization of the standards on hand which represent each class of isomers (the degree of chlorination, monochloro- dichloro-, etc., defines the "class" of isomers) known to comprise the group of compounds being determined. In the case of the PCDD/PCDF, 210 separate isomers are possible, distributed among eight classes (mono- through octachlorinated) of PCDD and eight classes of PCDF. The following procedure for characterization of the isomers was developed and applied.

Obtain mass spectra for each available isomer standard, noting in particular the relative abundances of the m^+ and $m+2^+$ peaks. (The isomer standards used for calibration purposes in the present study are listed later in this chapter.)

Determine gas chromatographic retention times for each isomer standard available (or a reasonable number of these).

Based on the above data, select the ion masses appropriate for detecting and quantifying the isomers in the class of interest, and select the gas chromatographic retention time window appropriate for all isomers in the class (that is, the time interval during which the earliest-eluting and latest-eluting isomers of a particular class elute from the gas chromatographic column and enter the mass spectrometer source—this must be estimated in cases where not all isomers of a given class are available). This retention time window must also include the retention time for the internal standard employed in quantifying the isomers of a particular class. Since a limited number of stable isotopically labeled internal standards are available, the following ^{37}Cl-labeled internal standards were utilized to represent the PCDD/PCDF indicated:

1. tetra- and penta-CDD and -CDF: $^{37}Cl_4$-2,3,7,8-tetra-CDD;
2. hexa- and hepta-CDD and -CDF: $^{37}Cl_4$-1,2,3,4,6,7,8-hepta-CDD; and
3. octa-CDD and -CDF: $^{37}Cl_8$-octa-CDD.

Based on the gas chromatographic retention time data and the mass spectral data obtained as indicated above, the procedures listed in Table 8.1 were developed and applied in the analysis of tetra- through octachlorinated PCDD/PCDF. Further details of the configuration and operational parameters for the GC/MS/DS are:

1. Conditions for the gas chromatograph were:

- column: 50-m WCOT (OV-101) fused-silica capillary,
- carrier gas: hydrogen, 30 lb head pressure,
- column temperature: programmed, see Table 8.1 for details,
- interface temperature: 250°C, and
- injector: split at 50:1.

2. Conditions for the mass spectrometer were:

- ionizing voltage: 70 eV,
- accelerating voltage: 4 kV,
- source temperature: 250°C, and
- multiple ion monitoring mode: see Table 8.1 for details.

3. Data Acquisition/Handling: The DS55SM Data System and associated peripherals were employed to acquire both analog (total ion current plots, specific ion current monitoring plots) and digital (peak areas) data, from which the quantities of PCDD/PCDF present in the samples were calculated.

4. Construction of Calibration Curves, Quantification: The general approach to quantifying PCDD/PCDF native to each of the samples (that is, the PCDD/PCDF present in the as-received sample) was to determine the ratios of the mass spectral responses obtained for the ions characteristic of the native PCDD/PCDF to those of the appropriate internal standards that had been added to the sample in known concentrations. The concentrations of the native PCDD/PCDF were determined by comparing the above data with calibration curves, prepared by plotting the analytical results obtained for a series of standards, each of which contained a fixed concentration of internal standard, but varying concentrations of representative PCDD and/or PCDF. More specifically, the ratio of the ion

Table 8.1. Sequence of Operations in GC/MS (MS-25) Analyses of PCDD and PCDF in Sample Extract

Elapsed Time (min)	*Event*	*GC Column Temperature (°C)*	*Temperature Program Rate (°C/min)*	*Ions Monitored by Mass Spectrometer (m/z)*	*Compounds Monitored*
0.00	Injection, splitless	190			
1.50	Turn on split valve	190			
2.00	Begin temp. program to 220°C	190	5		
5.00	Open column flow to mass spectrometer	205	5		
6.50	Start PROGRAM 2; sweep = 200 ppm;	220		235.980	Di-CDF
	time on each mass = 0.12 sec			237.977	Di-CDF
				251.974	Di-CDD
				253.972	Di-CDD
10.50	Stop PROGRAM 2	220			
11.00	Start PROGRAM 3; sweep = 200 ppm;	220		269.941	Tri-CDF
	time on each mass = 0.15 sec			271.938	Tri-CDF
				285.935	Tri-CDD
				287.932	Tri-CDD
15.00	Stop PROGRAM 3	220			
15.50	Start PROGRAM 4; sweep = 100 ppm;	220		258.930	Tetra-CDD
	time on each mass = 0.15 sec			303.902	Tetra-CDF
				305.899	Tetra-CDF
				319.897	Tetra-CDD
				321.894	Tetra-CDD
				327.885	$^{37}Cl_4$-Tetra-CDD
23.50	Stop PROGRAM 4	220			
24.00	Start PROGRAM 5; sweep = 750 ppm;	220		337.863	Penta-CDF
	time on each mass = 0.2 sec			339.860	Penta-CDF
				353.858	Penta-CDD
				355.855	Penta-CDD

Table 8.1. Sequence of Operations in GC/MS (MS-25) Analyses of PCDD and PCDF in Sample Extract (*continued*)

Elapsed Time (min)	*Event*	*GC Column Temperature (°C)*	*Temperature Program Rate (°C/min)*	*Ions Monitored by Mass Spectrometer (m/z)*	*Compounds Monitored*
25.00	Begin temp. program to 230°C	220	5		
34.50	Stop PROGRAM 5				
35.00	Start PROGRAM 6; sweep = 150 ppm; time on each mass = 0.25 sec	230		373.821	Hexa-CDF
				375.818	Hexa-CDF
				389.816	Hexa-CDD
				391.813	Hexa-CDD
50.00	Begin temp. program to 235°C		5		
53.50	Stop PROGRAM 6	235			
54.00	Start PROGRAM 7; sweep =750 ppm; time on each mass = 0.35 sec	235		407.782	Hepta-CDF
				409.779	Hepta-CDF
				423.777	Hepta-CDD
				425.774	Hepta-CDD
				431.765	$^{37}Cl_4$-Hepta-CDD
70.00	Stop PROGRAM 7				
85.00	Start PROGRAM 8; sweep = 750 ppm; time on each mass = 0.50 sec	235		441.732	Octa-CDF
				443.740	Octa-CDF
				457.738	Octa-CDD
				459.735	Octa-CDD
				471.717	$^{37}Cl_8$-Octa-CDD
95.00	Stop PROGRAM 8	235			
130.00	Return to initial temp.				

intensity recorded for the native PCDD/PCDF to the ion intensity obtained for the ^{37}Cl-labeled internal standard was plotted as a function of the ion intensity obtained for the ^{37}Cl-labeled internal standard. Since the ^{37}Cl-labeled internal standard was added at the beginning of the sample preparation/analysis scheme, and, as indicated in the procedures outlined above, the internal standard is quantified at the same time as the native analyte(s), it is clear that the quantitative result obtained for the internal standard reflects losses of PCDD/PCDF incurred during the course of sample handling and analysis. However, the internal standard may not correct for the effects of the sample matrix; that is, the added ^{37}Cl-labeled internal standard may not penetrate the sample and become sorbed or chemisorbed in the same fashion as, for example, the PCDD/PCDF that are incorporated in the particulates that resulted from combustion. However, the use of stable isotopically labeled internal standards as described above represents the best current approach for assessing the efficiency of the overall analytical procedure (including the extraction efficiency). In addition, as stated previously, the use of a stable isotopically labeled isomer to represent the several isomers in the same class or other classes is based on the assumption that the chemical/physical properties of the labeled isomer essentially parallel the properties of the other isomers being quantified.

During the analyses, it was necessary to establish criteria governing the identifications of PCDD/PCDF. These criteria are as follows:

1. The mass chromatographic peaks produced by the unknown component must exhibit appropriate GC retention times; that is, they must fall within a retention time "window" established for a particular class of PCDD or PCDF (e.g., tetra-CDD). These "windows" are established by injecting available standards for each class of PCDD and PCDF, and, in some instances, complex mixtures of standards containing PCDD and/or PCDF isomers known to be of a specific chlorinated class, but for which the actual isomeric structures have not been established.
2. Mass chromatographic peaks produced by the component must exhibit the appropriate response for at least two major m/z ratios characteristic (that is, are known to appear in the mass spectrum) of the particular PCDD or PCDF class being monitored. The ratio of relative intensities of the two ions monitored as indicators must correspond to that resulting from injection of an appropriate calibration standard within ±30%.
3. In cases where the sample matrix causes shifts in the GC rentention times of the components of interest, as compared to the retention times of the corresponding standards determined from a separate injection of standards, and as indicated by analogous shifts in the retention times for the internal standards added to the sample, the identification of a specific PCDD and/or PCDF must be confirmed by coinjection of the sample with an added quantity of the PCDD or PCDF isomer in question. Enhancement of a given mass chromatographic peak upon such coinjection leads to tentative assignment of the unknown peak as a specific PCDD or PCDF isomer.

PCB

The general approach for quantification of PCB present in the extracts of the samples was the same as that outlined above for the PCDD/PCDF. However, in this case, the internal standard employed was D_6-3,3',4,4'-tetra-CBP. Here, again, it was assumed that the extraction, chromatographic and mass spectrometric behavior of all PCB parallels that of the D_6-3,3',4,4'-tetra-CBP employed as an internal standard. The representative PCB standards listed below were employed in calibrating the GC/MS to determine the gas chromatographic retention time "windows" appropriate for each class of PCB as well as to determine mass spectral characteristics of each class. As in the case of the PCDD/PCDF, the assumption was made that the PCB employed in the calibration of the GC/MS were indeed representative of all of the compounds comprising each of the ten classes of PCB. This assumption could only be verified by exhaustive calibration employing all of the available PCB isomers; certified standards of all 209 isomers are not as yet available.

The GC/MS/DS operational parameters employed for the quantification of PCB are:

1. Conditions for the gas chromatograph were:

- column: 50-m WCOT (OV-101) fused-silica capillary;
- carrier gas: hydrogen, 30 lb head pressure;
- column temperature: programmed, see Table 8.2 for details.
- interface temperature: 250°C;
- injector: split at 50:1.

2. Conditions for the mass spectrometer were:

- ionizing voltage: 70 eV;
- accelerating voltage: 4 kV;
- source temperature: 250°C; and
- multiple ion monitoring mode: see Table 8.2 for details.

3. Data acquisition/handling was identical to that described for PCDD/PCDF.

Chlorobenzenes

The general approach for quantification of chlorobenzenes present in the extracts of the samples was the same as that outlined above for PCDD/PCDF. In this case $^{13}C_6$-hexachlorobenzene was employed as the internal standard; it was assumed that the behavior of this compound during extraction, chromatographic separation and mass spectrometric analyses parallels that of the other chlorobenzene isomers. As noted previously, the chlorinated benzenes (listed below) were

Table 8.2. Sequence of Operations in GC/MS (MS-25) Analyses of Sample Extracts for PCB

Elapsed Time (min)	*Event*	*GC Column Temperature (°C)*	*Temperature Program Rate (°C/min)*	*Ions Monitored by Mass Spectrometer (m/z)*	*PCB Monitored*
0.00	Injection, splitless	190			
1.50	Split valve on	190			
7.00	Open column flow to mass spectrometer; begin temp. program to 270°C	190			
	Start MID program: sweep width = 100; time on ion mass = 0.1 sec	190	5	188.039	Mono-
				190.036	Mono-
	Time on D_6-tetra-CBP ion masses = 0.1 sec			222.000	Di-
				223.997	Di-
				255.961	Tri-
				257.958	Tri-
				289.922	Tetra-
				291.919	Tetra-
				325.880	Penta-
				327.877	Penta-
				359.841	Hexa-
				361.839	Hexa-
				393.802	Hepta-
				395.800	Hepta-
				427.763	Octa-
				429.761	Octa-
				463.722	Nona-
				465.719	Nona-
				497.683	Deca-
				499.680	Deca-
				295.922	D_6-Tetra
				297.919	D_6-Tetra
23.00	Hold column temp.	270			
35.00	Stop MID program	270			
40.00	Cool column to initial temp.				

assumed to be representative of the various classes of chlorobenzenes; these standards were employed to determine the retention windows and ion masses appropriate for quantifying the various classes of halobenzenes. The GC/MS/DS operational parameters appropriate for quantification of the chlorobenzenes are:

1. Conditions for the gas chromatograph were:

- column: 50-m WCOT (OV-101) fused silica capillary;
- carrier gas: hydrogen, 30 lb head pressure;
- column temperature: programmed, 60°C starting temperature, hold for 2 min, program to 250°C at 8°C/min;
- injector: split at 50:1; and
- interface temperature: 250°C.

2. Conditions for the mass spectrometer were:

- ionizing voltage: 70 eV;
- accelerating voltage: 4 kV;
- source temperature: 250°C; and
- multiple ion monitoring mode: see Table 8.3 for details.

3. Data acqusition/handling was identical to that described for PCDD/PCDF.

Chlorophenols

The general approach for quantification of the CP present in the extracts of the samples was the same as described previously for PCDD/PCDF. In this case, $^{13}C_6$-penta-CP was employed as the internal standard; it was assumed that the extraction, chromatographic and mass spectrometric behavior of all CP parallels that of the $^{13}C_6$-penta-CP employed as the internal standard. The CP listed below were employed as calibration standards and were assumed to be representative of the various isomers comprising each of the classes of CP. Gas chromatographic retention time data and mass spectrometric data were obtained using the representative CP and the sequence of operations outlined in Table 8.4 was devised based on these data. The GC/MS/DS operational parameters appropriate for the quantification of CP are:

1. Conditions for the gas chromatograph were:

- column: 50-m WCOT (OV-101) fused-silica capillary;
- carrier gas: hydrogen, 30 lb head pressure;
- column temperature: programmed, 60°C starting temperature, hold for 2 min, program to 250°C at 12°C/min;
- injector: split at 50:1; and
- interface temperature: 250°C.

Table 8.3. Sequence of Operations in GC/MS (MS-25) Analyses of Sample Extracts for Chlorinated Benzenes

Elapsed Time (min)	*Event*	*GC Column Temperature (°C)*	*Temperature Program Rate (°C/min)*	*Ions Monitored by Mass Spectrometer (m/z)*	*Chlorinated Benzenes Monitored*
0.00	Injection, splitless	60			
1.50	Split valve on	60			
2.00	Begin temp. program to 250°C	60	8		
7.00	Start MID program; sweep = 500; time on ion mass = 0.1 sec; time of $^{13}C_6$-hexachlorobenzene ion masses = 0.1 sec	100	8	145.969	Di-
				147.966	Di-
				179.930	Tri-
				181.927	Tri-
				213.891	Tetra-
				215.888	Tetra-
				247.852	Penta-
				249.849	Penta-
				283.810	Hexa-
				285.807	Hexa-
				289.831	$^{13}C_6$-Hexa-
				291.828	$^{13}C_6$-Hexa-
22.00	Stop MID program	220			
26.00	Final temp.	250			
28.00	Cool column to initial temp.				

Table 8.4. Sequence of Operations in GC/MS (MS-25) Analyses of Sample Extractions for CP

Elapsed Time (min)	*Event*	*GC Column Temperature (°C)*	*Temperature Program Rate (°C/min)*	*Ions Monitored by Mass Spectrometer (m/z)*	*CP Monitored*
0.00	Injection, splitless	60			
1.50	Split valve on	60			
2.00	Begin temp. program to 250°C	60	12		
10.00	Start MID program; sweep = 500; time on ion mass = 0.1 sec; time on $^{13}C_6$-penta-CP ion mass = 0.1 sec.	156	12	195.887	Tri-
				197.884	Tri-
				229.876	Tetra-
				231.864	Tetra-
				263.847	Penta-
				256.844	Penta-
		252		273.861	$^{13}C_6$-Penta-
18.00	Column temp. on hold				
20.00	Stop MID program				
25.00	Cool column to initial temp.				

2. Conditions for the mass spectrometer were:

- ionizing voltage: 70 eV;
- accelerating voltage: 4 kV;
- source temperature: 250 °C; and
- multiple ion monitoring mode: see Table 8.4 for details.

3. Data acquisition/handling was identical to that described for PCDD/PCDF.

Quality Assurance

As is the case for all analytical programs conducted at our laboratory, all analytical measurements reported here were accomplished in accordance with good laboratory practice. An extensive quality assurance program is established and followed for all projects such as those described here. This program includes analyses of solvent and method blanks, analyses of internally spiked control samples, determinations to validate the efficacy of the analytical procedures applied, as already described. Quality control for the individual analyses reported here was also provided on a continuing basis because of the incorporation of known quantities of stable isotopically labeled internal standards into each sample before analysis.

Reagents and Chemicals

The following reagents and chemicals were used in the procedures outlined above:

1. Potassium hydroxide, anhydrous sodium sulfate, pyridine, acetic anhydride, hydrochloric acid and sulfuric acid were all reagent grade and were obtained from either J.T. Baker Chemical Co. (American Scientific Products, Obetz, Ohio) or Fisher Scientific Co. (Cincinnati, Ohio).
2. Methanol, hexane, methylene chloride, benzene, acetonitrile, toluene, isooctane were "Distilled in Glass" quality obtained from Burdick and Jackson (Muskegon, Michigan).
3. Woelm basic alumina (activity grade I) was obtained from ICN Pharmaceuticals (Cleveland, Ohio).
4. Silica (Bio-Sil A) was obtained from Bio-Rad (Rockville Centre, New York).
5. Doubly distilled water was obtained using the all-glass distillation apparatus in the Brehm Laboratory.
6. Prepurified nitrogen was obtained from Airco, Inc. (Montvale, New Jersey).
7. High-purity helium was obtained from Union Carbide Corp., Linde Division (Cleveland, Ohio).

The various chlorocarbon standards employed in this work were obtained from the sources listed below. Chlorinated dibenzo-*p*-dioxins:

1. 1,2,3,4-tetra- and octa-CDD from Analabs, Inc. (North Haven, Connecticut);
2. $^{37}Cl_4$-2,3,7,8-tetra-, 1,2,3,7,8-penta-, 1,2,3,4,6,7-hexa-, 1,2,3,4,6,7,8-hepta-, $^{37}Cl_4$-1,2,3,4,6,7,8-hepta- and $^{37}Cl_8$-octa-CDD from KOR Isotopes (Cambridge, Massachusetts);
3. 2,3,7,8-tetra-CDD and the 21 other tetra-CDD isomers from Dow Chemical Co. (Midland, Michigan) and H.R. Buser (Swiss Federal Research Station, Wadenswil, Switzerland); and
4. 1,2,3,4,7,8-hexa-CDD from A. Poland (University of Rochester, Rochester, New York).

The following chlorinated dibenzofurans were obtained from the U.S. Food and Drug Administration (Washington, D.C.):

- 1,2,4,8-tetra-CDF;
- 2,3,7,8-tetra-CDF;
- 2,3,6,8-tetra-CDF;
- 1,2,4,7,8-penta-CDF;
- 1,2,4,6,7,9-hexa-CDF;
- 1,2,3,4,6,8,9-hepta-CDF; and
- octa-CDF.

The following chlorinated biphenyls were obtained from Ultra Scientific, Inc. (Hope, Rhode Island):

- 4-CBP;
- 3,3′-di-CBP;
- 2,4′,5-tri-CBP;
- 3,3′,4,4′-tetra-CBP
- 2,2′,6,6′-tetra-CBP;
- 2,3,4,5,6-penta-CBP;
- 2,2′,4,5,5′-penta-CBP;
- 2,2′,4,4′,5,5′-hexa-CBP;
- 2,2′,3,4,4′,5′,6-hepta-CBP;
- 2,2′,3,3′,4,4′,5,5′-octa-CBP;
- 2,2′,3,3′,4,4′,5,6,6′-nona-CBP; and
- 2,2′,3,3′,4,4′,5,5′,6,6′-deca-CBP.

The D_6-3,3′,4,4′-tetra-CBP was supplied by KOR Isotopes (Cambridge,Massachusetts). The following chlorinated benzenes were obtained from Ultra Scientific Inc.:

- 1,4-dichlorobenzene;
- 1,3,5-trichlorobenzene;
- 1,2,3,5-tetrachlorobenzene;
- pentachlorobenzene; and
- hexachlorobenzene.

The $^{13}C_6$-hexachlorobenzene was obtained from KOR Isotopes (Cambridge, Massachusetts). The following CP standards were obtained from Ultra Scientific, Inc.:

- 2,3,4-tri-CP;
- 2,3,4,5-tetra-CP;
- 2,3,5,6-tetra-CP; and
- penta-CP.

The $^{13}C_6$-penta-CP was obtained from KOR Isotopes (Cambridge, Massachusetts).

RESULTS AND DISCUSSION

As mentioned previously, seven discrete samples resulted from each stack sampling effort or test in which the modified EPA Method V train was implemented to sample the incinerator stack effluents; two of these (the particulate filter and the toluene wash of the front of the sampling train) were composited to obtain the total particulates collected during the sampling. These samples and blanks are listed in Table 8.5. The data reported here resulted from analyses of these 21 samples plus one precipitator ash sample, obtained during three separate sampling events. The levels of the tetra- through octa-CDD and -CDF, and the chlorobenzenes, CP, and PCB in the samples, were determined using the analytical methodology described above. The analytical results obtained for each group of analytes are presented in separate sections below, followed by a concluding section, in which all data are discussed.

PCDD and PCDF

The concentrations of PCDD/PCDF found in the combustion effluent samples and associated blanks are listed in Tables 8.6 through 8.9; Table 8.10 contains the weights of particulates determined to be present in the particulate filters and probe wash samples resulting from each test. As seen in the tables, readily measurable quantities of all classes of PCDD and PCDF were determined to be present in the Method V train samples resulting from tests 3, 5 and 7. The total quantity of each class of PCDD and each class of PCDF is given in each table.

Table 8.5. Municipal Incinerator Stack Effluent Samples Obtained by Scott Environmental Services, Inc., during September 1981, and Analyzed during the Initial Phase of the Project

Test	*WSU Sample No.*	*Sample Description*
3[a]	SES-25, 30	Particulate filter and toluene front wash
	SES-31	Methanol front wash
	SES-26	XAD-2 trap
	SES-28	Toluene back wash
	SES-29	Methanol back wash
	SES-24	Ash
5[b]	SES-40, 45	Particulate filter and toluene front wash
	SES-46	Methanol front wash
	SES-41	XAD-2 trap
	SES-43	Toluene back wash
	SES-44	Methanol back wash
	SES-42	Impinger catch
7[c]	SES-55, 60	Particulate filter and toluene front wash
	SES-61	Methanol front wash
	SES-56	XAD-2 trap
	SES-58	Toluene back wash
	SES-59	Methanol back wash
	SES-57	Impinger catch
	SES-63	Toluene blank
	SES-64	Methanol blank
	SES-54	XAD-2 blank
	SES-4	Filter blank

[a]Three-hour sampling run, September 15, 1981.
[b]Three-hour sampling run, September 21, 1981.
[c]Five-hour sampling run, September 22, 1981.

Table 8.6. Quantities (ng/sample) of Tetra- through Octa-CDD and -CDF in Municipal Waste Incinerator Stack Effluent Samples: Test 3 (Conducted September 15, 1981)[a]

	Type of Sample[b]					
	1	*2,3*	*4*	*6*	*Total*[c]	*Ash*[d] (ng/g)
Tetra-CDD	14	500	83	147	744	474
Tetra-CDF	33	893	400	722	2048	419
Penta-CDD	15	438	79	120	652	349
Penta-CDF	23	359	172	239	793	107

Table 8.6. Quantities (ng/sample) of Tetra- through Octa-CDD and -CDF in Municipal Waste Incinerator Stack Effluent Samples: Test 3 (Conducted September 15, 1981)[a] *(continued)*

	Type of Sample[b]					
	1	*2,3*	*4*	*6*	*Total*[c]	*Ash*[d] (ng/g)
Hexa-CDD	50	474	103	101	728	615
Hexa-CDF	55	652	228	324	1259	263
Hepta-CDD	45	865	90	63	1063	370
Hepta-CDF	55	1140	225	230	1650	287
Octa-CDD	26	368	29	20	443	162
Octa-CDF	4	77	7	8	96	25

[a]Recoveries obtained for the ^{37}Cl-enriched internal standards added to each sample are: $^{37}Cl_4$-2,3,7,8-tetra-CDD, range = 60–95%, average = 77%; $^{37}Cl_4$-1,2,3,4,6,7,8-hepta-CDD, range = 70–100%, average = 85%; $^{37}Cl_8$-octa-CDD, range = 50–100%, average =74%.

[b]See text for explanation of sample types.

[c]The impinger catch sample was not analyzed due to funding limitations; therefore, the totals listed do not include the PCDD/PCDF present in the impinger liquid. No PCDD/PCDF were detected in sample type 5.

[d]Precipitator ash sample from test.

Table 8.7. Quantities (ng/sample) of Tetra- through Octa-CDD and -CDF in Municipal Waste Incinerator Stack Effluent Samples: Test 5 (Conducted September 21, 1981)[a]

	Type of Sample[b]						
	1	*2,3*	*7*	*4*	*5*	*6*	*Total*
Tetra-CDD	9	266	1.271	1,037	83	495	3,161
Tetra-CDF	74	1,069	6,658	4,607	425	1,920	14,653
Penta-CDD	23	361	1,564	1,674	89	441	4,152
Penta-CDF	33	521	1,885	2,109	131	538	5,217
Hexa-CDD	58	656	2,089	2,706	113	1,350	6,972
Hexa-CDF	77	519	2,068	2,625	124	908	6,321
Hepta-CDD	60	741	1,100	1,013	88	452	3,454
Hepta-CDF	69	667	700	796	94	332	2,658
Octa-CDD	25	376	192	220	24	73	910
Octa-CDF	3	48	34	37	4	18	144

[a]Recoveries obtained for the ^{37}Cl-enriched internal standards added to each sample are: $^{37}Cl_4$-2,3,7,8-tetra-CDD, range = 40–100%, average = 78%; $^{37}Cl_4$-1,2,3,4,6,7,8-hepta-CDD, range = 60–90%, average = 74%; $^{37}Cl_8$-octa-CDD, range = 60–85%, average =72%.

[b]See text for explanation of sample types.

Table 8.8. Quantities (ng/sample) of Tetra- through Octa-CDD and -CDF in Municipal Waste Incinerator Stack Effluent Samples: Test 7 (Conducted September 22, 1981)[a]

	Type of Sample[b]						
	1	*2*	*7*	*4*	*5*	*6*	*Total*
Tetra-CDD	14	410	951	628	57	502	2,562
Tetra-CDF	128	1,090	7,211	5,068	448	3,728	17,673
Penta-CDD	24	781	1,313	1,082	54	385	3,639
Penta-CDF	73	1,066	5,045	3,470	240	1,114	11,008
Hexa-CDD	55	2,546	1,122	1,706	68	302	5,799
Hexa-CDF	175	2,471	3,567	4,659	264	912	12,048
Hepta-CDD	155	11,586	680	1,185	58	272	13,936
Hepta-CDF	240	9,219	2,372	2,361	210	611	15,012
Octa-CDD	97	2,734	108	328	22	56	3,343
Octa-CDF	31	861	74	145	12	35	1,153

[a]Recoveries obtained for the ^{37}Cl-enriched internal standards added to each sample are: $^{37}Cl_4$-2,3,7,8-tetra-CDD, range = 40–100%, average = 78%; $^{37}Cl_4$-1,2,3,4,6,7,8-hepta-CDD, range = 70–100%, average = 81%; $^{37}Cl_8$-octa-CDD, range = 50–90%, average =73%.

[b]See text for explanation of sample types.

Table 8.9. Quantities (ng/sample) of Tetra- through Octa-CDD and -CDF in Procedural Blanks Corresponding to Tests 3, 5 and 7[a]

	Sample Description			
	Filter	*XAD-2 Resin*	*Toluene*	*Methanol*
Tetra-CDD	ND[b]	ND	ND	ND
Tetra-CDF	ND	ND	ND	ND
Penta-CDD	ND	ND	ND	ND
Penta-CDF	ND	ND	ND	ND
Hexa-CDD	ND	ND	ND	ND
Hexa-CDF	ND	ND	ND	ND
Hepta-CDD	ND	ND	ND	ND
Hepta-CDF	ND	ND	ND	ND
Octa-CDD	3	ND	ND	ND
Octa-CDF	ND	ND	ND	ND

[a]Recoveries obtained for the ^{37}Cl-enriched internal standards added to each sample are: $^{37}Cl_4$-2,3,7,8-tetra-CDD, range = 40–100%, average = 73%; $^{37}Cl_4$-1,2,3,4,6,7,8-hepta-CDD, range = 60–75%, average = 69%; $^{37}Cl_4$-octa-CDD, range = 50–72%, average =62%.

[b]ND = none detected; minimum detectable concentration = 2 ng/sample.

Table 8.10. Weights of Particulates Collected from Stack Effluents from Refuse Combustion during Scott Environmental Services Sampling

WSU Sample No.	*Sample Description*	*Weight of Particulate (g)*	*Total Weight of Particulate (g)*
SES-4	Filter blank	0.034	
	Test 3 (3 hr)		0.409
SES-25	Filter	0.347	
SES-30	Probe wash, toluene	0.062	
	Test 5 (3 hr)		2.149
SES-40A	Filter	0.714	
SES-40B	Filter	0.558	
SES-40C	Filter	0.398	
SES-45	Probe wash, toluene	0.479	
	Test 7 (5 hr)		2.147
SES-55A	Filter	0.615	
SES-55B	Filter	0.413	
SES-55C	Filter	0.606	
SES-60	Probe wash, toluene	0.513	

Typical multiple-ion plots obtained during GC/MS analyses of these samples are shown in Figures 8.2 through 8.6. These levels of PCDD/PCDF are clearly well above the quantities of PCDD/PCDF determined to be present in the various reagents employed during testing and analysis (see Table 8.9). In addition, it is important to note that the tetra-CDD and tetra-CDF constitute a major fraction of the total PCDD/PCDF detected. The implications of these findings from toxicological and epidemiological viewpoints cannot be assessed fully at this point because of the lack of definitive human toxicological data. However, the emission rates from incinerators of the type evaluated in the present study must be regarded as an important source of PCDD/PCDF. Of additional importance in assessing the environmental and toxicological implications of the above findings is the relative abundance of various isomers within each class, since it is known that the structures of various isomers markedly influence the toxicity. Although the OV-101 column employed for these analyses is probably not capable of effecting complete separation of all of the individual isomers that comprise a particular class of PCDD or PCDF, at least partial gas chromatographic separation of the individual isomers is clearly achieved.

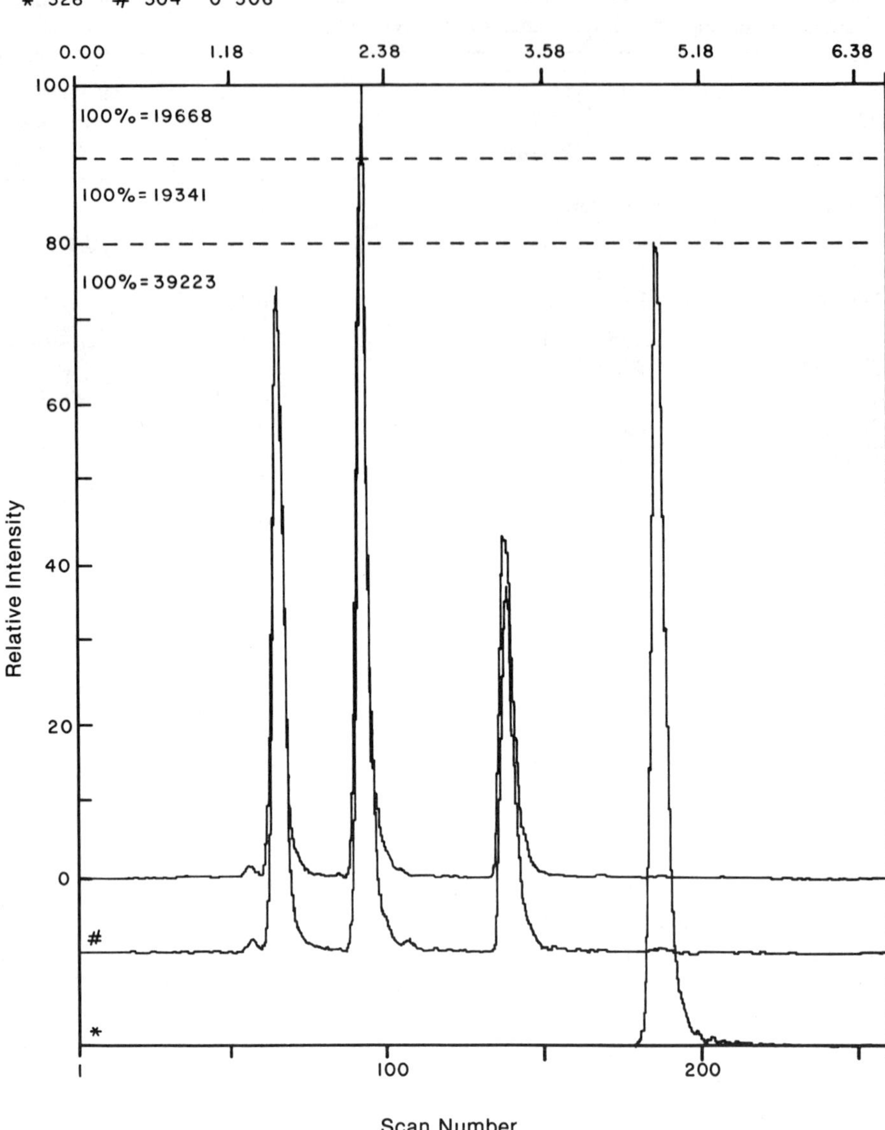

Figure 8.2. Selected-Ion Mass Chromatograms Obtained for Calibration Standards Containing 2,3,6,8-, 1,2,4,8-, and 2,3,7,8-Tetra-CDF, and $^{37}Cl_4$-2,3,7,8-Tetra-CDD.

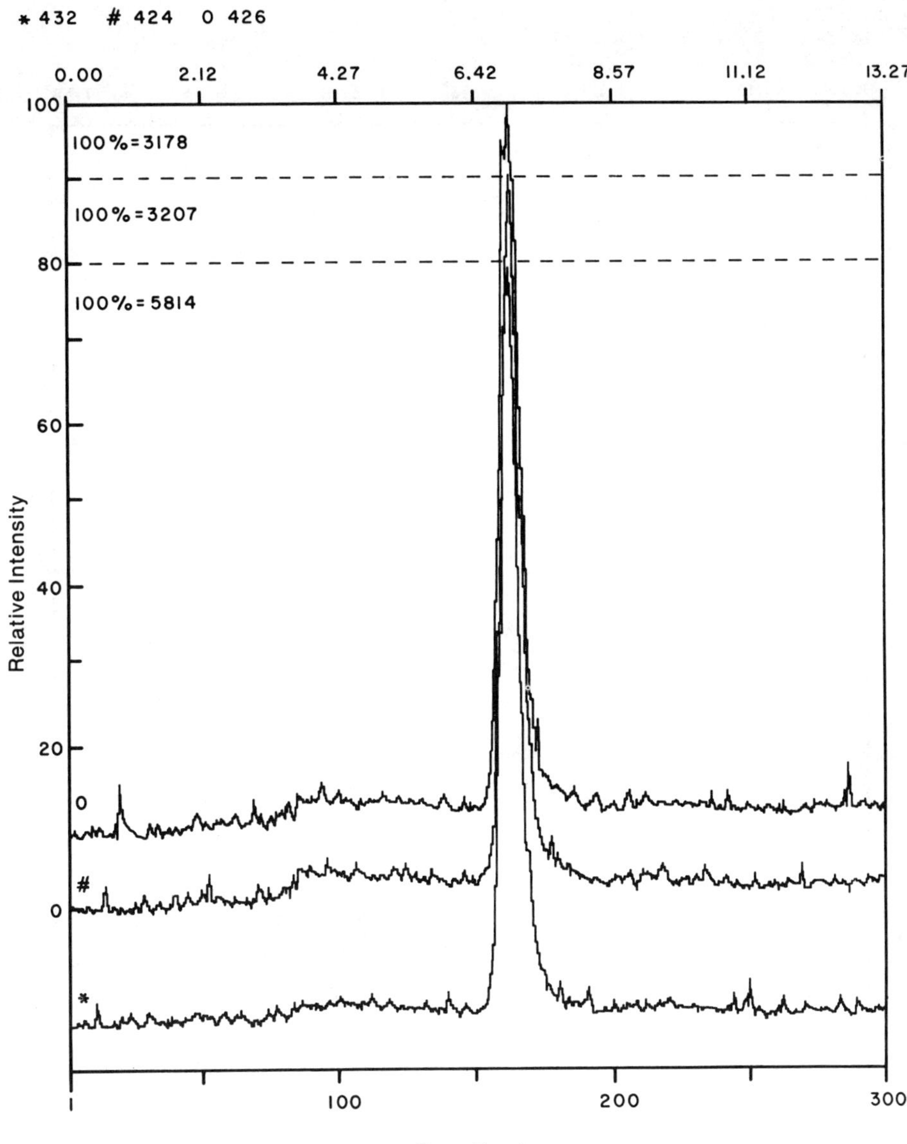

Figure 8.3. Selected-Ion Mass Chromatograms Obtained for Calibration Standard Containing 1,2,3,4,6,7,8- and $^{37}Cl_4$-1,2,3,4,6,7,8-Hepta-CDD.

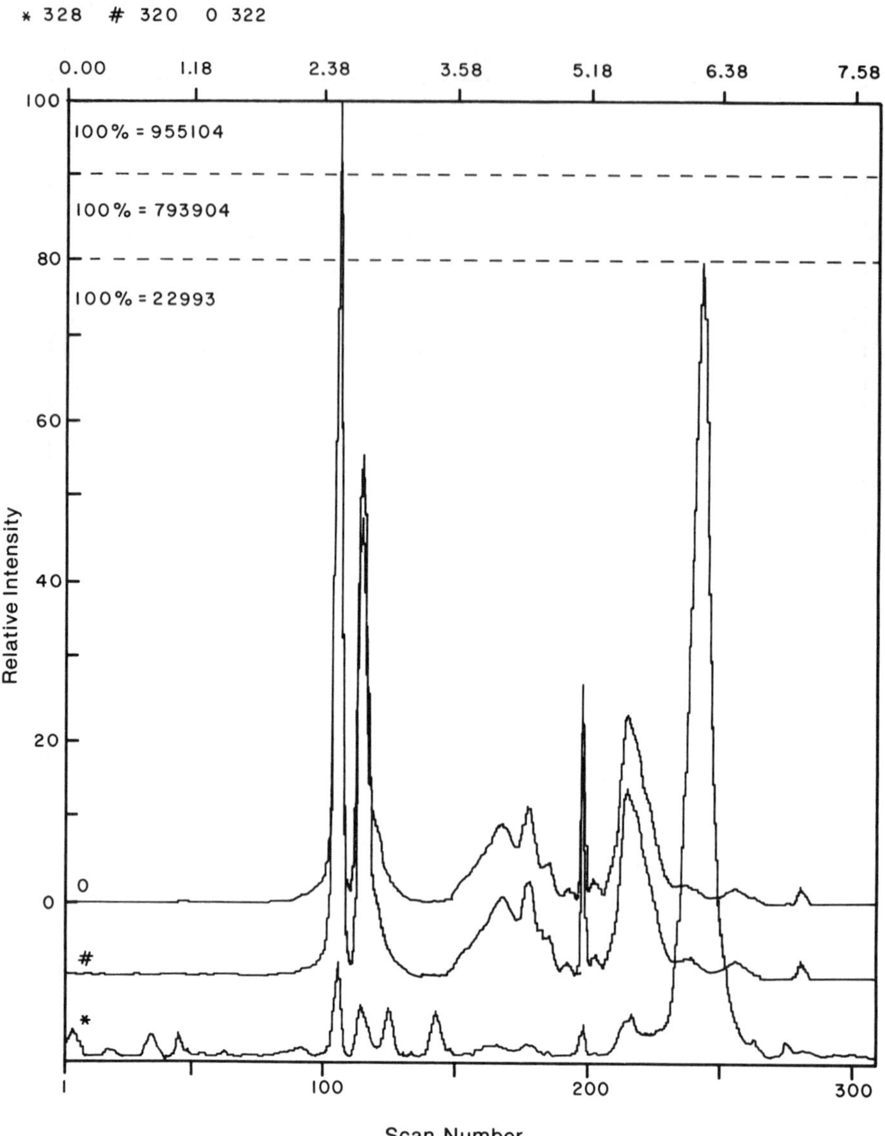

Figure 8.4. Selected-Ion Mass Chromatograms Obtained in Analysis of Extract of Particulate Filter/Toluene Front Wash Composite (test 7) for Tetra-CDD.

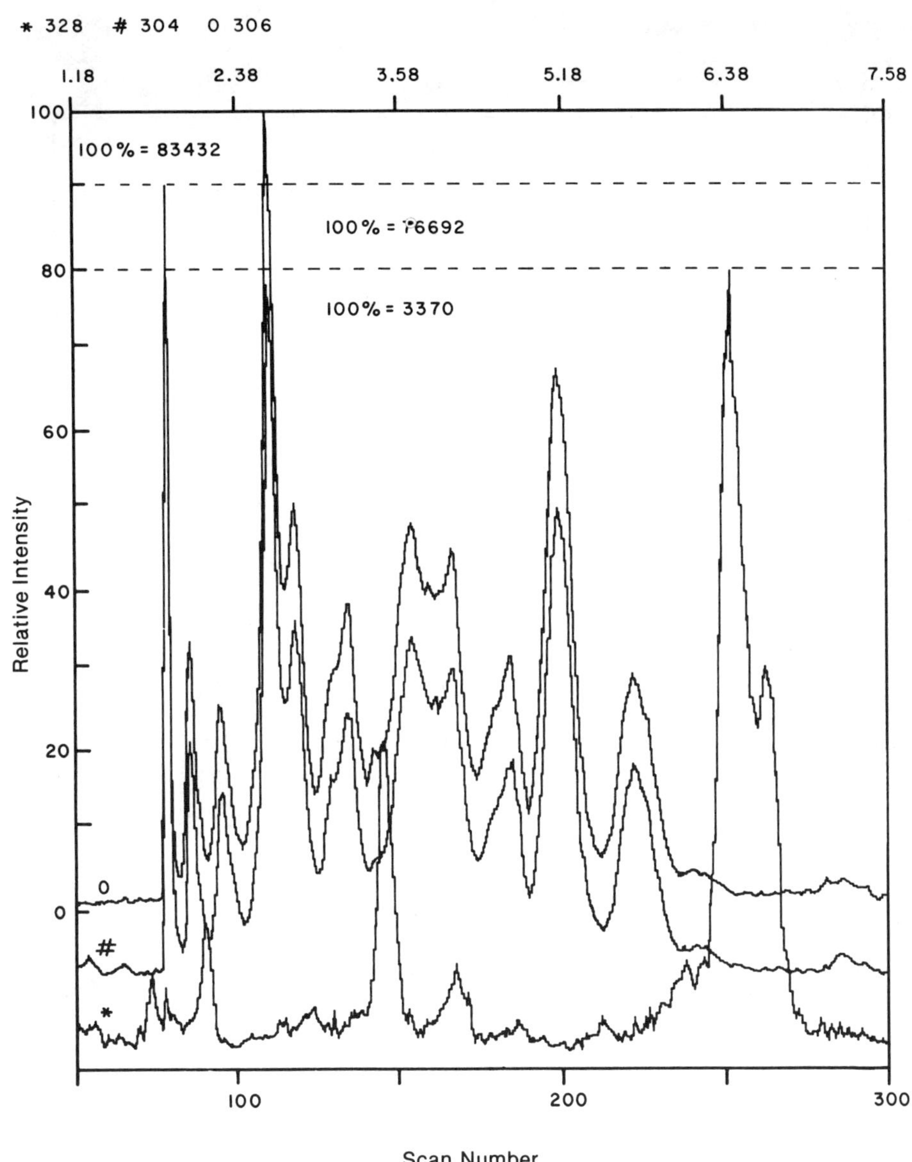

Figure 8.5. Selected-Ion Mass Chromatograms Obtained in Analysis of Extract of XAD-2 Resin Sample (test 5) for Tetra-CDF.

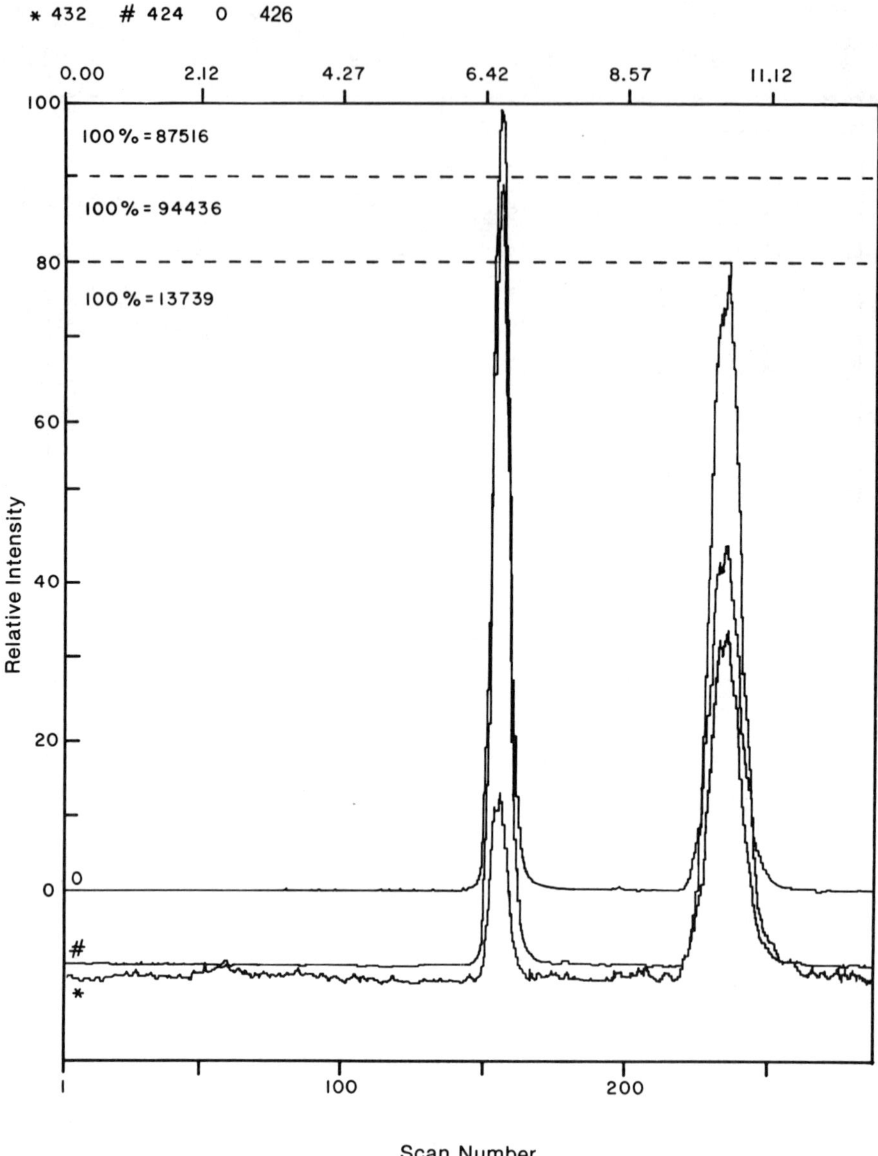

Figure 8.6. Selected-Ion Mass Chromatograms Obtained in Analysis of Extract of XAD-2 Resin Sample (test 5) for Hepta-CDD.

Table 8.11. Number of Apparent Isomers in Each Class of PCDD/PCDF Found in the Extract of Particulate Filter, Toluene Wash/Front of Sampling Train: Test 7 (September 22, 1981)[a]

PCDD/PCDF Class	*No. of Isomers Detected*
Tetra-CDDs	11
Tetra-CDFs	17
Penta-CDDs	8
Penta-CDFs	12
Hexa-CDDs	8
Hexa-CDFs	11
Hepta-CDDs	1
Hepta-CDFs	4
Octa-CDDs	1
Octa-CDFs	1

[a]Isomers were identified on the basis of gas chromatographic retention times and relative abundances of the two principal ion masses in the molecular ion cluster. Where possible, GC/MS data obtained for an extract were compared with similar data obtained for authentic standards available in the Brehm Laboratory, and identifications were made on the basis of these direct comparisons. Where specific isomer standards were not available, the identity of a peak in the mass chromatogram was inferred by comparisons with data for other closely related isomers. Until a laboratory possesses all isomers for a given class of CDDs/CDFs and demonstrates the ability to separate these, no unequivocal identification of specific isomers can be made.

Table 8.11 shows the results of counting the number of mass chromatographic peaks for each class of PCDD/PCDF observed in the specific ion plots obtained in analysis of the composite sample consisting of the particulate filter and the toluene wash of the front portion of the sampling train from test 7. These numbers represent the minimum number of isomers of each chlorinated class present in the sample. Obviously, there may be more isomers actually present if all the isomers are not resolved. As seen in Table 8.11, substantial numbers of apparent isomers of each class of PCDD/PCDF are present in this sample. For example, of the 22 possible tetra-CDD isomers, at least 11 appear to be present. Of the 38 possible tetra-CDF, at least 17 are present; 8 of the 14 possible penta-CDD are present; and 12 of the 28 possible penta-CDF are present. In the case of tetra-CDD, since all 22 tetra-CDD isomers are available in our laboratory, it is possible to obtain more definitive data on the tetra-CDD isomers present in the test 7 sample described above. These data are summarized in Table 8.12. As indicated by the information provided in the footnotes to Table 8.12, the OV-101 column employed in these analyses is not capable of separating all 22 tetra-CDD isomers; therefore, certain peaks in the mass chromatogram may result from more than one isomer. Calibration runs with all 22 tetra-CDD isomers provide the basis for the isomer identifications reported in Table 8.12. The relative abundances of the various isomers and isomer groups observed in this sample are similar to those observed in combustion product samples by other workers [16]. For example,

Table 8.12. Specific Tetra-CDD Isomers Identified in the Extract of Particulate Filter, Toluene Wash/Front of Sampling Train: Test 7 (September 22, 1981)

Tetra-CDD Isomer	*% of Total Tetra-CDD*
1,3,6,8-	20.97
1,3,7,9-	13.29
1,3,6,9-	5.17
1,3,7,8-/1,4,6,9-/1,2,4,7-/1,2,4,8-[a]	10.64
1,2,4,6-/1,2,4,9-	14.77
1,2,6,8-	7.09
1,4,7,8-	6.65
1,2,7,9-	2.07
1,2,3,4-/1,2,3,6-/1,2,6,9-[b]	11.08
1,2,3,7-/1,2,3,8-/2,3,7,8-[c]	3.84
1,2,7,8	4.43

[a]Using the OV-101 WCOT glass capillary column, 1,3,7,8-, 1,4,6,9-, 1,2,4,7- and 1,2,4,8-tetra-CDD are resolved from the other tetra-CDD isomers but are not resolved from each other. Therefore, the quantity of tetra-CDD identified here could arise from one or more of these four unresolved isomers.

[b]Using the OV-101 WCOT glass capillary column, 1,2,3,4-, 1,2,3,6- and 1,2,6,9-tetra-CDD are resolved from the other tetra-CDD isomers but are not resolved from each other. Therefore, the quantity of tetra-CDD could arise from one or more of these unresolved isomers.

[c]Using the OV-101 WCOT glass capillary column, 1,2,3,7-, 1,2,3,8- and 2,3,7,8-tetra-CDD are resolved from the other tetra-CDD isomers, but are resolved from each other. Therefore, the quantity of tetra-CDD identified here could arise from one or more of these three unresolved isomers.

1,3,6,8- and 1,3,7,9-tetra-CDD seem to be the more prominent isomers present, although the 1,2,4,6- and 1,2,4,9- isomers also account for 14.77% of the total tetra-CDD observed. The retention time for the three-isomer group that includes 2,3,7,8-tetra-CDD accounted for only 3.84% of the total tetra-CDD. Whether or not any of this small quantity is ascribable to the 2,3,7,8- isomer cannot be deduced from the present data, although this could be determined readily. Other capillary columns available in our laboratory are known to separate 2,3,7,8-isomer from all other tetra-CDD isomers. However, in the present project, the use of the OV-101 was considered more practical, to accomplish a survey of the various classes of PCDD/PCDF present in the sample extracts within a reasonable analytical period.

Chlorinated Benzenes

Tables 8.13 through 8.15 include the levels of chlorobenzenes found in the 22 individual samples analyzed. The quantities of the chlorobenzenes detected in the various tests generally parallel the duration of sampling. Thus, tests 3 and 5, which were of equal sampling duration, yielded samples containing similar levels of chlorobenzenes; in these tests, the predominant chlorobenzene observed was pentachlorobenzene. In test 7, where the duration of sampling was 67% longer

Table 8.13. Quantities (ng/sample) of Chlorobenzenes in Municipal Waste Incinerator Stack Effluent Samples: Test 3 (Conducted September 15, 1981)[a]

	Sample Type[b]								
Chlorobenzene	*1*	*2,3*	*4*	*5*	*6*	*Total*[c]	*Ash*	*XAD-2 Blank*	*Methanol Blank*
Di-	ND[d]	ND	13	ND	ND	13	ND	ND	ND
Tri-	ND	ND	13	ND	1,459	1,472	1	ND	ND
Tetra-	4	380	40	ND	7,673	8,097	2	1	ND
Penta-	65	3,000	43	ND	16,250	19,358	221	0.5	0.5
Hexa-	55	200	85	ND	5,514	5,854	746	1	1

[a]Recoveries obtained for the $^{13}C_6$-hexachlorobenzene internal standard added to each sample ranged 30–100% and averaged 58 %.

[b]See text for explanation of sample types.

[c]The impinger catch sample was not analyzed due to funding limitations; therefore, the values listed do not include chlorobenzenes present in the impinger fluid.

[d]ND = none detected. Minimum detectable concentration = 1 ng/sample.

Table 8.14. Quantities (ng/sample) of Chlorobenzenes in Municipal Waste Incinerator Stack Effluent Samples: Test 5 (Conducted September 21, 1981)[a]

	Type of Sample[b]						
Chlorobenzene	*1*	*2,3*	*7*	*4*	*5*	*6*	*Total*
Di-	100	40	ND[c]	1,210	35	1,285	2,670
Tri-	100	130	1,275	165	15	3,135	4,820
Tetra-	80	800	1,330	540	60	3,323	6,133
Penta-	184	1,634	17,200	575	30	2,850	22,473
Hexa-	120	491	1,210	655	160	5,688	8,234

[a]Recoveries obtained for the $^{13}C_6$-hexachlorobenzene internal standards added to each sample ranged 30–100% and averaged 58%.

[b]See text for explanation of sample types.

[c]ND = none detected. Minimum detectable concentration = 1 ng/sample.

Table 8.15. Quantities (ng/sample) of Chlorobenzenes in Municipal Waste Incinerator Stack Effluent Samples: Test 7 (Conducted September 22, 1981)[a]

	Type of Sample[b]						
Chlorobenzene	*1*	*2,3*	*7*	*4*	*5*	*6*	*Total*
Di-	ND[c]	ND	ND	ND	ND	30,000	30,000
Tri-	15	580	5	ND	ND	129,000	129,595
Tetra-	180	4,590	120	ND	ND	190,000	194,890
Penta-	237	5,737	1,950	75	ND	260,000	267,999
Hexa-	240	3,060	1,620	520	ND	71,600	77,040

[a]Recoveries obtained for the $^{13}C_6$-hexachlorobenzene internal standards added to each sample ranged 30–100% and averaged 58%.

[b]See text for explanation of sample types.

[c]ND = none detected. Minimum detectable concentration = 1 ng/sample.

(5 vs. 3 hr for tests 3 and 5), the total quantities of chlorobenzenes found in the samples were proportionately much higher (by a factor of 10), than those detected from tests 3 and 5. Of course, in assessing these data, many other variables must be taken into account, such as the various operating parameters of the incinerator that prevailed during the sampling period (temperatures, fuel, combustion efficiency, etc.). Overall, the quantities of chlorobenzenes detected in these combustion effluent samples are indeed substantial, and these compounds should reasonably be considered to be potential precursors of the PCDD/PCDF. Laboratory studies in which various chlorobenzenes and mixtures of these were pyrolyzed at 620°C indicated that both PCDD and PCDF were found as products, although the yields of PCDF were substantially greater than those of the PCDD [20].

Chlorinated Phenols

Tables 8.16 through 8.19 show the quantities of the various classes of CP present in the extracts of the stack effluent samples and blanks analyzed. For these data, the relationship between the quantities detected and the duration of each of the tests is not obvious. However, it is particularly interesting that large quantities of tri-CP are present in these samples, since it is well known that bimolecular condensation reactions involving tri-CP yield tetra-CDD [21]. Ample quantities of other phenols are also present; therefore, substantial quantities of the higher-chlorinated PCDD could also be derived from condensation reactions involving these compounds. Whether such reactions occur during combustion cannot, of course, be determined on the basis of these data. Further work is needed to clarify the origins of the PCDD/PCDF detected in such combustion effluents.

Table 8.16. Quantities (ng/sample) of CP in Municipal Waste Incinerator Stack Effluent Samples: Test 3 (Conducted September 15, 1981)[a]

	Type of Sample[b]						
	1	*2,3*	*4*	*5*	*6*	*Total*[c]	*Ash (ng/g)*
Tri-CP	ND[d]	14,140	580	2,280	40,500	57,500	500
Tetra-CP	ND	3,960	415	860	11,900	17,135	369
Penta-CP	ND	5,310	400	ND	4,900	10,610	193

[a]Recoveries obtained for the $^{13}C_6$-penta-CP internal standards added to each sample ranged 40–100% and averaged 87%.

[b]See text for explanation of sample types.

[c]The impinger catch sample was not analyzed due to funding limitations; therefore, the totals listed do not include CP present in the impinger fluid.

[d]ND = none detected. Minimum detectable concentration = 250 ng/sample.

Table 8.17. Quantities (ng/sample) of CP in Municipal Waste Incinerator Stack Effluent Samples: Test 5 (Conducted September 21, 1981)[a]

	Type of Sample[b]						
	1	*2,3*	*7*	*4*	*5*	*6*	*Total*
Tri-CP	ND[c]	6,550	137,500	9,800	490	145,000	299,340
Tetra-CP	ND	2,250	61,800	9,050	540	54,700	128,340
Penta-CP	ND	980	19,800	7,900	ND	10,100	38,780

[a]Recoveries obtained for the $^{13}C_6$-penta-CP internal standards added to each sample ranged 40–100% and averaged 87%.

[b]See text for explanation of sample types.

[c]ND = none detected. Minimum detectable concentration = 250 ng/sample.

Table 8.18. Quantities (ng/sample) of CP in Municipal Waste Incinerator Stack Effluent Samples: Test 7 (Conducted September 22, 1981)[a]

	Type of Sample[b]						
	1	*2,3*	*7*	*4*	*5*	*6*	*Total*
Tri-CP	ND[c]	15,600	694,000	4,800	730	164,000	879,130
Tetra-CP	ND	8,050	372,000	9,800	800	47,700	438,350
Penta-CP	ND	11,200	224,000	20,500	ND	20,700	276,400

[a]Recoveries obtained for the $^{13}C_6$-penta-CP internal standards added to each sample ranged 40–100% and averaged 87%.

[b]See text for explanation of sample types.

[c]ND = none detected. Minimum detectable concentration = 250 ng/sample.

Table 8.19. Quantities (ng/sample) of CP in Solvents and Filtration Media Analyzed as Blanks[a]

	Filter Blank	*XAD-2 Resin Blank*	*Toluene Blank*	*Methanol Blank*
Tri-CP	ND[b]	ND	120	240
Tetra-CP	10	10	200	200
Penta-CP	230	100	960	3500

[a]Recoveries obtained for the $^{13}C_6$-penta-CP internal standards added to each sample ranged 40–100% and averaged 87%.

[b]ND = none detected. Minimum detectable concentration = 250 ng/sample.

Polychlorinated Biphenyls

Tables 8.20 through 8.22 contain the quantities of the various classes of PCB present in the extracts of the stack effluent samples and blanks analyzed. The quantities of PCB found in these samples are quite low, in comparison with the other halogenated hydrocarbons quantified in the study. In view of the ubiquity of PCB in the environment, these comparatively low levels are somewhat surprising. Possibly, these species are not efficiently produced from refuse fuel at the temperatures and conditions prevailing in this combustor. Alternatively, it is possible that these compounds are formed, but that even at the relatively mild conditions prevailing in the incinerator, they are readily converted to "oxidized" species, such as PCDD/PCDF and CP and hence are not observed in the effluents. It has already been shown that under certain pyrolysis conditions, PCB give rise to PCDD/PCDF [22–25].

Table 8.20. Quantities (ng/sample) of PCB in Municipal Waste Incinerator Stack Effluent Samples: Test 3 (Conducted September 15, 1981)[a]

	Type of Sample[b]						
Chlorobiphenyl	*1*	*2,3*	*4*	*5*	*6*	*Total*[c]	*Ash*[d]
Mono-	ND[e]	ND	ND	ND	ND	ND	ND
Di-	ND	ND	ND	ND	ND	ND	ND
Tri-	ND	ND	80	ND	3300	3380	60
Tetra-	ND	ND	100	ND	1660	1760	20
Penta-	ND	ND	60	ND	10	70	30
Hexa-	ND	ND	ND	ND	15	15	ND
Hepta-	ND	ND	ND	ND	ND	ND	ND
Octa-	ND	ND	ND	ND	ND	ND	ND
Nona-	ND	ND	ND	ND	ND	ND	ND
Deca-	ND	ND	ND	ND	ND	ND	ND

[a]Recoveries obtained for the D_6-3,3′,4,4′-tetra-CBP internal standards added to each sample ranged 40–100% and averaged 85%.

[b]See text for explanation of sample types.

[c]The impinger catch sample was not analyzed due to funding limitations.

[d]Ash from test 2.

[e]ND = none detected. Minimum detectable concentration = 10 ng/sample.

Table 8.21. Quantities (ng/sample) of PCB in Municipal Waste Incinerator Stack Effluent Samples: Test 5 (Conducted September 21, 1981)[a]

	Type of Sample[b]						
Chlorobiphenyl	*1*	*2,3*	*7*	*4*	*5*	*6*	*Total*
Mono-	ND[c]	ND	ND	ND	ND	ND	ND
Di-	ND	ND	ND	ND	ND	ND	ND
Tri-	ND	ND	ND	ND	ND	ND	ND
Tetra-	ND	ND	ND	ND	ND	ND	ND
Penta-	ND	ND	ND	ND	ND	ND	ND
Hexa-	ND	ND	ND	ND	ND	ND	ND
Hepta-	ND	ND	ND	ND	ND	ND	ND
Octa-	ND	ND	ND	ND	ND	ND	ND
Nona-	ND	ND	ND	ND	ND	ND	ND
Deca-	ND	ND	ND	ND	ND	ND	ND

[a]Recoveries obtained for the D_6-3,3′,4,4′-tetra-CBP internal standards added to each sample ranged 40–100% and averaged 85%.

[b]See text for explanation of sample types.

[c]ND = none detected. Minimum detectable concentration = 10 ng/sample.

Table 8.22. Quantities (ng/sample) of PCB in Municipal Waste Incinerator Stack Effluent Samples: Test 7 (Conducted September 22, 1981)[a]

	Type of Sample[b]										
Chlorobiphenyl	*1*	*2,3*	*7*	*4*	*5*	*6*	*Total*	*XAD-2 Blank*	*Filter Blank*	*Methanol Blank*	*Toluene Blank*
Mono-	ND[c]	ND	ND[d]	ND	ND	ND[d]	ND	ND	ND	ND	ND
Di-	ND	ND	ND[d]	ND	ND	ND[d]	ND	ND	ND	ND	ND
Tri-	ND	70	40	40	ND	200	350	10	ND	40	5
Tetra-	ND	100	120	130	ND	160	510	20	2	60	5
Penta-	ND	110	220	150	ND	80	560	11	3	70	7
Hexa-	ND	ND	180	120	ND	30	330	ND	ND	20	ND
Hepta-	ND	ND	ND	ND	ND	ND	ND	ND	ND	ND	ND
Octa-	ND	ND	ND	ND	ND	ND	ND	ND	ND	ND	ND
Nona-	ND	ND	ND	ND	ND	ND	ND	ND	ND	ND	ND
Deca-	ND	ND	ND	ND	ND	ND	ND	ND	ND	ND	ND

[a]Recoveries obtained for the D_6-3,3′,4,4′-tetra-CBP internal standards added to each sample ranged 40–100% and averaged 85%.

[b]See text for explanation of sample types.

[c]ND = none detected. Unless otherwise indicated, minimum detectable concentration = 10 ng/sample.

[d]Minimum detectable concentration = 50 ng/sample, because of presence of unknown compounds that interfered with quantification.

CONCLUSIONS

The steam boiler from which the combustion effluent samples were obtained for the present study must, under the conditions prevailing during the periods when samples were obtained, be regarded as a source of significant quantities of PCDD, PCDF and other chlorinated aromatic hydrocarbons. In addition, it is noteworthy that a major portion of the total PCDD/PCDF observed in the effluent samples are tetrachlorinated compounds. In the case of the PCDF, the tetrachlorinated isomers were the most abundant of the PCDF detected in each of the three tests. Of course, the threat to human health posed by these findings cannot presently be estimated. Data such as these must be treated using established dispersion models to estimate the concentrations of airborne PCDD/PCDF that would result in the vicinity of exposed populations adjacent to the incinerator. These exposure levels can then be evaluated in light of current toxicological information.

It is not known whether the refuse-fueled boiler sampled in this study was operating under optimum combustion conditions. However, the typical operating temperatures mentioned previously in this paper are lower than those of other incinerators previously evaluated [26] by our laboratory. The quantities of CP and chlorinated benzenes detected in the combustion effluent samples were surprisingly high, while, as already noted, the quantities of PCB were unexpectedly low. Of course, the PCB, CP and chlorobenzene concentrations, if any, in the refuse being combusted, or the quantities of materials such as polyvinyl chloride (PVC) plastics and lignin that may generate PCB, CP or chlorobenzenes are unknown. Other workers [27] have reported that PVC plastics give rise to chlorinated aromatic compounds when combusted in laboratory experiments. It is clear from the foregoing discussion that the PCDD/PCDF observed in the present study could be formed by a variety of mechanisms, and that the data reported here do not permit a clear choice between these. Probably, several types of reactions occur simultaneously in the combustion plasma; the comparatively large quantities of CP and chlorinated benzenes detected in the combustion effluents strongly suggest that these compounds are involved in the formation of PCDD/PCDF. Further studies are needed, possibly including carefully controlled, laboratory-scale combustion experiments, to define further the factors contributing to the evolution of PCDD and PCDF from combustion processes. The results obtained here are important, however, since they represent the first reported measurements of such a broad group of chlorinated compounds in the flue gas effluents from a refuse-fueled incinerator.

REFERENCES

1. Lee, D.H.K., and H.L. Falk, Eds. *Environ. Health Persp.* Experimental Issue 5 (1973).
2. Blair, E., Ed. *Chlorodioxins—Origin and Fate* (Washington, D.C.: American Chemical Society, 1973).

3. "Proceedings of the National Conference on Polychlorinated Biphenyls," U.S. EPA, Washington, D.C. (1976).
4. Nicholson, W.J., and J.A. Moore, Eds. *Ann. N.Y. Acad. Sci.* 320:1 (1979).
5. Esposito, M.P., T.O. Tiernan and F.E. Dryden. "Dioxins," U.S. EPA, Cincinnati, OH (1980).
6. Kimbrough, R.D., Ed. *Halogenated Biphenyls, Terphenyls, Naphthalenes, Dibenzodioxins and Related Products* (Amsterdam: Elsevier/North Holland Biomedical Press, 1980).
7. Khan, M.A.Q., and R.H. Stanton, Eds. *Toxicology of Halogenated Hydrocarbons—Health and Ecological Effects* (New York: Pergamon Press, 1981).
8. Hutzinger, O., R.W. Frei, E. Merian and F. Pocchiari, Eds. *Chlorinated Dioxins and Related Compounds—Impact on the Environment* (New York: Pergamon Press, 1982).
9. Kimbrough, R.D. *Arch. Environ. Health* 32:77 (1977).
10. Kurastsune, M., T. Yoshimura, J. Matsuzaka and A. Yamaquchi. *Environ. Health. Persp.* 1:119 (1972).
11. Chen, P.H., J.M. Gaw, C.K. Wong and C.J. Chen. *Bull. Environ. Contam. Toxicol.* 25:325 (1980).
12. Zack, J.A., and R.R. Suskind. *J. Occup. Med.* 22:11 (1980).
13. Bonaccorsi, A., R. Fanelli and G. Tognoni. *Ambio* 1:234 (1978).
14. Poland, A., and A. Kende. *Fed. Proc.* 35:2404 (1976).
15. Olie, K., P.L. Vermeulen and O. Hutzinger. *Chemosphere* 6:455 (1977).
16. Buser, H.R., H.P. Bosshardt, C. Rappe and R. Lindahl. *Chemosphere* 7:419 (1978).
17. Bumb, R.R., W.B. Crummett, S.S. Cutie, J.R. Glendhill, R.H. Hummel, R.O. Kagel, L.L. Lamparski, E.V. Luoma, D.L. Miller, T.J. Nestrick, L.A. Shadoff, R.A. Stehl and J.S. Woods. *Science* 207:59 (1980).
18. Lustenhouwer, J.W.A., K. Olie and O. Hutzinger. *Chemosphere* 9:501 (1980).
19. Tiernan, T.O., M.L. Taylor, J.G. Solch, G.F. VanNess and J.H. Garrett. *Resources Conserv.* 9:343 (1982).
20. Buser, H.R. *Chemosphere* 6:415 (1979).
21. Nestrick, T.J., L.L. Lamparski and R.H. Stehl. *Anal. Chem.* 51:2273 (1979).
22. Buser, H.R., and C. Rappe. *Chemosphere* 8:157 (1979).
23. Buser, H.R., H.-P. Bosshardt, C. Rappe and R. Lindahl. *Chemosphere* 7:419 (1978).
24. Buser, H.R., H.-P. Bosshardt and C. Rappe. *Chemosphere* 7:109 (1978).
25. Morita, M., J. Nakagawa and C. Rappe. *Bull. Environ. Contam. Toxicol.* 19:665 (1977).
26. Tiernan, T.O., M.L. Taylor, J.G. Solch, G.F. VanNess, J.H. Garrett and M.D. Porter, in *Detoxication of Hazardous Waste*, J. Exner, Ed. (Ann Arbor: Ann Arbor Science Publishers, 1982).
27. Ahling, B., A. Bjørseth and G. Lunde. *Chemosphere* 7:799 (1978).

9

Determination of 2,3,7,8-Tetrachlorodibenzo-*p*-dioxin in Industrial and Municipal Wastewaters, Method 613: Development and Detection Limits

A.S. Wong, M.W. Orbanosky, P.A. Taylor, Carl R. McMillin, Roy W. Noble, D. Wood, James E. Longbottom, D.L. Foerst and Raymond J. Wesselman

On June 7, 1976, the U.S. Environmental Protection Agency (EPA) agreed to a Consent Decree establishing a list of toxic compounds for which industrial and municipal wastewater effluent limitations were to be developed and promulgated by June 30, 1982. These specific toxic compounds, known as the "priority pollutants," included 114 organic compounds [1]. The final regulations concerning effluent limitations will be applied to point sources through the National Pollutant Discharge Elimination System (NPDES). As such, analytical methods for the 114 organic compounds had to be developed and made available for the NPDES program in a relatively short period of time.

In 1977 the EPA Environmental Monitoring and Support Laboratory (EMSL) located in Cincinnati, Ohio, began a contract program to develop analytical methods for the priority pollutants. The 114 organic compounds were subdivided into 13 groups of chemically related compounds. Contracts for the development of analytical methods were awarded to five laboratories, which eventually produced Methods 601 through 613. These methods were published in the *Federal Register* [2] as proposed rulemaking under section 304(h) of the Clean Water Act.

Method 613 for 2,3,7,8-tetrachlorodibenzo-*p*-dioxin (2,3,7,8-tetra-CDD)

is unique in that it alone contains one target compound and requires the use of high-resolution gas chromatography/low-resolution mass spectrometry (HRGC/LRMS) during the assay step. As a result of public comment and recent advances in analytical procedures, the final version of Method 613 is different from the initial version that appeared in the *Federal Register.* The following discussions will detail the development of both the initial and current versions of Method 613.

INITIAL METHOD DEVELOPMENT

Literature Review

The contractor for the development of Method 613, California Analytical Laboratories, Inc., Sacramento, California, performed a literature review at the beginning of the contract in spring 1978 to determine the state of the art for 2,3,7,8-tetra-CDD analyses. The literature review found four papers relevant to the determination of 2,3,7,8-tetra-CDD in herbicide and pesticide formulations [3–6]. Common to those procedures was the use of alumina columns during the cleanup step to separate 2,3,7,8-tetra-CDD from free acid and ester herbicides or chlorophenols [3–5]. All four papers described the use of packed-column gas chromatography during the assay step. Electron capture detectors (ECD) were used in the earlier procedures [3,4]. A flame ionization detector (FID) [5] and mass spectrometry techniques [6] were used in the more recent work in which instrumental detection limits of 30 and 10 ppb, respectively, were reported.

The use of glass capillary columns was prevalent in the more recent papers, which included reports concerning the utility of OV-101, OV-17 [7] and Silar-10C [8] liquid phases. Other recent advances in the mode of detection for detecting 2,3,7,8-tetra-CDD included the use of chemical ionization (CI) and negative ion chemical ionization (NICI) mass spectrometry [9,10].

The method to be developed for the NPDES program had restrictions; namely, the method should not involve techniques or equipment not routinely found in or readily available to a commercial analytical laboratory. This ruled out requiring the use of CI or NICI and high-resolution mass spectrometry (HRMS) during the assay step. Glass capillary columns coated with OV-101 and OV-17 or their equivalents (SP-2100 and SP-2250) were commercially available, but columns coated with Silar-10C were not.

The restrictions applied to methods development permitted ECD screening. Any positive extract would be submitted for confirmation and quantification by GC/MS. This was done to keep costs in line with the other NPDES methods being developed simultaneously. The screening approach did not permit adding isotopically labeled 2,3,7,8-tetra-CDD to the original sample; thus, the state of the art reflected in the early version of Method 613 was somewhat compromised by these restrictions.

Table 9.1. Extraction Efficiency: Recovery vs. Solvent System and pH

	Solvent System		*Percent Recovery*[a]		
Dosing Level (ng/L)	*% Hexane*	*% Methylene Chloride*	*pH 2*	*pH 7*	*pH 10*
100	85	15	93 ± 3.7	100 ± 6.8	95 ± 7.7
100	0	100	92 ± 6.2	92 ± 5.5	91 ± 5.5

[a]Average of three replicates.

EXPERIMENTAL

Extraction

The pH of the aqueous sample and choice of extracting solvent can affect the recovery through the extraction step. Therefore, a series of experiments was performed to determine if the traditional solvent system for pesticide analysis (15% methylene chloride in hexane) would be favored over the more convenient solvent system of 100% methylene chloride.

The efficiency of each solvent system was evaluated in triplicate at three pH values: 2, 7 and 10. One-liter aliquots of reagent water were adjusted to the desired pH by adding 0.1 *N* HC1 or 0.1 *N* NaOH as appropriate. The aliquots were then spiked with 100 ng of 2,3,7,8-tetra-CDD and serially extracted with three 60-mL portions of the extracting solvent. The organic phases were dried over sodium sulfate and concentrated to 10 mL using Kuderna-Danish (KD) techniques. Toluene was added (1.0 mL), and the samples were concentrated to 5.0 mL using KD and finally concentrated to 1.0 mL using nitrogen blowdown. The results are summarized in Table 9.1. Neither the solvent system nor the pH of the original sample exhibited a deleterious effect on recovery. Recoveries were uniformly above 90% in all cases. Fewer emulsion problems were encountered using 100% methylene chloride compared to using 15% methylene chloride in hexane. Since the organic phase is on the bottom of the separatory funnel when 100% methylene chloride is used, the potential for exposure is reduced because the number of manipulations involved is less, compared to using 15% methylene chloride in hexane. These considerations led to choosing 100% methylene chloride as the extracting solvent. No pH adjustment was judged necessary to optimize recovery.

Removal of Acidic and Basic Coextractants

The literature review indicated a need for washing the organic extract with aqueous base and acid before further concentration. Thus, no further investigation of this step was performed. After the extraction from water, the method specified extracting the 180-mL methylene chloride phase twice with 100 mL of 1.0 *N* sodium hydroxide, twice with 100 mL of 1.0 *N* sulfuric acid and three times with 100 mL of reagent water. These manipulations involve relatively cumbersome processing techniques because the less dense aqueous phase was to be discarded.

Concentration and Solvent Exchange

Both methylene chloride and hexane solutions of 2,3,7,8-tetra-CDD are concentrated at various steps of the procedure. Table 9.2 summarizes the recovery of 2,3,7,8-tetra-CDD when different solvent systems were concentrated from 200 to 1.0 mL using KD procedures followed by nitrogen blowdown. The recovery of 2,3,7,8-tetra-CDD was quantitative for 100% hexane, 15% methylene chloride in hexane and 100% methylene chloride.

Sample Preservation

The effluents from treatment processes are often chlorinated as a final polishing step in the treatment process. Chlorine, which is a strong oxidizer, can react with analytes when they are stored even for short periods of time. The temperature and pH during storage can also affect the stability of the sample before extraction. The effect of the four variables—chlorination, pH, temperature and time—was studied with one series of experiments. Sufficient sodium hypochlorite was added to reagent water to give a 10-ppm level of free residual chlorine. The pH was adjusted to pH 2, 7, or 10. Samples were stored in glass containers for 14 days at 4 or 25 °C. The spiking level was 1000 ng/L of 2,3,7,8-tetra-CDD. The results are summarized in Tables 9.3 and 9.4. The GC/MS system was used during the assay step in these series of experiments, since the chlorinated samples gave interference problems with the packed-column/ECD system.

The chlorinated water samples gave uniformly lower recoveries of 2,3,7,8-tetra-CDD after 14 days, regardless of the temperature at which the samples were stored. There was also some loss exhibited for the nonchlorinated samples. A further series of experiments was performed to investigate the effect in chlorinated water. A 1000-ng/L solution of 2,3,7,8-tetra-CDD at pH 7 was divided into 90-mL aliquots in separate 100-mL volumetric flasks, and sodium hypochlorite was added to give 10 ppm of free residual chlorine. The flasks were capped, stored at 25 °C and extracted with 1.0 mL of hexane on days 0, 3, 6, 9, 12 and 15. A plot of recovery time is given in Figure 9.1. The rate of disappearance was not as large as that found in the earlier study, and showed a 10% loss on the sixth day and 22% loss on the fifteenth day. By requiring that samples be analyzed by the seventh day after collection, the studies indicated there is no need to add a chemical preservative to guard against a serious loss of 2,3,7,8-tetra-CDD due to chlorination.

Table 9.2. Concentration Efficiency: Recovery vs. Solvent System

Dosing Level (ng)	*Solvent System*		*Percent Recovery,*[a] *200 to 1.0 mL*
	% Hexane	*% Methylene Chloride*	
100	100	0	97 ± 5.8
100	15	85	103 ± 2.5
100	0	100	104 ± 7.8

[a] Average of three replicates.

Table 9.3. Recovery from Chlorinated Water: pH vs. Time vs. Temperature

Dosing Level (ng/L)	*pH*	*Percent Recovery*[a]		
			Day 14	
		Day 0	4°C	25°C
1000	2	91 ± 10	44 ± 6.2	51 ± 6.0
1000	7	82 ± 8.2	50 ± 5.5	38 ± 8.6
1000	10	84 ± 5.7	43 ± 8.2	32 ± 8.6

[a] Average of three replicates.

Table 9.4. Recovery from Nonchlorinated Water: pH vs. Time vs. Temperature

Dosing Level (ng/L)	*pH*	*Day 0*	*Percent Recovery*[a]	
			Day 14	
			4°C	25°C
1000	2	81 ± 9.0	66 ± 8.0	78 ± 10.9
1000	7	91 ± 9.1	68 ± 7.6	77 ± 5.9
1000	10	98 ± 8.1	74 ± 4.6	79 ± 7.2

[a] Average of three replicates.

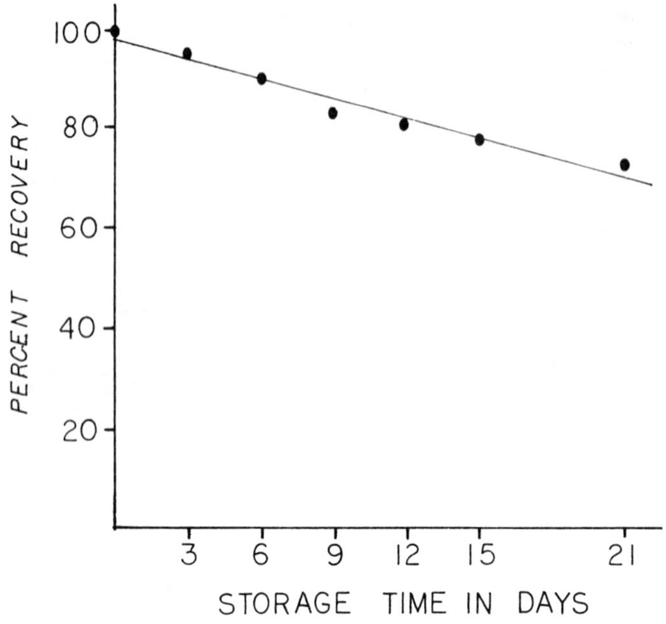

Figure 9.1. Recovery of 2,3,7,8-Tetra-CDD from Chlorinated Water.

Sample Cleanup

Four open-column, gravity-fed chromatographic systems were evaluated for separating 2,3,7,8-tetra-CDD from coextractants in complex samples. The systems tested were alumina, silica gel, Florisil and silica gel/charcoal.

Alumina cleanup columns were prepared from Fisher grade A 540 80/200 mesh neutral alumina that was activated at 130°C for 24 hr. A 30-cm × 10-mm i.d. glass column was filled to a height of 15 cm and topped with 1 cm of sodium sulfate. After equilibration with 50 mL of benzene, the 1-mL hexane solution of 100 ng of 2,3,7,8-tetra-CDD was transferred quantitatively to the column, which was then eluted with 50 mL of 3% methylene chloride in hexane, 50 mL of 20% methylene chloride in hexane and 50 mL of 100% methylene chloride. 2,3,7,8-tetra-CDD was only found in the 20% methylene chloride fraction and exhibited an average recovery of 86 ± 9% (Table 9.5.). This column is useful for separating polychlorinated biphenyls (PCB), 2,4,5-trichlorophenol and hexachlorophene from 2,3,7,8-tetra-CDD.

The silica gel cleanup column was adapted from that described earlier [6]. Silica gel (14 g of 100/200 mesh Davidson 923) was packed into a 30-cm × 10-mm i.d. glass column and equilibrated with 50 mL of 20% benzene in hexane. A 1.0-mL aliquot containing 100 ng of 2,3,7,8-tetra-CDD was then added to the

Table 9.5. Efficiency of Various Cleanup Procedures

Column	*Spike Level (ng)*	*Fraction Collected*	*Percent Recovery*
Alumina	100[a]	20% methylene chloride in hexane	86 ± 9
Silica Gel	100[a]	20% benzene in hexane	90 ± 8
Florisil	100[b]	6% diethyl ether in hexane	30 ± 3
		15% diethyl ether in hexane	62 ± 5
Silica Gel/Charcoal	10[a]	100% benzene	46 ± 4
	100[a]	100% benzene	43 ± 5

[a] Average of three replicates.
[b] Average of five replicates.

column. Another 40 mL of 20% benzene in hexane was collected, concentrated and analyzed. The mean recovery was 90 ± 8% for the triplicate experiments (Table 9.5). This column is useful for separating 2,3,7,8-tetra-CDD from chlorophenoxy acid herbicides, since these herbicides are retained by the silica gel.

EPA procedures for pesticides generally recommend Florisil column cleanup procedures. Therefore, the Florisil technique in Method 608, "The Determination of Chlorinated Pesticides and PCBs," was evaluated. Of the 100 ng of 2,3,7,8-tetra-CDD, 30% eluted from the Florisil column in the 6% diethyl ether/hexane fraction and 62% eluted in the 15% diethyl ether/hexane fraction (Table 9.5). This procedure does not afford a useful separation. Most of the pesticides listed in Method 608, including all PCB, elute in the 6% diethyl ether/hexane fraction. This procedure was not included in Method 613.

California Analytical Labs was also developing methodology for determining 2,3,7,8-tetra-CDD in sludges and sediments. Some of the environmental samples on hand provided an exceptional challenge for the standard cleanup procedures. The extract was still a dark, oily intractable mixture after sequential alumina and silica gel cleanup was performed. A silica gel/charcoal cleanup procedure was developed in response to these samples and applied to aqueous samples. The adsorbent was a mixture of 1 g of 50/200 mesh activated coconut charcoal (Fisher Scientific 5-690-A) and 140 g of 100/200 mesh silica gel (Davidson 923). A 5-mm-i.d. disposable Pasteur pipette was packed to give a 5-cm bed of the mixed adsorbent. After equilibrating this microcolumn with 5 mL of hexane, the 1.0-mL aliquot of 10 or 100 ng of 2,3,7,8-tetra-CDD was added to the column. The first fraction, 10 mL of hexane, was discarded. 2,3,7,8-Tetra-CDD eluted in the next 10 mL of benzene. Recovery was low (40–50%), but the precision was adequate (Table 9.5). This procedure was a great aid in simplifying the sediment and sludge extracts, so it was included in the early version of Method 613. It was to be used only after alumina and silica gel procedures were used and the extract was still intractable and needed further cleanup.

Chromatography

The packed glass columns and glass capillary columns that were evaluated were chosen because of availability and applicability to other EPA procedures. The packed glass columns were

- 1.5% OV-17 + 1.95% QF-1 on 100-200 mesh Gas Chrom Q, 1.8 m × 4 mm i.d.;
- 3% OV-17 on 100/120 mesh Gas Chrom Q, 1.8 m × 4 mm i.d.; and
- 1% SP-2250 on 100/120 mesh Supelcoport, 1.2 m × 2 mm i.d.

All were adequate and gave instrumental detection limits in the range of 10–20 pg per injection using either ECD or SIM MS. The mixed-phase column was recommended because it is the primary analytical column in Method 608 and should be readily available in laboratories performing routine pesticide analyses.

The capillary columns evaluated were 30-m × 0.25mm-i.d. borosilicate glass columns coated with SP-2250 or SP-2100, the equivalent of the columns cited by Buser [7]. SP-2250 was recommended because it gave better resolution between 2,3,7,8- and 1,2,3,4-tetra-CDD. The latter isomer could be used as a surrogate compound as an internal standard. Figure 9.2 shows the performance expected for the SP-2250 capillary column after a splitless injection of 20 pg of 2,3,7,8-tetra-CDD when the MS was operated in the SIM mode.

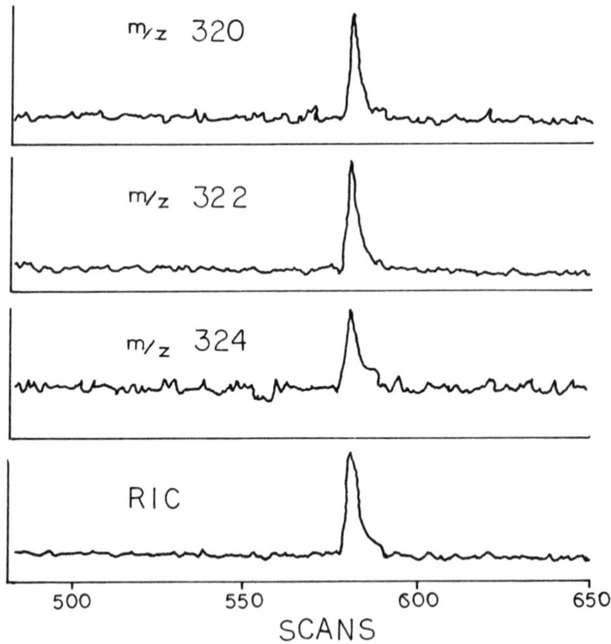

Figure 9.2. SP-2250-Capillary Column Performance for 2,3,7,8-Tetra-CDD.

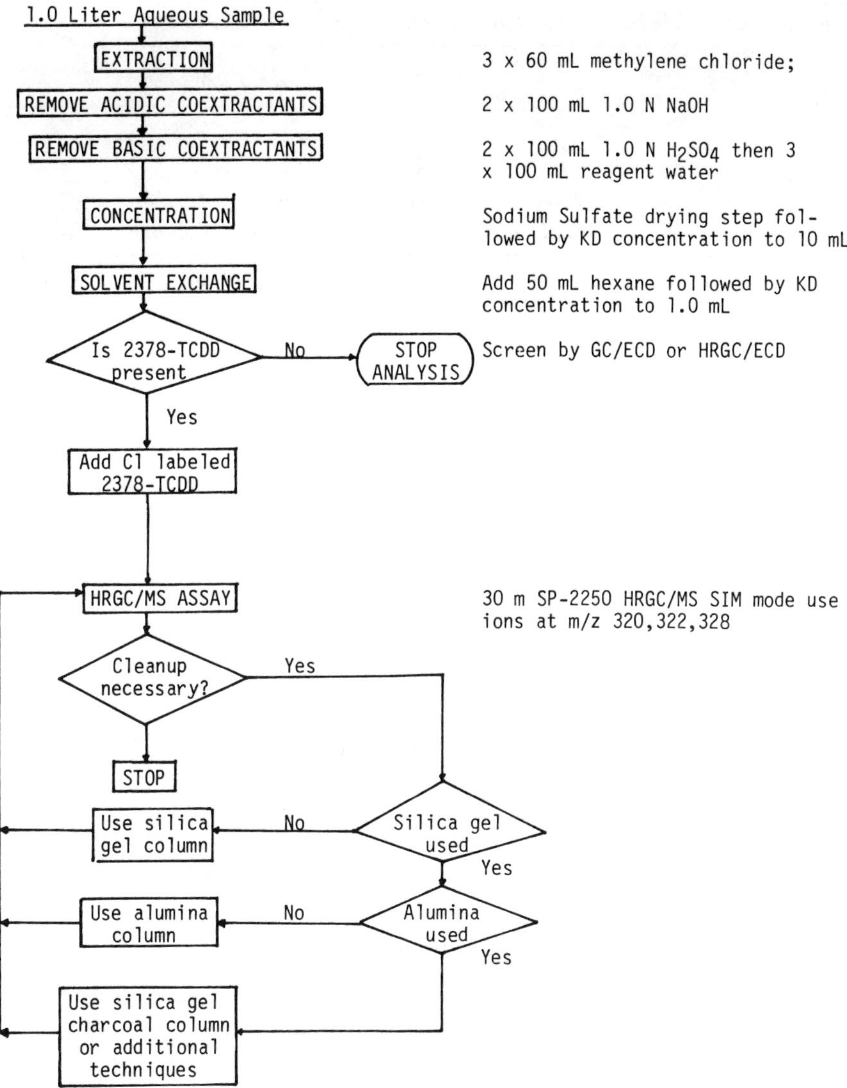

Figure 9.3. Flowchart of Initial Method.

Evaluation of the Initial Method

The initial Method 613 that appeared in the *Federal Register* is outlined as a flowchart in Figure 9.3. After screening by GC/ECD, the presence of 2,3,7,8-tetra-CDD is confirmed and the concentration measured using HRGC/MS.

Table 9.6. Initial Method Validation: Reagent Water and Wastewater

			Percent Recovery			
Matrix[a]	*Spike Level (ng/L)*	*Number of Replicates*	*No Cleanup Column*	*After One Cleanup Column*[b]	*After Two Cleanup Columns*[c]	*After Three Cleanup Columns*[d]
Reagent Water	100	5	93 ± 2.6[e]	85 ± 2.3[e]	77 ± 12[e]	38 ± 8.2[e]
Industrial	20	1		63		
Wastewater	20	1		71		
	40	1		70		
	80	1		75		
	100	3				42 ± 6.8
Municipal	15	7		85 ± 5.1[f]		
Wastewater	150	7		82 ± 4.2[f]		

[a]No 2,3,7,8-tetra-CDD found in any control samples at a detection limit of 3 ng/L.
[b]Silica gel column.
[c]Silica gel then alumina column.
[d]Silica gel then alumina then silica gel/charcoal column.
[e]HRGC/ECD results.
[f]Alumina column only.

The following criteria were to be achieved to confirm the presence of 2,3,7,8-tetra-CDD.

1. The retention time of the native 2,3,7,8-tetra-CDD is the same as that of the authentic standard.
2. The isotope ratio (320/322) must agree to within ±10% of the authentic standard.
3. m/z 320 and 322 must exhibit a simultaneous maximum.

The contractor evaluated this method using 1.0-L aliquots of reagent water, industrial wastewater and municipal wastewater that were spiked with 2,3,7,8-tetra-CDD over the range of 15–150 ng/L. The recoveries were above 70%, except when the silica gel/charcoal cleanup procedure was used (Table 9.6). The reagent water and wastewater recovery data were nearly identical. The contractor's final report is available through NTIS [11]. The detection limit of 3 ng/L is based on a signal at 5 times the noise level assuming a 5-μL injection of a 1-mL extract from a 1-L sample that exhibits no recovery losses.

PUBLIC COMMENT

The method was published as a proposed test procedure with Methods 601–612, 624 and 625 in the December 3, 1979, *Federal Register*, and as such was subject to a period of public comment to proposed rulemaking. Only single-laboratory data were available at that time. All the proposed methods were to undergo an interlaboratory validation study involving 15–20 laboratories before the methods were published in the final rulemaking in fall 1982. The interlaboratory results for Method 613 are the topic of another paper [12].

The initial public comment period to the proposed rulemaking was extended to April 28, 1980. On January 13, 1981, at the request of the Chemical Manufacturing Association, the public comment period was reopened until February 2, 1981. The first round of public comment questioned the need for screening by GC/ECD. The scientific community stated that the procedure should use only GC/MS during any assay step and that the internal standard should be added to the original sample. This would make all samples potentially hazardous, because each sample would contain the labeled 2,3,7,8-tetra-CDD.

The second round of public comment reflected the advances in the state of the art for 2,3,7,8-tetra-CDD analyses that occurred in summer and fall 1981. Use of reverse- and normal-phase HPLC was shown to give isomer-specific separation of 2,3,7,8-tetra-CDD [13]. This high-resolution cleanup procedure was reported to be effective in determining nanogram-per-liter amounts of 2,3,7,8-tetra-CDD in human milk [14] and particulate samples [15]. Achieving isomer-specific separation before the assay step obviates the need for capillary column GC, which still had not demonstrated isomer-specific separation of 2,3,7,8-tetra-CDD. This high-resolution HPLC cleanup technique appears too labor-intensive for use in a routine commercial laboratory.

During this time a comprehensive method for determining 2,3,7,8-tetra-CDD at the parts-per-trillion level in human milk, fish, water, soil and sediment was reported [16]. The original sample was always fortified with labeled 2,3,7,8-tetra-CDD in this method. Basic and acidic coextractants were removed using 1 *N* KOH and concentrated sulfuric acid. Woelm alumina (neutral grade, acitivity I) and Florisil were used for cleanup. Extracts were taken just to dryness and dissolved in 60 μL of benzene for HRGC/HRMS analysis. A 30-m $\times$ 0.25-mm glass capillary column coated with SE-30 was used for chromatographic separation and was interfaced to a high-resolution mass spectrometer that differentiated any other ions that were 30×10^{-3} mass units larger or smaller than molecular ions of 2,3,7,8-tetra-CDD. The selectivity of the SE-30 capillary column was judged equivalent to OV-101 capillary columns, which had previously been shown to coelute at least two other tetra-CDD isomers with 2,3,7,8-tetra-CDD [17]. The requirements for achieving confirmation of the presence of 2,3,7,8-tetra-CDD included those listed above for Method 613, plus two additional criteria:

1. The signals must be greater than 2.5 times the noise level.
2. Correct response must be obtained after all extracts are fortified with authentic 2,3,7,8-tetra-CDD and internal standard.

The latter criterion demands two HRGC/HRMS runs per extract (the raw extract and the extract after fortification). The extremely small volume of the final extract and the requirement for HRMS removes the final steps of this method from the repertoire of the routine commercial analytical laboratory.

The most serious fault in the early version of Method 613 was a lack of isomer specificity. A Silar-10C capillary column (55m × 0.25 mm i.d.) was reported to give unique separation of 2,3,7,8-tetra-CDD from the other 21 tetra-CDD isomers [18]. Some of the public comment deprecated the utility of this column. Because of temperature restrictions, it was apparent that no penta-, hexa-, hepta- or octa-CDD would elute from this highly polar column. This constraint is irrelevant for a method that contains a single specific isomer as the only target compound.

FINAL METHODS DEVELOPMENT

The interlaboratory methods validation studies mentioned above are conducted by the Quality Assurance Branch of EMSL. Monsanto Research Corporation was awarded the contract for the Method 613 validation study in spring 1980. The method detection limit (MDL) and analytical curve for Method 613 were determined by Monsanto [19]. Below 100 ng/L the initial version of Method 613 performed very well and a MDL of 6 ng/L was established (Table 9.7). The higher dosing levels used to determine the analytical curve (working range) demonstrated a trend toward 50% recovery as the concentration approached 1000 ng/L. At high concentrations, there may not be a sufficient volume of extracting solvent to remove all the 2,3,7,8-tetra-CDD from the aqueous sample. A brief study indicated that at a high dosing level there was a significant amount of 2,3,7,8-tetra-CDD left in the aqueous phase after extracting with three 60-mL aliquots of methylene chloride. This problem still needs more exploration; however, this conclusion is not consistent with the results in Table 9.4.

Table 9.7. Initial Method: Reagent Water MDL and Analytical Curve

Spike Level (ng/L)	*Number of Replicates*	*Mean Result (ng/L)*	*(ng/L)*	*MDL*[a]
15	7	14.3 ± 1.89		5.9
3	2	Not detected		
60	2	60.5 ± 3.5		
600	2	310 ± 71		
1000	2	495 ± 7.1		

[a] Requires at least seven replicates MDL = std. dev. × Students' t (α = 0.001).

Revision of the Method

The sample processing steps of the initial method were cumbersome. After the preliminary extraction, the analyst had to manipulate 200 mL of the organic phase. During the acidic and basic extractions, the methylene chloride phase, which was to be retained, was at the bottom of the separatory funnel. In view of the public comment and at the advice of Monsanto Research Corporation, the sample processing steps were revised. The GC/ECD screening step and the charcoal/silica gel cleanup column were deleted. The charcoal/silica gel cleanup is appropriate for sediment and soil samples, but not for water samples. The sample processing was revised to include:

1. addition of the internal standard to every sample before extraction;
2. a solvent exchange to hexane immediately after the initial extraction;
3. washing with 10 *N* NaOH and concentrated H_2SO_4; and
4. using a final extract volume of 0.2–1.0 mL.

The criteria for qualitative identification were expanded to include using a 60-m Silar-10C (or equivalent) capillary column to obtain isomer specificity and requiring that all peaks be greater than 2.5 times the noise level. The quality assurance (QA) requirements were expanded to include calculating the recovery of the internal standard and flagging the results as suspect if the internal standard exhibited less than 50% recovery at the assay step. This is a check on extreme matrix effects and errors during sample processing. The final version of the method is presented as a flowchart in Figure 9.4 and is the method used during the interlaboratory validation study. The data summarized in Tables 9.1 through 9.5 are still appropriate to the revised method because the data describe performance at individual steps in the method. The steps are the same, only the order in which they are performed was revised.

During the interlaboratory study, 2 of the 11 laboratories experienced a total loss of 2,3,7,8-tetra-CDD at the alumina column cleanup step [10]. This indicates that the activity of the specified alumina may not be uniform between batches and lots and emphasizes the importance of using laboratory control samples (LCS). The method specifies that any laboratory using the method must analyze four LCS samples and demonstrate acceptable proficiency before analyzing any environmental samples.

The entire revised method was validated in reagent water, an industrial wastewater and municipal wastewater (Table 9.8). The dosed reagent water samples exhibited a MDL of 2 ng/L based on the standard deviation of seven replicates dosed at 5 ng/L and processed through the method. This is reasonable compared with the not-detected results for the 3-ng/L spike in Table 9.7 because the final volume is now 0.2 mL, not 1.0 mL. The industrial wastewater gave results equivalent to reagent water. The municipal wastewater results showed acceptable accuracy (−8% bias) and precision (20% RSD) even though a matrix

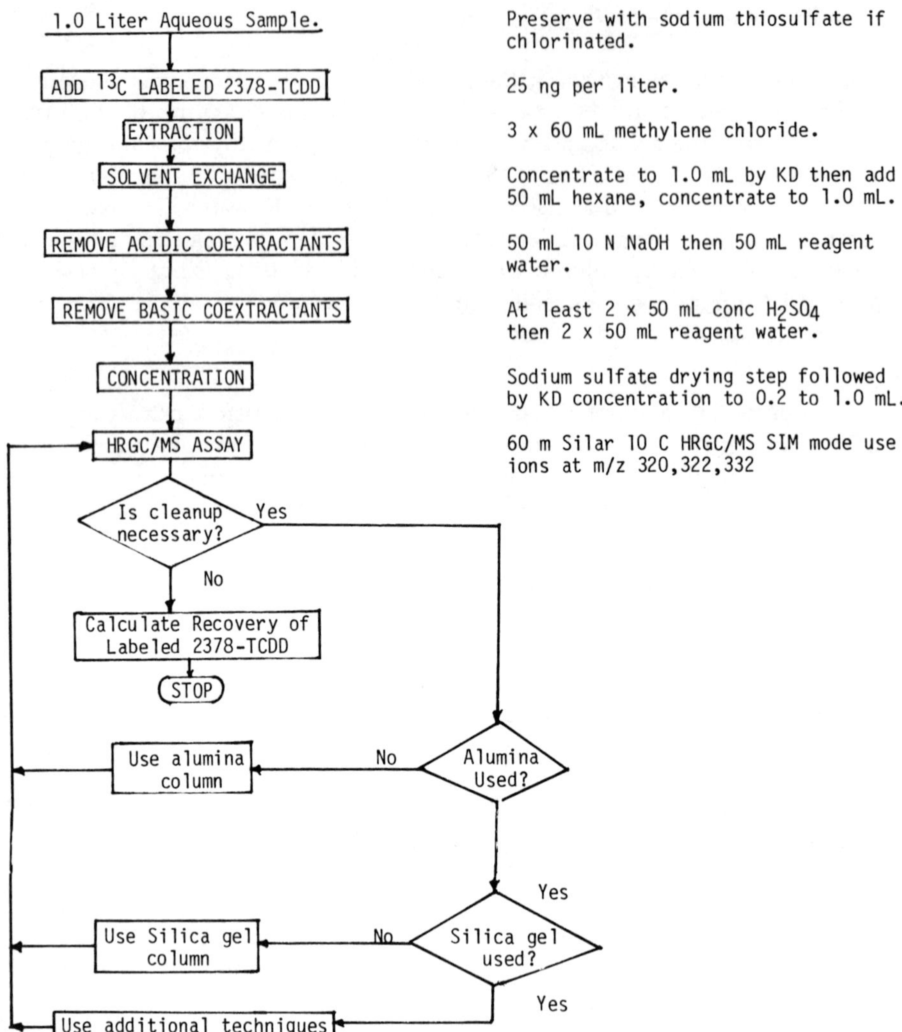

Figure 9.4. Flowchart of Final Method.

effect was evident. The small amount of bias would have been much larger had the internal standard been added after the extraction step as was specified in the earlier version of Method 613. The 50% criterion for the percent recovery of the internal standard may be too rigorous a cutoff for a QA validation check. In this case, the recovery of internal standard averaged 14%, which requires that the results be flagged as suspect, even though the accuracy showed only a −8% bias.

Table 9.8. Revised Method Validation: Reagent Water and Wastewater

Matrix	*Spike Level (ng/L)*	*Number of Replicates*	*Mean Result (ng/L)*	*MDL[a] (ng/L)*	*% Recovery of Internal Std.*
Reagent Water	5	7	4.77 ± 0.51	1.6	87 ± 23
Industrial Wastewater	5	3	4.29 ± 0.03		104 ± 25
Municipal Wastewater	25	3	23.1 ± 4.7[b]		14 ± 6

[a] Requires at least seven replicates: MDL = std. dev. × Students' t (=0.001).
[b] Recovery of labeled internal standard less than 50%.

SUMMARY

The final version of Method 613 is somewhat typical of the 600 series of methods (604–611) for solvent extractables, except that the labeled internal standard is added to every sample. The aqueous sample is extracted with methylene chloride, which is then exchanged to hexane, which is then washed with 10 *N* NaOH and concentrated H_2SO_4 to remove acidic and basic coextractants and any neutral compounds that react easily with concentrated sulfuric acid. The assay step is performed using a capillary column that gives a unique isomer-specific separation for 2,3,7,8-tetra-CDD [12] and is coupled to a MS detector operating in the SIM mode. The sensitivity requirement demands that the exit end of the capillary column be positioned as close to the ion source as practical. Two cleanup procedures are provided to help separate 2,3,7,8-tetra-CDD from interferences. Criteria for qualitative confirmation of the presence of 2,3,7,8-tetra-CDD are:

1. The GC column is isomer-specific.
2. The native 2,3,7,8-tetra-CDD and labeled 2,3,7,8-tetra-CDD internal standard coelute.
3. The chlorine isotope ratio (m/z 320/322) is within ±10% of the calibration standard.
4. All peaks are greater than 2.5 times the noise level.

DISCLAIMER

This chapter has been reviewed in accordance with EPA peer and administrative review policies and approved for presentation and publication. Mention of trade names or commercial products does not constitute endorsement or recommendation for use.

REFERENCES

1. Keith, L.H., and W.A. Telliard. *Environ. Sci. Technol.* 13:416 (1979).
2. *Federal Register* 44:69464–69575 (1979).
3. Woolson, E.A., R.F. Thomas and P.D.J. Ensor. *J. Agric. Food Chem.* 20:351 (1972).
4. Edmunds, J.W., D.F. Lec and C.M.L. Nickel. *Pest. Sci.* 4:101 (1973).
5. Vogel, H., and R.D. Weeren. *Anal. Chem.* 48:9 (1976).
6. Ramstad, R., N.H. Nahle and R. Matalon. *Anal. Chem.* 49:386 (1977).
7. Buser, H.R. *Anal. Chem.* 48:1553 (1976).
8. Passivirta, J., J.E. Enquist and S. Raeisaenen. *Chemosphere* 6:355 (1977).
9. Stalling, D.L. Proceedings of the International Conference on Environmental Sensing and Assessment, Volume 1 (1976), p. 42.
10. Hunt, D.R., T.M. Harvey and J.W. Reissel. *Anal. Chem.* 47:995 (1975).
11. Longbottom, J.E., and A.S. Wong. "Determination of TCDD in Industrial and Municipal Wastewaters," PB82-196822, National Technical Information Service, Springfield, VA.
12. McMillin, C.R., F.D. Hileman, D.E. Kirk, T. Mazer, B.J. Warner, R.J. Wesselman and J.E. Longbottom. "Determination of 2,3,7,8-Tetrachlorodibenzo-*p*-dioxin in Industrial and Municipal Wastewaters, Method 613: Performance Evaluation and Preliminary Method Study Results," Chapter 10, this volume.
13. Nesterick, T.J., L.L. Lamparski and R.H. Stehl. *Anal. Chem.* 51:2273 (1979).
14. Langhorst, M.L., and L.A. Shadoff. *Anal. Chem.* 52:2037 (1980).
15. Lamparski, L.L., and T.J. Nesterick. *Anal. Chem.* 52:2045 (1980).
16. Harless, R.L., E.O. Oswald, M.K. Wilkinson, A.E. Dupuy, Jr., D.D. McDaniel and H. Tai. *Anal. Chem.* 52:1239 (1980).
17. Buser, H.R. *Anal. Chem.* 49:918 (1977).
18. Buser, H.R. and C. Rappe. *Anal. Chem.* 52:2257 (1980).
19. Glaser, J.A., D.L. Foerst, G.D. McKee, S.A. Quave and W.L. Budde. *Environ. Sci. Technol.* 15:1426 (1981).

10

Determination of 2,3,7,8-Tetrachlorodibenzo-*p*-dioxin in Industrial and Municipal Wastewaters, Method 613: Performance Evaluation and Preliminary Method Study Results

Carl R. McMillin, Fred D. Hileman,
David E. Kirk, Thomas Mazer,
Beverly J. Warner, Raymond J. Wesselman
and James E. Longbottom

In November 1979 the U.S. Environmental Protection Agency (EPA) exercised an option on Monsanto Research Corporation (MRC) contract EPA-68-03-2856 to conduct an interlaboratory validation study for 2,3,7,8-tetrachlorodibenzo-*p*-dioxin (2,3,7,8-tetra-CDD) using EPA Method 613.

The original Method 613, as published in the December 3, 1979, *Federal Register*, was evaluated and a detection limit of 5.9 ppt was found in interference-free water. It was also determined that the ampules of 2,3,7,8-tetra-CDD spiking solution were stable for more than 90 days. Requests for proposals for Method 613 validation study were sent to more than 250 potential laboratories. The participating laboratories were required to conduct capillary column gas chromatography/mass spectrometry (GC/MS) analysis of 2,3,7,8-tetra-CDD, using a capillary column that, at the time, was only available in glass (Silar-10C). When fewer than the desired number of responses were received, it was determined that the reasons were primarily either a lack of capillary GC/MS capability or a concern for the safety and handling of 2,3,7,8-tetra-CDD.

A number of changes to Method 613, which are described by Wong et al.

[1], made the sample preparation portion of the method easier to conduct. It was decided that MRC would supply the dioxin standards, so that high-hazard facilities would not be required by the participating laboratories, and that the wastes generated during the study would be returned to MRC for disposal. A total of 29 laboratories responded to a request for proposals. The responses were ranked technically [experience, facilities, quality assurance/quality control (QA/QC) and personnel], and arrangements were made with 18 laboratories of high technical rating to participate in the method study, contingent on acceptable analyses of the performance evaluation (PE) samples. Two additional laboratories received the PE samples for analyses by Method 613 as a quality assurance check on other projects. Their data are also included in the inter-laboratory statistics for the PE samples.

PERFORMANCE EVALUATION SAMPLES

The three objectives of the PE study were (1) to determine if the laboratories could separate the 2,3,7,8-tetra-CDD from the other potentially interfering isomers; (2) to determine if the laboratories had adequate sensitivity for the study; and (3) to generate intralaboratory reproducibility data for potential use in evaluating both the laboratories and Method 613. The laboratories were requested to spike four one-L samples of water from one vial of spiking solution and to analyze the samples using Method 613 with the alumina column cleanup.

At the time the PE sample was synthesized, only Silar-10C had been shown to give resolution of 2,3,7,8-tetra-CDD from the other 21 tetra-CDD isomers [2]. EPA Method 613 specified that if a positive peak is found, it must be confirmed using a Silar-10C column or its equivalent. A PE sample was synthesized that included approximately 50 ppt of the 2,3,7,8-, 1,2,3,4- and 1,4,7,8-tetra-CDD-isomers. The latter two isomers elute immediately before and immediately after the 2,3,7,8-tetra-CDD on Silar-10C [2] and are therefore the most difficult isomers to separate from 2,3,7,8-tetra-CDD with this column.

Synthetic routes for various tetra-CDD isomers had previously been cited in the literature [3]. Based on this literature, 2,3,6- and 2,4,5-trichlorophenol were reacted with potassium hydroxide to yield the respective trichlorophenates, which were then subjected to a condensation reaction in a sealed tube at 300°C to yield a mixture of primarily the 1,4,7,8-, 1,2,6,9- and 1,2,7,8-tetra-CDD isomers. 2,3,7,8-tetra-CDD was synthesized similarly from 2,4,5-trichlorophenate.

By combining the tetra-CDD produced from the condensation reactions in the appropriate ratios with 1,2,3,4-tetra-CDD obtained from KOR Isotopes (Cambridge, Massachusetts), a PE mixture was ultimately prepared containing approximately 50 ppt of 1,4,7,8-, 1,2,3,4- and 2,3,7,8-tetra-CDD.

The PE kits, sent to the laboratories in September 1981, included the PE sample vial, a vial containing a stated quantity of ^{13}C-labeled 2,3,7,8-tetra-CDD, and a vial containing a stated quantity of unlabeled 2,3,7,8-tetra-CDD, all to be dissolved in acetone. Also included was a helper mix of the PE isomers to be dissolved in tetradecane and used to assist in the determination of the optimum chromatographic conditions. Vials of spiking solutions were produced at MRC by dispensing set quantities of the standard solutions into screw-top septum vials and evaporating the solvent so they could be shipped dry.

Before sending the PE samples to the participating laboratories, MRC evaluated a 60-m SP-2330 glass capillary column (Supelco, Bellefonte, Pennsylvania) for its applicability to the separation of the tetra-CDD isomers since SP-2330 is a highly polar liquid phase with McReynolds constants very close to those of Silar-10C. The SP-2330 column was found to give superior chromatographic resolution and peak shape to the Silar-10C column used previously.

The resolution achieved for a mixture of the 22 tetra-CDD isomers (provided by C. Rappe, University of Umeå, Sweden) and its specificity for 2,3,7,8-tetra-CDD, as shown in Figures 10.1 through 10.3, was comparable to those achieved by Buser and Rappe [2], promoting approval for its use as a column equivalent to Silar-10C. Sample chromatograms of the PE mix (Figure 10.4) were supplied to the participating laboratories along with the PE samples.

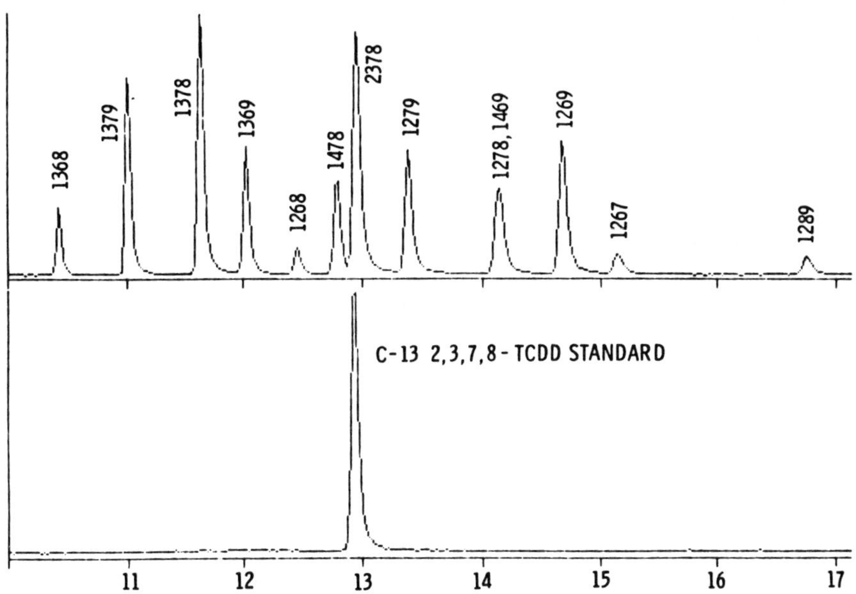

Figure 10.1. Tetra-CDD Isomers with 2:2 Chlorine Substitution Analyzed on a Glass SP-2330 Column.

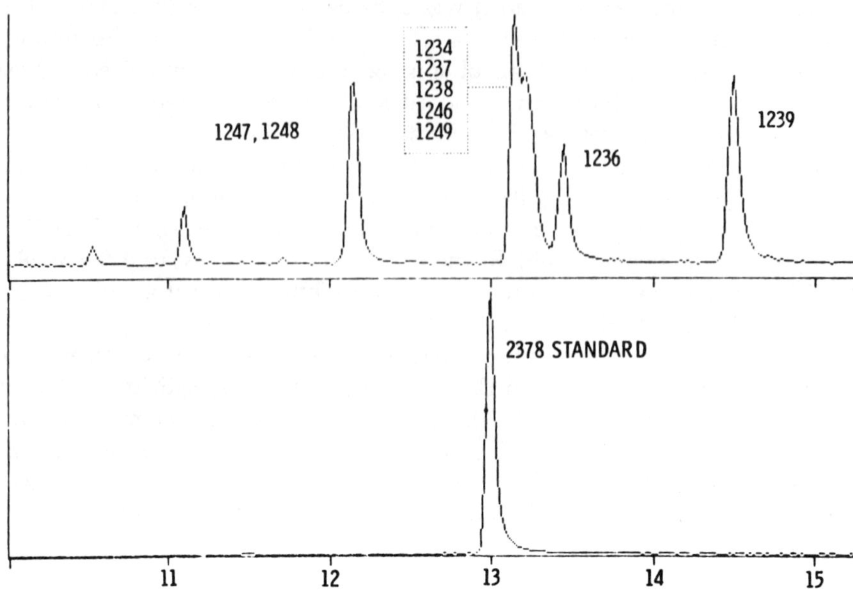

Figure 10.2. Tetra-CDD Isomers with 1:3 Chlorine Substitution Analyzed on a Glass SP-2330 Column.

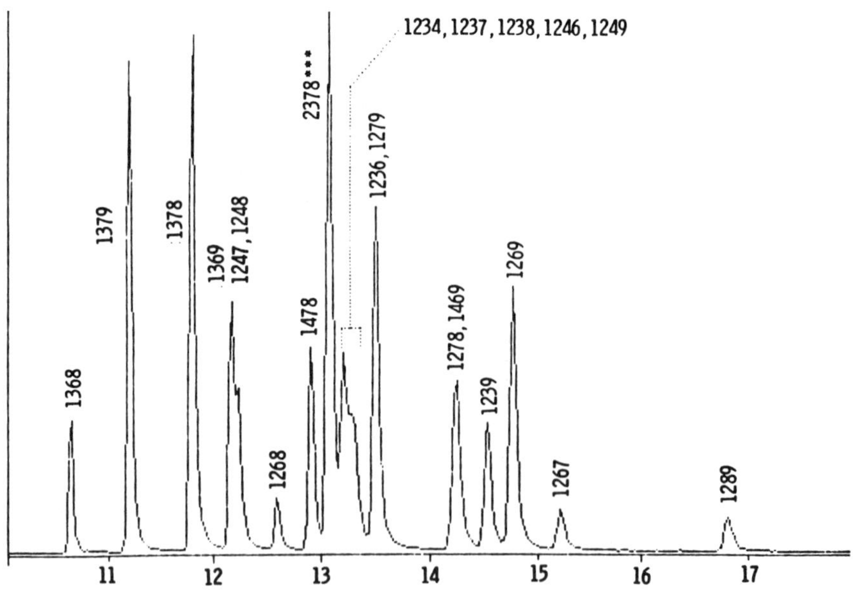

Figure 10.3. All 22 Tetra-CDD Isomers Analyzed on a Glass SP-2330 Column.

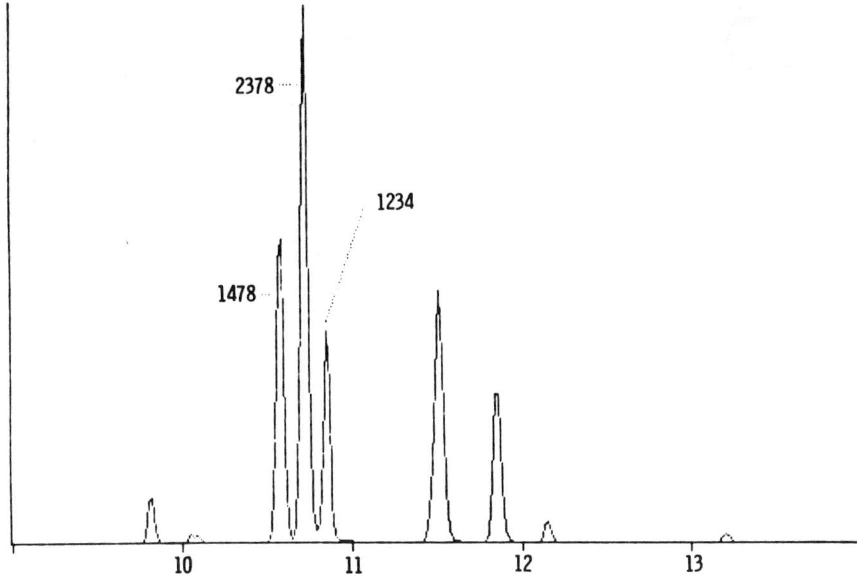

Figure 10.4. MRC PE Sample Analyzed on a 54-m Glass SP-2330 Column with Hydrogen Carrier Gas and Temperature-Programmed from 200 to 265°C.

PE STUDY RESULTS

Of the twenty laboratories, fourteen submitted numerical results, including chromatograms from their analyses of the PE samples. Of the other six laboratories, two were able to achieve separation of the isomers as demonstrated by chromatograms but were unable to analyze the PE samples because of sensitivity problems. Four laboratories dropped out of the study because of problems with sensitivity, the glass capillary interface, scheduling or personnel. Of the sixteen laboratories that supplied chromatographic results, seven used a polar SP-2330 (Silar-10C equavalent) column; nine used less polar DB-5, Carbowax 20M, SE-30 or SE-54 fused-silica columns. One of these laboratories switched to Carbowax 20M after encountering difficulties with SP-2330. One of the 14 laboratories did not use Method 613 for the analysis. The laboratories using SP-2330 columns achieved good chromatographic separation of the 2,3,7,8-tetra-CDD from the other isomers, as shown in Figure 10.5.

One problem encountered in the analysis of the PE sample in two laboratories was the deterioration of the SP-2330 column. In one laboratory this occurred while workers were attempting to deactivate the column. At the other laboratory the deterioration occurred after analysis of three or four water extracts, perhaps due to a leak in the system. No problems were encountered with the glass SP-2330 column in the other laboratories using this column except their general dis-

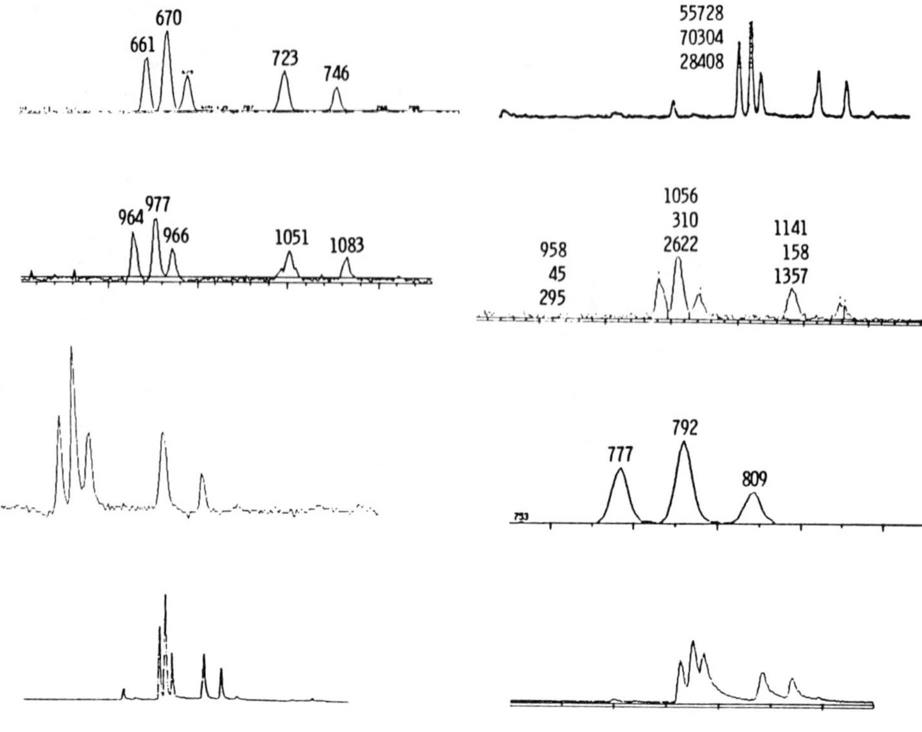

Figure 10.5. Chromatograms from Eight Laboratories of the PE Mixture Using SP-2330 Columns.

pleasure with having to use a glass column instead of the more rugged fused silica. A glass SP-2330 column has been used for about ten months at MRC for dioxin and dibenzofuran analyses with no problems.

The analytical results for the PE samples from the 13 laboratories using Method 613 showed an average 2,3,7,8-tetra-CDD value of 89 ±38 ppt (one standard deviation) and an average recovery of 70 ±27% (one standard deviation). The PE sample was analyzed four times by each laboratory. The average intralaboratory variability was ±12 ppt (average standard deviation). The average intralaboratory variability for recovery was ±19%.

A comparison of the data obtained from the analyses of the PE samples in which the SP-2330 column was used to the data from the other columns showed little difference in the accuracy, precision, variability, or recovery, as seen in Figure 10.6. This is not surprising, since the other columns used in the analyses readily resolved the test mix. The test mix included the two most difficult isomers for polar columns (such as Silar-10C or SP-2330) to separate from 2,3,7,8-tetra-CDD, but did not include interfering isomers for nonpolar columns. In retro-

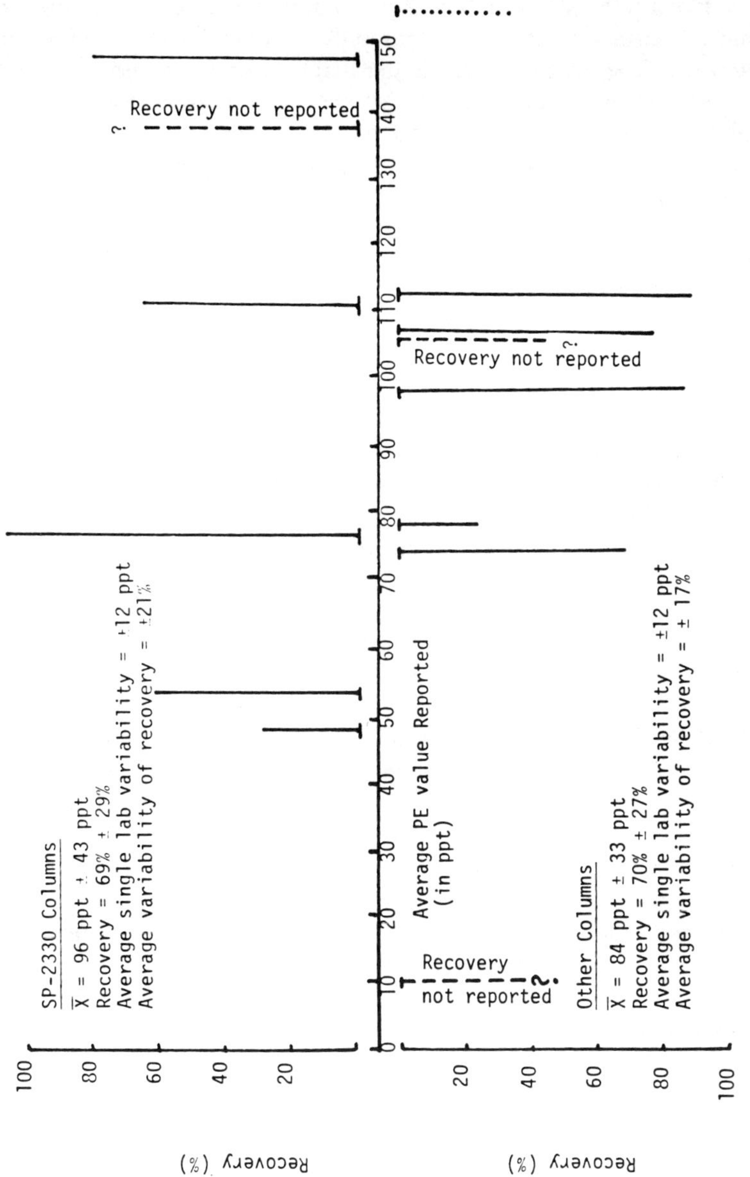

Figure 10.6. Summary of PE Results for 2,3,7,8-Tetra-CDD.

spect, either the PE analysis instruction should have demanded, rather than implied, use of the SP-2330 or Silar-10C column, or other interfering isomers should have been added to ensure that only a column capable of separating 2,3,7,8-tetra-CDD from the other 21 tetra-CDD isomers could be used (e.g., 1,2,3,8-tetra-CDD as an OV-101 interference, and so forth for SE-54, DB-5, SE-30 and other potential columns). The total ion chromatograms were examined for each of the laboratories and those 13 successfully completing the PE study were sent method study samples.

METHOD 613 INTERLABORATORY VALIDATION STUDY

Three target levels were chosen for the study: 20, 80 and 200 ppt. The low-level target value was selected to be about 5–10 times the distilled water method detection level, or about 25 ppt. Note that the original Method 613 was found to have a method detection limit (MDL) of 6 ppt, while the revised Method 613 was found to have a MDL of 1.5 ppt. Both were determined on a low-resolution quadrupole GC/MS system, since most participating laboratories would be using this type of instrument.

When the analytical curve for Method 613 was determined, a potential extraction problem was encountered for samples spiked at 600 ppt and higher. The quantity of methylene chloride was apparently insufficient to completely extract the spiked dioxin at these levels. The high level target value was therefore selected to be 200 ppt for the method study. A middle range target value of about 80 ppt was selected between the high and low values. In the method study, Youden pair concentrations were selected at about 5% above and below each of the low, middle and high target values, yielding a total of six concentrations.

The six concentrations of 2,3,7,8-tetra-CDD were then to be spiked into distilled water, tap water and a surface lake or pond water acquired by each of the participants and into three different wastewaters supplied to the participants. The wastewaters were collected so that they would be representative of sample matrices in which dioxins might be found. One of the wastewaters was particularly challenging, because it readily formed emulsions (wastewater A). The study consisted of six concentrations of 2,3,7,8-tetra-CDD to be analyzed in six different water types (totaling 36 analyses) plus the blanks for each water, a total of 42 analyses. The 42 labeled vials (including blank vials) were sent to the participants along with a vial of ^{13}C-labeled 2,3,7,8-tetra-CDD, a vial of unlabeled 2,3,7,8-tetra-CDD, and the one-L bottles of wastewater.

Later, newly available 60-m fused-silica SP-2330 and SP-2340 columns from Supelco were evaluated and found to give similar separation of the 2,3,7,8-tetra-CDD from the other 21 isomers when compared to the 60-m glass SP-2330 as shown in Figures 10.7 through 10.9. As a result of this, laboratories that desired to use these new columns for the method study were given approval to do so. A recent *Supelco Reporter* also presented these results to a wider audience [4].

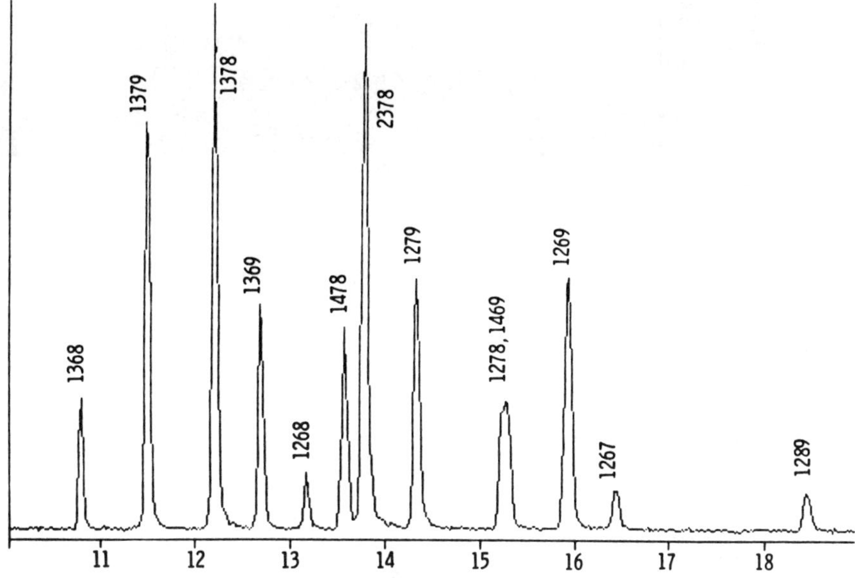

Figure 10.7. Tetra-CDD Isomers with 2:2 Chlorine Substitution Analyzed on a Fused-Silica SP-2340 Column.

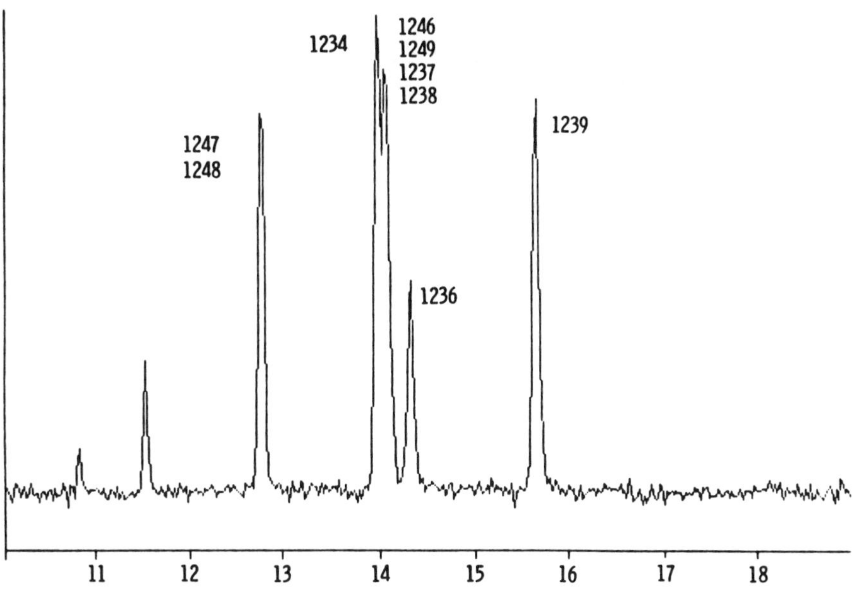

Figure 10.8. Tetra-CDD Isomers with 3:1 Chlorine Substitution Analyzed on a Fused-Silica SP-2340 Column.

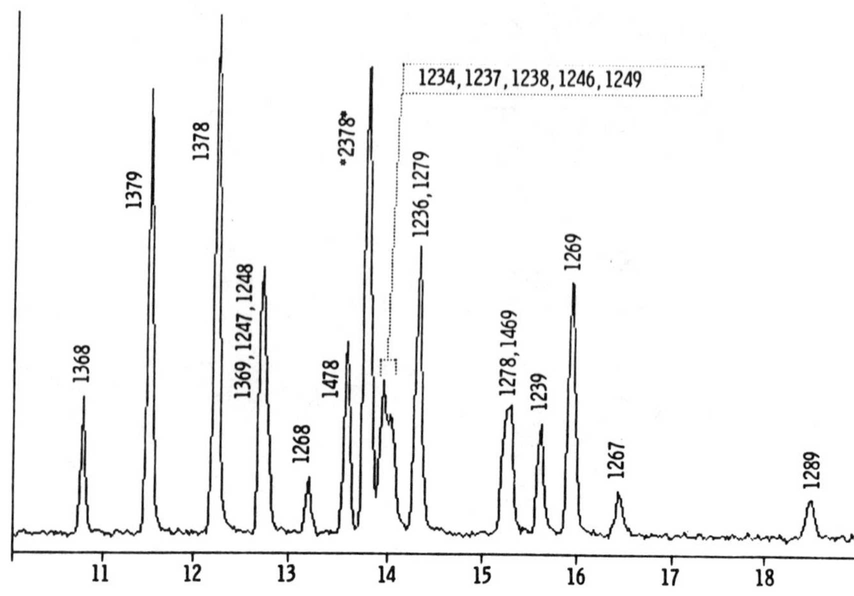

Figure 10.9. All 22 Tetra-CDD Isomers Analyzed on a Fused-Silica SP-2340 Column.

METHOD STUDY RESULTS

The detailed review and complete statistical analysis of the data have not been completed. Accuracy and precision equations will be computed and included in later versions of the method. Inspection of the data with a few statistical calculations does, however, give a preliminary indication of the overall quality of the data and the types of problems that were encountered.

Of the thirteen laboratories that were sent samples, one dropped out for economic reasons and two others had sensitivity problems that precluded low-parts-per-trillion analysis. Of the ten remaining laboratories, two encountered losses of the analyte in some samples when using the alumina column for sample cleanup. In the method, the sample is introduced into the alumina column followed by 50 mL of 3% methylene chloride in hexane that is discarded. If the alumina is not properly activated, the 2,3,7,8-tetra-CDD may be eluted in this fraction and lost. Next, 50 mL of 20% methylene chloride in hexane is used to elute the 2,3,7,8-tetra-CDD from the alumina column. In some cases, this solution failed to elute the 2,3,7,8-tetra-CDD from the alumina. Therefore, further work is needed to define specific alumina activation conditions and then to establish that the alumina is performing properly through the analysis of check standards.

A preliminary summary of some of the results of the method-validation study is shown in Table 10.1. Of these laboratories, nine used glass SP-2330

Table 10.1. Preliminary Review of Some Method 613 Study Data[a]

High-Concentration Vials		*Medium-Concentration Vials*		*Low-Concentration Vials*	
202[b]	*182*[b]	*81*[b]	*73*[b]	*26*[b]	*21*[b]
260[c]			104 ± 16[d]	31[c]	12 ±7.1[d]
174 ± 11[e]	152 ± 0.8[e]	75 ± 3.1[e]	65 ±1.4[e]	24 ±1.5[e]	20 ±0.5[e]
275 ±34[f,g]	230 ±11[h]	168 ± 21[h]	167 ± 22[h]	30 ± 4.5[h]	21 ± 9.0[g,h]
145 ± 10	163 ± 12	55 ±13	49 ± 2.6	26 ±4.4	24 ± 9.5
177 ± 14	166 ± 33	71 ± 6.0	67 ± 5.0	24 ± 1.9	24 ± 4.8
183 ± 11	191 ± 14[f]	67 ± 2.3	60 ± 2.7	22 ± 6.3	19 ± 2.7
213 ± 48[g,i]	137 ± 12[g,i]	67 ± 3.6[g,i]	51 ± 9.0[g,i]	24 ± 2.1[d]	18 ± 2.1[g,i]
112 ± 15	121 ± 48	55 ± 12	54 ± 16	23 ± 5.3	22 ± 5.8
203 ± 23[g]	212 ±49[g]	75 ± 7.5[f]	60 ± 9.7[f]	26 ±7.8[f]	19 ± 3.4[h]
213 ± 24	161 ± 7.7	79 ±12	71 ± 5.6	30 ± 6.1[f]	26 ± 6.3
111 ±61[f]	93 ± 79[f]	105[c]	48[c]	—	—
188 ± 29[j]	163 ± 26[j]	82 ± 8.9[j]	72 ± 9.0[j]	26 ± 4.4[j]	20 ± 5.1[j]

[a]Each value presented is an average of the six analyses for that vial at a laboratory in each of six different water types, unless otherwise noted (ppt ± one standard deviation).

[b]Calculated true value.

[c]Based on one value.

[d]Based on two values.

[e]Values corrected for a documented calculational error.

[f]Based on five values.

[g]Includes data points questioned by laboratory (e.g., low recovery).

[h]Based on four values.

[i]Based on three values.

[j]Average values.

capillary columns for the analyses; fused-silica SP-2330 was used by one laboratory. The carrier gas used was helium in nine laboratories and hydrogen in one laboratory.

Inspection of the data shown in Table 10.1 shows that the current version of Method 613 is capable of generating reasonable results by a variety of qualified laboratories. The correlation of the average analyzed values with the calculated true values is very good. Some fall-off at the higher concentrations is observed, but this may be related to the fact that these concentrations approach the solubility limits of 2,3,7,8-tetra-CDD in water.

ACKNOWLEDGMENTS

This project has been funded by the U.S. Environmental Protection Agency under contract number 68-03-2856. The content of this publication does not necessarily reflect the views or policies of the U.S. Environmental Protection Agency, nor does mention of trade names, commercial products or organizations imply endorsement by the U.S. government.

REFERENCES

1. Wong, A.S., M.W. Orbanosky, P.A. Taylor, C.R. McMillin, R.W. Noble, D. Wood, J.E. Longbottom, D.L. Foerst and R.J. Wesselman. "Determination of 2,3,7,8-Tetrachlorodibenzo-*p*-dioxin in Industrial and Municipal Wastewaters, Method 613: Development and Detection Limits," Chapter 9, this volume.
2. Buser, H.R., and C. Rappe. *Anal. Chem.* 52:2257–2262 (1980).
3. Nestrick, T.J., L.L. Lamparski and R.H. Stehl. *Anal. Chem.* 51:2273–2281 (1979).
4. "Rapid Separation of 2,3,7,8-TDD from Other TCDD Isomers," *Supelco Reporter* 4th ed. 1(4) (1982).

11

Occurrence of Tetrachlorodibenzo-*p*-dioxin in Environmental Samples from Southwest Missouri

Robert D. Kleopfer, William W. Bunn,
Kenneth T. Yue and Daniel J. Harris

BACKGROUND

Although polychlorinated dibenzo-*p*-dioxins (PCDD) have never been manufactured intentionally, they are found quite frequently in the environment. This chapter describes the nature and extent of tetrachlorodibenzo-*p*-dioxin (tetra-CDD) contamination resulting from chemical manufacturing activities near Springfield, Missouri.

The environmental problems associated with those activities first surfaced in 1971, when three horse arenas in Missouri were sprayed for dust control with a mixture of used oil and chemical waste [1]. In the ensuing three years, approximately 90 horses died (or were destroyed) along with numerous cats, dogs, rodents, chickens, birds and insects. Several people also became ill as a result of contact with the toxic contaminant, which was identified in 1974 as 2,3,7,8-tetra-CDD. Investigations traced the source of the substance to the defunct Northeastern Pharmaceutical and Chemical Company (NEPACCO), which had leased manufacturing facilities at a chemical plant in Verona, Missouri. The plant produced pharmaceutical-grade hexachlorophene, which involved the intermediate production of 2,4,5-trichlorophenol (tri-CP). The process required the distillation of tri-CP, which resulted in a still bottom waste stream with tetra-CDD concentrations of several hundred parts per million. Between approximately April 1970 and January 1972, a reported 328 batches of 2,4,5-tri-CP were distilled for hexachlorophene production. Before 1970 the facilities had been used to produce Herbicide Orange for the military. However, that process resulted in a technical-grade herbicide that reportedly, permitted impurities, such as tetra-CDD, to go forward into the final product.

U.S. ENVIRONMENTAL PROTECTION AGENCY (EPA) INVESTIGATIONS

More recent EPA investigations of past waste disposal practices began in 1979 as a result of an anonymous phone call. Those investigations resulted in the identification of three potential sites, which included a municipal park, a farm site and trenches on plant property. Subsequent publicity led to tips that resulted in the identification of additional waste sites, including other farm sites and a water/wastewater technical school at Neosho, Missouri [2]. As of mid-1982, 24 sites of potential tetra-CDD contamination had been identified. To date, the investigations have accounted for an estimated 110 lb of still bottom-derived tetra-CDD, plus relatively minor amounts of tetra-CDD in other waste streams.

The Spring River Basin encompasses most of the known disposal sites in southwestern Missouri. Since tetra-CDD contamination of fish, livestock and wildlife had been reported in several areas where an environmental release of tetra-CDD has occurred [3], a preliminary sampling survey was conducted in 1981 to determine if contamination of the Spring River had occurred. The results of these investigations are discussed here.

REMEDIAL ACTION

Among the various contaminated wastes produced by NEPACCO was 4300 gal of tri-CP still bottom wastes stored in a tank on plant property at Verona, Missouri. The entire plant had been acquired by Syntex Agribusiness, Inc., shortly after NEPACCO leased the hexachlorophene manufacturing area from Hoffman-Taff, Inc. During summer 1980 Syntex began a process to destroy the estimated 12 lb of tetra-CDD present in the storage tank. The process, which was successful, involved photolysis with ultraviolet (UV) light in the presence of a proton donor [4]. During the process EPA monitored various process waste streams as well as ambient air for the presence of tetra-CDD.

In 1981 Syntex initiated a second cleanup effort to remove tetra-CDD from a drum disposal area on a farm near Verona. This project involved construction of two microbiological degradation basins for destruction of tetra-CDD in soil. Additional cleanup projects are in the early stages.

ANALYTICAL METHODOLOGY

Several different laboratories were involved in the various analyses associated with these investigations. A brief discussion of the methodologies used is presented here.

Water (EPA)

A modified EPA method 613 [5] was used for analyses of groundwater and wastewater samples. A 1-L sample of water was spiked with an internal standard of isotopically labeled 2,3,7,8-tetra-CDD. The spiked sample was then extracted with methylene chloride using separatory funnel techniques. The extract was concentrated to a volume of 1.0 mL or less. Analysis was by high-resolution gas chromatography/low-resolution mass spectrometry (HRGC/LRMS).

Fish

Columbia National Fisheries Research Laboratory (CNFRL)/EPA

This procedure was developed at CNFRL [6]. A ground tissue sample was mixed with anhydrous sodium sulfate and extracted with cyclohexane: methylene chloride (1:1) using column extraction. In the same process, the eluate passed through (1) potassium silicate, (2) silica gel, (3) cesium silicate, (4) silica gel, and (5) carbon/glass adsorbent. The tetra-CDD was recovered from the carbon adsorbent by reverse elution with toluene. The concentrated toluene extract was exchanged into hexane and then applied to a second series of adsorbents containing (1) cesium silicate and sulfuric acid–impregnated silica gel and (2) activated alumina. The desired fraction was reduced in volume and analyzed by HRGC/LRMS-multiple ion detection (MID).

Both CNFRL and EPA Region VII laboratories used a Finnigan Model 4000 GC/MS with an Incos Data System. A 30-m x 0.25-mm (i.d.) DB-5 fused-silica capillary column was used with helium carrier gas. A MID program was used to monitor the nominal mass values corresponding to the moelcular ions (320 and 322) for tetra-CDD. Quantification was based on the use of isotopically enriched internal standard $^{13}C_{12}$-tetra-CDD, which was added to each sample before extraction. Nominal masses 332 and 334 were monitored for the internal standard.

University of Nebraska—Lincoln (UNL)

This procedure was used by the Midwest Center for Mass Sepctrometry at UNL [7]. The ground tissue sample was fortified with the isotopically enriched $^{13}C_{12}$-tetra-CDD internal standard. After saponification with ethanol and aqueous potassium hydroxide, the solution was extracted with hexane. The hexane extract was cleaned with concentrated H_2SO_4 before extract concentration. Isomer-specific analysis was by high-resolution GC/high-resolution MS. A Carlo-Erba gas chromatograph equipped with a SE-54 fused-silica capillary column (0.25 mm i.d. by 30 m) was used for the analyses. The ions at masses 319.8965, 321.8936 and 333.9339 were monitored for the tetra-CDD with a MS resolving power of 7500.

National Center for Toxicological Research (NCTR)

Two fish samples were analyzed by NCTR [8]. The samples were prepared by the UNL lab by the previously described procedure. Analysis was by combined capillary GC/atmospheric pressure ionization MS. The ions monitored were 176, 178 and 182, which correspond to the 4,5-dichloro-1,2-benzoquinone radical anion (176, 178 for the native isotopes, 182 for the $^{13}C_6$ isotope). This analytical procedure is selective for tetra-CDD isomers having the 2:2 substitution pattern.

Soil

Wright State University

Aliquots (typically 1–10 g) of air-dried, finely pulverized soil were dosed with $^{37}Cl_4$-2,3,7,8-tetra-CDD and extracted by shaking for 20 min with petroleum ether. The extract was washed consecutively with water, 20% KOH, water, concentrated H_2SO_4 and water. The resulting dried, concentrated extract was subjected to additional cleanup by HPLC on a Du Pont Zorbax ODS column [9]. LRGC/HRMS analyses were performed using a Varian 3740 GC coupled through an AEI silicone membrane separator to an AEI Double-Focusing, Double-Beam MS-30 mass spectrometer at a static resolution of 1:12,500. The ions monitored were m/e 319.8966, 325.8805 and 327.846.

University of Nebraska—Lincoln

A 5 to 10-g sample was dosed with $^{37}Cl_4$-tetra-CDD, treated with ammonium chloride solution, and extracted by stirring with 1:1 hexane/acetone. The extract was washed consecutively with 1*N* KOH, water, concentrated and then subjected to liquid chromatography on alumina. Analysis was by LRGC/HRMS using a Kratos MS-5076 mass sepctrometer at a resolving power of 10,000. Masses 321.8936 and 327.848 were monitored initially. Confirmation was accomplished by monitoring masses 321.8936 and 319.8965.

Soil/Sediment (EPA)

Aliquots (50 g) of soil/sediment were air-dried, dosed with $^{13}C_{12}$-tetra-CDD and mixed with anhydrous sodium sulfate before extraction with acetone/hexane on a column or toluene in a Soxhlet apparatus. The concentrated extract was cleaned up by liquid chromatography on silica gel, acidified silica gel and alumina. Analysis was by HRGC/LRMS using the equipment and conditions described for fish extract analysis.

Ambient Air (EPA)

Air samples were collected by pumping a measured amount of air (10–200 L) through an adsorption tube containing Florisil. The inlet tube and Florisil were desorbed with methylene chloride. The extracts were exchanged into isooctane and analyzed by HRGC/EC. The methodology yielded a detection limit of 1 ng/m^3 with recoveries of 92 ± 10% (mean and standard deviation for seven data points) for vapor phase dosing of TCDD.

ANALYTICAL RESULTS

The highest values from each of the matrices sampled are summarized in Table 11.1. Although tetra-CDD was detected in sediment and fish samples from the Spring River, it has never been found in any of the ambient air or water samples. Concentrations from the various NEPACCO waste materials ranged up to several hundred parts per million.

Table 11.2 summarizes the results of analyses of fish and sediment from the Spring River for samples collected during November and December 1981. The sampling locations are depicted in Figure 11.1.

Selected samples were analyzed under conditions such that 2,3,7,8-tetra-CDD could be separated from the other 21 tetra-CDD isomers. Except for a small amount of 1,3,7,8-tetra-CDD (less than 3% of the total tetra-CDD) in the horse arena samples, all of the tetra-CDD was in the form of the 2,3,7,8- isomer.

Table 11.1. Summary of Analytical Results for Tetra-CDD (Highest Values Found)

Sample	*Results (ppb)*
Storage Tank, TCP still bottoms	343,000
Ambient Air	ND (1 ng/m^3)[a]
Soil from Farm Site	72
Fish from Spring River	0.055
Groundwater	ND (0.002)
Drum Samples from Farm Site	319,000
Sediment from Spring River	0.012
Hexachlorophene, Filter Cake Material	8
Feed Lot Samples	0.4
Neosho Digester	60
Horse Arena Soil	33,000

[a] ND = none detected (detection limit).

Table 11.2. Summary of Spring River Fish/Sediment

Stream Mile[a]	*Fish Species (Number in Composite)*	*Range of TCDD in Fish (ng/kg)*	*Sediment (ng/kg)*
0	Rainbow trout (5)	ND (1)[b]	ND (5)
0.7	White sucker (5)	15–25	ND (10)
	Hog sucker (2)	16	
2.9	Location of former producer of hexachlorophene		
3.2	White/hog sucker (4)	37–55	12
4.3	White sucker (3)	15[c]	
5.6	White sucker (5)	36–39	ND (20)
	White sucker (5)	17[c]	
11.9	White sucker (5)	6[c]	ND (10)
	Hog sucker (2)	5[c]	
	Creek chub (3)	ND (1)	
36	White sucker (4)	ND (1)	
	Smallmouth bass (7)	6.2	
46	Spotted sucker (6)	ND (1)	
	Largemouth bass (9)	2.5, 1.4[c]	
	Shorthead redhorse (4)	1.1	
69	Northern hog suckers (3)	ND (1)	
	Shorthead/river redhorse (3)	0.8	
96	Carp (5)	ND (3)	

[a]The headwaters of the Spring River are at a trout hatchery (stream mile 0). Locations are indicated in stream miles downstream from that point.

[b]ND = none detected (detection limit).

[c]Fillets. All other data are for whole fish.

DISCUSSION

The results of these investigations reveal a rather widespread distribution of 2,3,7,8-tetra-CDD in the environment of southwest Missouri due primarily to inappropriate waste disposal practices used some 10 years previously. Although remedial cleanup actions have been highly successful, accurate estimates of the total volumes of tetra-CDD wastes are not possible, and the whereabouts of any remaining material is uncertain. In 1982 Missouri residents were cautioned about consumption of fish taken from the upper portions of the Spring River. Additional sampling and investigative work are planned to identify additional sources of contamination and to protect public health.

In spite of the complexity of tetra-CDD analyses, the interlaboratory comparisons of split samples have been excellent. One particular fish sample was analyzed by four different laboratories using three different procedures with

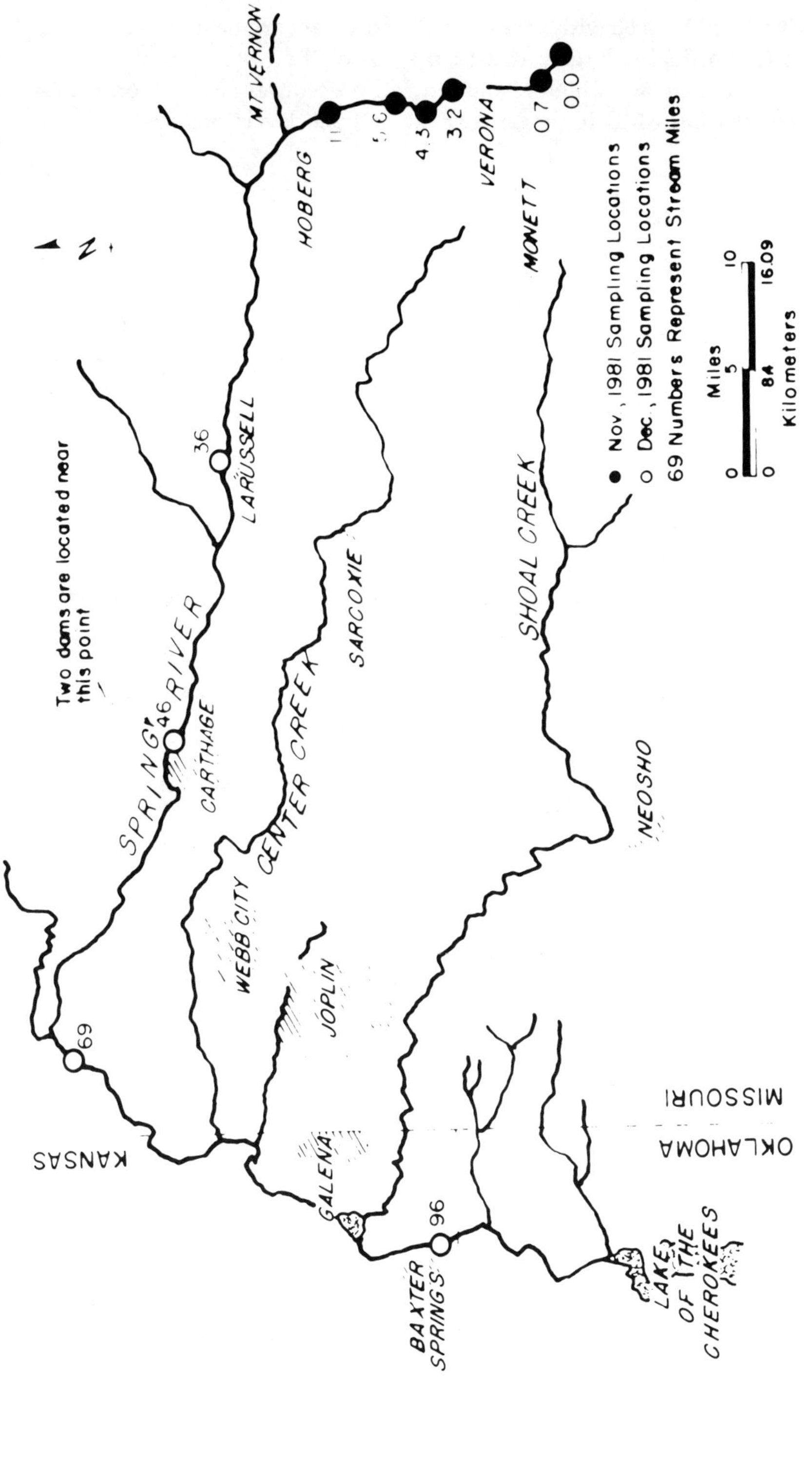

Figure 11.1. Spring River Sampling Locations.

results ranging from 37 to 55 with a mean of 47 and a standard deviation of 8 ppt. This success is, no doubt, largely due to the use of the isotope dilution procedures. The agreement with high-resolution MS procedures also demonstrates the utility of HRGC/LRMS for measurement of trace levels of tetra-CDD in ambient samples.

SUMMARY

Manufacturing activities and waste disposal practices that occurred over 10 years earlier have resulted in rather widespread tetra-CDD contamination in southwest Missouri. Although cleanup efforts are underway, investigations and monitoring of the problems are expected to continue for several years to come.

ACKNOWLEDGMENTS

The following laboratories provided analytical support for these investigations:

1. Center for Disease Control, Atlanta, Georgia;
2. Columbia National Fisheries Research Laboratories, Columbia, Missouri;
3. EPA Region VII Laboratory, Kansas City, Kansas;
4. I.T. Enviroscience, Knoxville, Tennessee;
5. National Center for Toxicological Research, Jefferson, Arkansas;
6. Syntex Agribusiness, Inc., Springfield, Missouri;
7. University of Nebraska, Midwest Center for Mass Spectrometry, Lincoln, Nebraska; and
8. Wright State University, Brehm Laboratory, Dayton, Ohio.

REFERENCES

1. Kimbrough, R.D., C.D. Coleman, J.A. Liddle, R.E. Kline and P.E. Phillips. *Arch. Environ. Health* 28:77 (1977).
2. Harris, D.J. Paper presented at the National Conference of Uncontrolled Hazardous Waste Sites, Washington, DC, October 1981.
3. Environmental Secretariat, National Research Council of Canada. "Polychlorinated Dibenzo-*p*-dioxins: Criteria for Their Effects on Man and His Environment," NRCC No. 18574 (1981).
4. *Waste Age,* (October 1981), pp. 60–63.
5. *Federal Register* 44(233):694664 (1979).
6. Stalling, D.C., L.M. Smith, J.D. Petty, J.W. Hogan, J.L. Johnson, C. Rappe and H.R. Buser. Paper presented at the Second International Conference on TCDD and Related Compounds, Washington, DC, 1981.

7. Gross, M.L., T. Sun, B.A. Lyon, S.F. Wejinsk and D.R. Hiker. *Anal. Chem.* 53: 1902 (1981).
8. Mitchum, R.K., W.A. Korfmacher, G.F. Moler and D.L. Stalling. *Anal. Chem.* 54: 719 (1981).
9. Nestrick, T.J., L.L. Lamperski and R.H. Stehl. *Anal. Chem.* 51:2273 (1979).

12

Composition of Polychlorinated Dibenzofuran and Dibenzo-*p*-dioxin Residues in Sediments of the Hudson and Housatonic Rivers

J.D. Petty, L.M. Smith, P.-A. Bergqvist, J.L. Johnson, D.L. Stalling and C. Rappe

Polychlorinated dibenzofurans (PCDF) are a series of tricyclic aromatic compounds similar to the polychlorinated dibenzo-*p*-dioxins (PCDD) in chemical, biological and toxicological properties [1,2] (Figure 12.1). Both PCDD and PCDF have many positional isomers, among which the 2,3,7,8-, 1,2,3,7,8- and 2,3,4,7,8-substituted isomers have been shown to be toxic to a variety of mammals (including rats, mice, guinea pigs and nonhuman primates) [1,3].

Polychlorinated biphenyls (PCB) have been shown to contain traces of PCDF, and pyrolysis of PCB can produce percent quantities of PCDF [4,5]. Chlorinated phenols and certain phenoxy acid herbicides have been recognized as sources of PCDD and PCDF [6,7]. Particular attention was directed to assessing the presence of 2,3,7,8-tetra-CDD in 2,4,5-tri-CP) or materials prepared from this phenol such as hexachlorophene or 2,4,5-trichlorophenoxyacetic acid herbicides

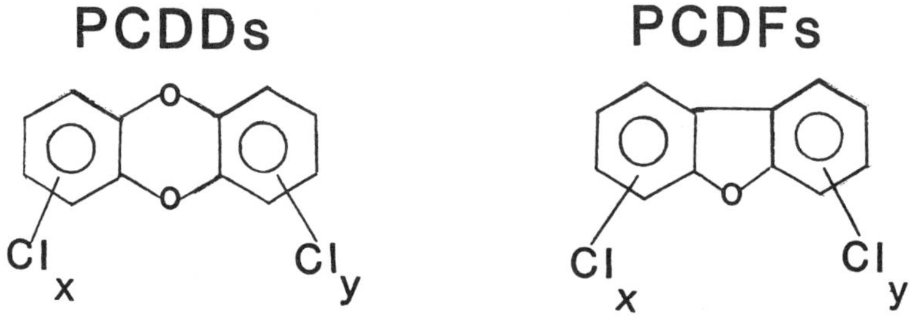

PCDDs

PCDFs

Cl_x Cl_y Cl_x Cl_y

Figure 12.1. Structures of PCDD and PCDF.

[6]. Another source of PCDD and PCDF contamination is known to be from waste incinerators [6]; however, the overall significance of this latter source has not been fully assessed. Knowledge of the distribution of the toxic 2,3,7,8- chlorine-substituted components of these two groups in the environment is expanding, primarily because of improved analytical methodology capable of measuring their presence at low part-per-trillion concentrations [6,8,9]. Unfortunately, information necessary for the assessment of the potential hazard posed by the presence of these materials in the biota is very sparse, especially in aquatic ecosystems.

EXPERIMENTAL

The experimental procedures used for preparation of the various adsorbents, chromatographic conditions, contaminant enrichment processes and the analytical conditions for gas chromatography/mass spectrometry (GC/MS) of PCDD and PCDF have been reported [8,9]. Sediment samples (10 g) were acid treated [10] and subsequently mixed and blended with 20 g anhydrous sodium sulfate (Mallinckrodt, Inc., St. Louis, Missouri) for Soxhlet extraction. Immediately before extraction, the mixture was spiked with 500 pg each of uniformly labeled (UL) ^{13}C-2,3,7,8-tetra-CDD (R. Mitchum, U.S. Department of Agriculture) and UL $^{37}C1$-2,3,7,8-tetra-CDF (F. Hileman, Monsanto Research Corporation). The sediment/sodium sulfate mixture was Soxhlet-extracted for 48 hr with 100 mL of toluene (Burdick and Jackson, Muskegon, Michigan), and the extract was reduced by rotary evaporation to dryness. The PCDF and PCDD residues were enriched as previously described [9].

A second 10-g portion of each sediment was prepared as above and subsequently subjected to column extraction and enrichment [9].

RESULTS AND DISCUSSION

The occurrence and composition of residues of PCDF and PCDD in sediments from the Housatonic and Hudson Rivers were examined following a survey in this laboratory for these residues in various freshwater fisheries [8,9]. Among other findings, that survey revealed the presence of a tetra-CDF having analytical characteristics identical to 2,3,7,8-tetra-CDF at approximately 1 ppb and other PCDF congeners tentatively identified as having 2,3,7,8- substitution in whole fish samples from the Housatonic River (Woods Pond, Massachusetts).

Two extraction methods for recovery of PCDF and PCDD residues in sediments were compared by analyzing duplicate sediment samples from the Hudson River. The first method employed Soxhlet extraction of a mixture of acid-treated sediment and sodium sulfate; the second method involved column extraction of a mixture of sediment and sodium sulfate. The extracts were then subjected to an enrichment procedure using sequential columns of cesium silicate/silica gel and carbon dispersed on glass fibers. Analyses of the samples yielded variable concentrations of PCDF and PCDD, even though internal standards (UL ^{13}C-tetra-CDD and UL ^{37}Cl-tetra-CDF) were added to the samples before extraction. In general, the recovery of the internal standards was good (Tables 12.1 and 12.2).

Table 12.1. Residues of PCDD (pg/g) in Hudson River Sediment

	Cl_4	Cl_5	Cl_6	Cl_7	Cl_8	*Total*
Soxhlet 1	7	70	1,263	6,240	14,600	22,180
Column 1 (182%)[a]	ND[b]	27	91	726	3,298	4,140
Soxhlet 2 (63%)	33	65	249	3,775	6,311	10,430
Column 2 (81%)	36	87	214	4,000	11,377	15,710
Soxhlet 3 (120%)	44	235	392	6,340	3,420	10,430
Column 3 (132%)	36	30	480	1,620	5,813	7,980
Soxhlet 4 (104%)	ND	ND	1,335	18,950	15,500	35,790
Column 4 (143%)	ND	ND	396	3,130	15,300	18,830
Soxhlet 5 (114%)	ND	ND	ND	270	4,170	4,440
Column 5 (105%)	ND	ND	178	1,647	9,100	10,930
Soxhlet 6 (96%)	26	ND	ND	ND	1,990	2,020
Column 6 (145%)	16	89	187	3,320	7,530	11,130
Blank, Soxhlet (104%)	ND	ND	ND	ND	10	10
Blank, Column (110%)	ND	ND	ND	ND	3	3

[a]Recovery of UL ^{13}C-2,3,7,8-tetra-CDD internal standard.
[b]ND = not detected, <3 pg/g.

Table 12.2. Residues of PCDF (pg/g) in Hudson River Sediment

	Cl_4	Cl_5	Cl_6	Cl_7	Cl_8	*Total*
Soxhlet 1 (19%)[a]	11	60	257	1,410	460	2,200
Column 1 (125%)	31	47	35	630	297	1,040
Soxhlet 2 (84%)	52	60	79	714	325	1,230
Column 2 (90%)	38	75	129	1,010	448	1,700
Soxhlet 3 (88%)	48	69	270	1,910	720	3,020
Column 3 (140%)	155	193	180	2,436	846	3,810
Soxhlet 4 (66%)	200	182	377	1,700	1,010	3,470
Column 4 (83%)	ND[b]	ND	53	1,200	536	1,790
Soxhlet 5 (149%)	62	ND	ND	ND	118	180
Column 5 (74%)	28	30	105	796	455	1,410
Soxhlet 6 (96%)	48	ND	ND	ND	ND	48
Column 6 (126%)	26	57	108	664	300	1,160
Blank, Soxhlet (104%)	ND	ND	ND	ND	10	10
Blank, Column (110%)	ND	ND	ND	ND	3	3

[a]Recovery of UL-^{13}Cl-2,3,7,8,-tetra-CDF internal standard.
[b]ND = not detected, <3 pg/g.

Table 12.3. Residues of PCDF and PCB in Sediment and Fish from Woods Pond, Massachusetts

	PCDF Concentration (ng/g)		
	Sediment	*Yellow Perch*	*Aroclor 1260*
Tetra-CDF	<0.005	1.06	290
Penta-CDF	0.009	0.64	1330
Hexa-CDF	0.15	0.44	1810
Hepta-CDF	0.92	<0.005	780
Octa-CDF	0.27	<0.005	29
Total CDF	1.35	2.14	4240
PCB (g/g)	60	170	Neat
PCDF/PCB Ratio	22.5×10^{-6}	12.6×10^{-6}	4.24×10^{-6}

The variable results obtained with the two methods indicate the need for additional work in this area; however, it is quite apparent from the data presented that these Hudson River sediment samples contain residues of both PCDD and PCDF. Further, as shown in Tables 12.1 and 12.2, these residues are comprised of the higher-chlorinated congeners of both PCDD and PCDF. The concentration of hepta- and octa-CDD ranged from 5–15 ng/g (ppb), with the octa-CDD concentration, in many instances, accounting for more than half of the total PCDD residue. As with the PCDD residues, the hepta- and octa-CDF residues occurred at the highest levels, approaching 1 ppb.

In contrast, major differences exist between PCDD and PCDF residue composition observed in sediments and fish samples [9,11,12]. While the residue profiles observed in sediment samples from the Hudson and Housatonic Rivers were similar in that the higher chlorinated congeners predominated, the fish samples were dominated by the tetra- and penta- congeners (Table 12.3). However, comparison of PCB (presumably a major source of PCDF) residue profiles in a composite fish sample and a sediment sample, both from Woods Pond, revealed an almost identical PCB pattern.

The observed differences between residues in sediment and fish could result from the lower solubility of the hepta- and octa- congeners relative to the tetra- and penta- congeners and/or from large differences in adsorption on sediments. At the present time, insufficient data exist to address fully the factors responsible for the observed differences in residue profiles in fish and sediment.

REFERENCES

1. Poland, A., W.F. Greenlee and A.S. Kende. *Ann. N.Y. Acad. Sci.* 320:214 (1979).
2. Rappe, C., H.R. Buser and H.P. Bosshardt. *Ann. N.Y. Acad. Sci.* 320:1 (1979).

3. McConnell, E.E., and J.A. Moore. *Ann. N.Y. Acad. Sci.* 320:138 (1979).
4. Bowes, G.W., M.J. Mulvihill, B.R.T. Simoneit and A.L. Burlingame. *Nature* 256:305 (1975).
5. Buser, H.R., H.P. Bosshardt and C. Rappe. *Chemosphere* 7:109 (1978).
6. Rappe, C. and H.R. Buser. In: *Halogenated Biphenyls, Terphenyls, Naphthalenes, Dibenzodioxins, and Related Products,* R.D. Kimbrough, Ed. (Amsterdam: Elsevier/North-Holland Biomedical Press, 1980), pp. 41–76.
7. Mieure, J.P., O. Hicks, R.G. Kaley and P.R. Michael. *J. Chromatog. Sci.* 15:275 (1977).
8. Stalling, D.L., L.M. Smith and J.D. Petty. In: *Measurement of Organic Pollutants in Water and Wastewater,* C.E. VanHall, Ed., ASTM STP 686 (Philadelphia: American Society for Testing and Materials, 1979), pp. 302–323.
9. Stalling, D.L., J.D. Petty, L.M. Smith, C. Rappe and H.R. Buser. In: *Chlorinated Dioxins and Related Compounds,* O. Hutzinger, Ed. (New York: Pergamon Press, 1982), pp. 77–85.
10. Kooke, R.M.M., J.W.A. Lustenhouwer and H. Hutzinger. *Anal. Chem.* 53:461 (1981).
11. Kuehl, D.W., R.C. Dougherty, Y. Tondeur, D.L. Stalling, L.M. Smith and C. Rappe. In: *Environmental Health Chemistry,* J.D. McKinney, Ed. (Ann Arbor, MI: Ann Arbor Science Publishers, Inc. 1981), pp. 245–261.
12. Doughtery, R.C., M.J. Whitaker, L.M. Smith, D.L. Stalling and D.W. Kuehl. *Environ. Health Perspec.* 36:103 (1980).

IV

Analytical

13

Analytical Chemistry of Polychlorinated Dibenzo-*p*-dioxins and Dibenzofurans: A Review of the Current Status

Thomas O. Tiernan

The extraordinary toxicities of the polychlorinated dibenzo-*p*-dioxins (PCDD) and dibenzofurans (PCDF) that have been demonstrated by animal tests and, to some extent, by accidental exposure of humans to these compounds, have prompted extensive efforts to quantitatively measure these chemicals in various chemical, biological and environmental media [1,2]. These toxicity considerations have further dictated the requirement to analyze for these chemicals at concentrations ranging as low as parts-per-trillion. Development of an entirely new realm of analytical chemical capability has been required to achieve reliable quantitative determinations at the part-per-trillion level; this has only recently been realized successfully. Even when the sample medium contains higher concentrations of PCDD/PCDF, analysis for these compounds can be a formidable task, because of the large numbers and quantities of other chlorinated hydrocarbons frequently present in many samples of interest, which can complicate or even preclude accurate determination of the PCDD/PCDF if appropriate analytical methodology is not used.

In the present survey, the current state of the art in terms of analytical procedures for the determination of PCDD/PCDF is reviewed. While this review is comprehensive, it is not exhaustive in coverage. An attempt has been made, however, to discuss most of the analytical procedures currently in use for measurement of PCDD/PCDF. The evolution of the analytical methods is presented in a somewhat chronological sequence, since most of the reported analytical procedures have been developed for specific types of sample media in response to the need to assess various environmental problems encountered over the past several years. Obviously the state of the art of these methods has advanced considerably since the inception of many of these procedures.

SCOPE OF THE ANALYTICAL PROBLEM

The generalized molecular structures of PCDD and PCDF are shown in Figure 13.1. Each of these structures actually represents a whole series of discrete compounds having between one and eight chlorine atoms attached to the ring. Moreover, since the chlorine atoms on each molecule can occupy any of the eight available ring positions, there is a group of distinct positional isomeric compounds for each chlorinated class of PCDD/PCDF. The total number of PCDD/PCDF isomers as a function of the number of chlorine substituents is shown in Table 13.1. It can be seen that in all there are 75 PCDD isomers and 135 PCDF isomers. The magnitude of the analytical task associated with the determination of PCDD/PCDF in complex samples thus becomes apparent, especially when one recognizes the difficulty of separating and identifying a large number of isomers within a given chlorinated class, all of which are quite similar structurally, and therefore, also similar in chemical and physical properties. This task is further complicated by the fact that many PCDD/PCDF isomers have not even been prepared in pure form and, therefore, are not available for use in developing and calibrating analytical methodology.

Several synthetic routes for preparing PCDD [3-9] and PCDF [6,10-15] have been reported; various PCDD and PCDF isomers have been produced by these reactions. The entire sets of the isomers of tetra-, hexa-, hepta- and octa-CDD have been synthesized by one laboratory [8,9,16]. However, considerably fewer PCDF isomers have been synthesized [15]. While the PCDD and PCDF isomers prepared by the various investigators cited here have been characterized using several methods, including gas and liquid chromatography, mass spectrometry, proton and ^{13}C nuclear magnetic resonance, and photolytic dechlorination coupled with pattern recognition techniques [2], few, if any (except 2,3,7,8-tetra-CDD), have been confirmed structurally by single-crystal X-ray diffraction (XRD), the most definitive method. This is due in part to the fact that only quite small quantities of the synthesized isomers have been available, and also to the fact that single-crystal XRD requires considerable time and expense. The PCDD and PCDF isomers prepared by the investigators cited above have obviously not been widely available to laboratories seeking these for use as analytical standards. However, a limited number of PCDD and PCDF isomers, including several ^{13}C- and ^{37}Cl-labeled compounds, have been prepared in sufficient quantities for sale to numerous investigators by KOR Isotopes (Cambridge, Massachusetts). Even in laboratories that have synthesized or acquired PCDD and PCDF standards, there are no reported instances in which the quantities of these have been sufficient to evaluate and calibrate thoroughly the complex analytical procedures to be described later in this chapter for entire sets of the isomers of the several chlorinated PCDD and PCDF classes.

Despite these limitations, there has been widespread interest in the determination of specific PCDD and PCDF isomers. This interest stems largely from the fact that the toxicities of these compounds depend markedly on both the number and position of the chlorine substituents. 2,3,7,8-tetra-CDD is clearly the

Figure 13.1. Generalized Structures of PCDD and PCDF.

Table 13.1. Numbers of CDD and CDF Isomers as a Function of the Number of Chlorine Substituents

Number of Chlorine Atoms	*Number of CDD Isomers*	*Number of CDF Isomers*
1	2	4
2	10	16
3	14	28
4	22	38
5	14	28
6	10	16
7	2	4
8	1	1
Total	75	135

most toxic of these compounds, followed by 1,2,3,7,8-penta-, 1,2,3,6,7,8-hexa- and 1,2,3,7,8,9-hexa-CDD [1,2]. The toxicities of the PCDF isomers generally parallel those of PCDD [2,17]. It has been reported that several of the more toxic PCDF isomers, including 2,3,6,8-tetra-, 2,3,7,8-tetra-, 1,2,4,7,8-penta-, 2,3,4,7,8-penta- and 1,2,3,4,7,8-hexa-CDF, were retained in the bodies of deceased humans who had ingested rice oil contaminated with polychlorinated biphenyls (PCB) and PCDF [18]. The toxicology of PCDD and PCDF is discussed much more extensively elsewhere [1,2], but it is clear from the foregoing that there is a need to accomplish isomer-specific analyses of PCDD and PCDF in many types of sample media, and that simply determining these compounds by chlorinated class does not generally provide sufficient information to adequately evaluate hazards. The extent to which such analyses can actually be accomplished at present, however, is still rather limited for the reasons discussed above.

Because of the extraordinary toxicity of 2,3,7,8-tetra-CDD, which has been associated primarily with 2,4,5-trichlorophenol (tri-CP) and certain herbicide derivatives of this chlorophenol (all widely used at one time), and because this PCDD isomer was one of the earliest available in sufficient quantity to use as an analytical standard, many of the early analytical studies focused on determination of 2,3,7,8-tetra-CDD in various biological and environmental samples. Indeed, much of the current analytical methodology for PCDD and PCDF stems from that originally developed to detect tetra-CDD (and later, 2,3,7,8-tetra-CDD). Some of this methodology was gradually adapted, modified and expanded to accommodate analyses of other PCDD and PCDF, although, as will be discussed later, entirely new methods have also been developed. Among the many types of samples analyzed for PCDD and/or PCDF are [1,2]:

1. Animal tissues: adipose, muscle, liver, blood and whole animal (bovine, fish, deer, elk, turtle, rat, rabbit and birds);
2. human tissues: adipose, blood and liver;
3. human and animal milk;
4. soils and sediments;
5. water: rivers, streams, lakes, wells and leachates;
6. chemical formulations and wastes (2,4-D, 2,4,5-T, PCB, CP, other herbicides and industrial chemicals);
7. combustion products: ash, stack effluents, water and other process fluids from combustion of municipal waste, coal, wood, chemical wastes, PCB, and penta-CP–treated materials and wastes;
8. industrial hygiene samples: wipes and air samples (solid sorbents); and
9. hazardous waste site samples: complex chemical mixtures, landfill, sludges and leachates.

It should be mentioned that the level of sophistication applied in these analyses and the objectives thereof varied widely. In some cases, the objectives have been primarily to screen samples, in a generally qualitative or semiquantitative manner, to identify classes of PCDD and/or PCDF (that is, tetra-, penta- and so on). In other instances, the objective of the analysis has been to accurately quantify specific PCDD and/or PCDF isomers in such samples. Clearly, both types of analyses can be useful, depending on the purposes for which the analytical results are to be used.

Various criteria can be applied for judging the efficacy of specific analytical procedures for the determination of PCDD/PCDF, depending on the objectives of the analysis. Among the criteria that have usually been considered are:

1. A standard practice applied by most competent laboratories analyzing for PCDD/PCDF involves the addition of isotopically labeled PCDD and/or PCDF isomers to the sample in known quantity before analysis, and determination of these labeled compounds along with the native or unlabeled PCDD or PCDF. The percent recovery of the isotopically labeled standard thus provides a measure of the efficacy of the analytical procedures. Typically, ^{37}Cl- or C^{13}-CDD or -CDF have been used for this purpose. However, since only a few such standards have been prepared and are generally available (certain labeled isomers of tetra-, hepta- and octa-CDD, and labeled 2,3,7,8-tetra-CDF are commercially available), the extent to which these procedures can be used is limited. Addition

of a labeled PCDD/PCDF to a sample is also useful in that the added compound can be used as a true internal standard for purposes of quantifying the native PCDD/PCDF in the sample. Such an approach presumes that losses of the added internal standard incurred in the course of processing the sample are proportionately the same as the losses of the native PCDD/PCDF in the sample. Since the internal standard is added in an accurately known quantity, by measuring the native compound relative to the labeled internal standard it is possible to quantify reliably the native compound in the sample.

2. Since PCDD/PCDF have been detected at concentrations spanning a relatively wide range in various sample matrices (typically parts-per-trillion to parts-per thousand-levels), it is desirable to use analytical methodology that can detect (and if desired, quantitatively measure) PCDD/PCDF over this entire range.

3. If large numbers of samples are to be analyzed, as is frequently the case in environmental assessments, the analytical methodology should be amenable to efficient and reliable processing of such large numbers of samples in a reasonable time period. As will be seen from the ensuing discussion, the complexity of reliable analytical procedures for determining PCDD/PCDF is such that even the best methods currently available are quite labor- and resource-intensive. These analyses are therefore necessarily quite expensive, and a single analysis often requires 1–2 days, although economies can obviously be realized by processing of samples in batches.

4. As far as possible, acceptable analytical procedures for determining PCDD/PCDF should be applicable to analysis of a wide variety of different sample matrices (such as those listed above). In fact, the separation and instrumental detection procedures that have been developed for such analyses are rather similar for many types of samples, as will be seen from the later discussion.

SAMPLING METHODS

The procedures used in obtaining a sample to be analyzed for PCDD/PCDF can affect markedly the results obtained. A detailed discussion of sampling methods is beyond the scope of this review, but it should be noted that this aspect of PCDD/PCDF analyses is perhaps one of the most neglected areas. Obviously, the dictates of good analytical practice should be applied in sampling, and such factors as the homogeneity of the sample matrix, collection of a truly representative sample, statistical considerations with respect to replicate analyses, and the resultant effects of these factors on precision and accuracy must be taken into account. In cases where samples are acquired by simple grab sampling, such considerations can usually be applied. In other instances, such as the collection of flue gas samples from incineration sources, the sampling procedures are far less standardized, and their efficiencies and limitations have not generally been adequately evaluated. A discussion of some of the sampling procedures applicable for collection of the flue gas samples to be analyzed for PCDD/PCDF is presented by Tiernan et al. [20]. A more detailed discussion of sampling requirements and guidelines for use in environmental chemical measurements has recently been prepared by the American Chemical Society [21]. Sampling methodology is also discussed in a Canadian government report on analytical methodology applicable to the PCDD [22].

ANALYTICAL METHODOLOGY FOR DETERMINATION OF PCDD/PCDF

The analytical methodology developed for determination of PCDD/PCDF in various sample matrices generally entails a sequence of three operations:

1. extraction of PCDD/PCDF from the sample matrix, which involves in some cases digestion or destruction of the matrix;
2. preliminary separation of PCDD/PCDF from major matrix constituents and other major chlorinated residues and chemical contaminants; and
3. detection of PCDD/PCDF in the cleaned up sample extract.

Procedures to accomplish these operations have been developed and applied in several different laboratories; these procedures are similar in many respects, although they differ in various details, such as the choice of liquid chromatography columns and the digestion or extraction reagents. It is not possible in a review of this type to discuss in detail all of the reports in the literature, many of which discuss the application of rather similar analytical methodology, varying only in minor aspects. Instead, presented below is a representative listing of several methods currently in use at various laboratories that analyze or have analyzed in the past substantial numbers of samples for PCDD and/or PCDF. Still other variations of such methods reported by others are discussed and evaluated in the Canadian report [22]. All of the methods listed below entail initial addition of internal standards, followed by destruction of the sample matrix (where practical) or extraction with an organic solvent, partitioning the PCDD/PCDF into an organic phase, and the polar constituents into an aqueous phase and discarding of the latter; this is followed by liquid chromatographic cleanup [and in some cases high-performance liquid chromatography (HPLC) cleanup] of the extracts. The cleanup steps are specifically intended to remove matrix components (such as lipids in the case of biological samples), as well as most of the chlorinated hydrocarbon residues other than PCDD/PCDF present in the sample (for example, PCB, DDE and other common chlorocarbon residues), which may interfere with subsequent detection and quantification of PCDD/PCDF. Albro and Parker have discussed the types of chlorinated hydrocarbons that may constitute interferences in such analyses, and fractionation procedures that are aimed at elimination of these [23]. Some of the more widely used extraction and cleanup procedures will now be listed, followed by a discussion of detection methods.

Extraction and Cleanup Procedures Applied in Analyses for PCDD/PCDF

U.S. Environmental Protection Agency (EPA) [24]

Acid Extraction. This method has been applied to fish, adipose tissue, milk, water, soil and sediment:

1. Spike 10–20 g of sample with $^{37}Cl_4$-2,4,7,8-tetra-CDD standard.
2. Reflux with KOH.
3. Extract with hexane.
4. Remove and extract hexane fraction with concentrated H_2SO_4.
5. Dry on sodium carbonate.
6. Chromatograph on neutral alumina column using CCl_4 and CH_2Cl_2 to elute.

Neutral Extraction. This method has been applied to fish.

1. Blend 15–20 g of sample with sodium sulfate and dry ice:
2. Spike resultant powder with $^{37}Cl_4$-2,3,7,8-tetra-CDD standard.
3. Extract with acetonitrile.
4. Remove acetonitrile fraction and extract with acetonitrile-saturated hexane.
5. Discard hexane.
6. Concentrate exetonitrile and replace with hexane.
7. Chromatograph on Florisil column, then on neutral alumina column, eluting sequentially with 100% hexane, 10% CH_2Cl_2 in hexane, and 25% CH_2Cl_2 in hexane.

U.S. Food and Drug Administration (FDA) [25,26]

This method has been applied to chemical formulations and fish:

1. Dissolve 20 g of sample in alkaline solution by agitating at room temperature for 2–3 hr.
2. Extract with hexane.
3. Remove hexane and extract with concentrated acid.
4. Dry on sodium carbonate.
5. Chromatograph on neutral alumina, eluting with 20% CCl_4 in hexane, then CH_2Cl_2.
6. HPLC on Zorbax-ODS, 40°C, methanol eluant.

U.S. Fish and Wildlife Service [27]

This method has been applied to fish:

1. Blend sample with sodium sulfate to free-flowing powder.
2. Pack powder into glass column and extract with CH_2Cl_2 (200 mL/g fish).

3. Evaporate solution to 50 mL.
4. Add methanol and benzene to yield solution containing 20% CH_3OH and 5% benzene.
5. Pass solution through carbon-glass fiber sorbent, wash sorbent with $CH_2Cl_2/CH_3OH/C_6H_6$ (75/20/5), then elute with toluene.
6. Chromatograph on composite column on potassium silicate or cesium silicate over H_2SO_4–silica gel.

Dow Chemical Company

Acid Extraction [28,29]. This method has been applied to fish and milk:

1. Spike 10–20 g of sample with ^{13}C-tetra-CDD internal standard (and/or other PCDD standards).
2. Shake with concentrated HCl for 1 hr.
3. Extract with hexane (overnight shaking plus additional 3 hr).
4. Pass hexane extract through combined column and silica, concentrated H_2SO_4 on silica, and 1 M KOH on silica (22% H_2SO_4 and 44% H_2SO_4 on silica columns and NaOH on silica columns are used alternatively).
5. Pass hexane extract through second dual column of silver nitrate on silica and basic alumina.
6. Clean up dioxin fractions using normal-phase silica (Zorbax-SIL) HPLC, then reverse-phase HPLC (Zorbax ODS, methanol solvent).

Neutral Extraction [30]. This method has been applied to particulates (fly ash, dust, urban particulates, municipal sludge, and soils):

1. Spike 0.05–100 g of sample with ^{13}C-tetra-CDD, ^{13}C-hexa-CDD and ^{13}C-octa-CDD internal standards.
2. Soxhlet extract sample with benzene for at least 16 hr.
3. Concentrate extract using Snyder column.
4. Subject extract to liquid chromatographic and HPLC cleanup similar to that described in Acid Extraction (above).

Brehm Laboratory, Wright State University

Chemical Wastes, Aqueous Effluents and Soils [31,32]. This procedure has been applied to soils:

1. Spike 1–10 g of soil with ^{37}Cl-tetra-CDD internal standard (and/or other internal standards).
2. Extract spiked soil in bottle with petroleum ether or hexane-methanol by shaking for a period of 20 min to overnight.
3. Alternatively, extract spiked soil in a Soxhlet apparatus with methylene chlorine or benzene followed by concentration of extract on a Snyder column.

4. Wash the organic phase successively with 20% KOH, water, concentrated H_2SO_4 and water, discarding the wash solutions.
5. Dry organic extract over sodium sulfate.
6. Pass extract through basic alumina column.
7. Cleanup dioxin fractions using normal-phase silica (Zorbax-SiL) and reverse-phase (Zorbax-ODS) HPLC.

This procedure has been applied to chemical wastes:

1. Spike 1–10 g of the sample with ^{37}Cl-tetra-CDD internal standard (and/or other labeled standards).
2. Extract spiked sample in bottle with methanol and petroleum ether by shaking for period of 20 min to overnight.
3. Continue with cleanup as in points 4–7 above.

This procedure has been applied to aqueous effluents:

1. Spike 10–20 g of sample with ^{37}Cl-tetra-CDD internal standard (and/or other internal standards)..
2. Extract spiked sample in bottle with petroleum ether by shaking for 20 min.
3. Continue with cleanup as in points 4–7 above.

Procedure for Particulates and Other Samples from Flue Gas Stack Sampling Trains [20,33].

1. Spike appropriate quantities of organic and aqueous liquids, particulates, XAD-2 resins and other sorbents with ^{37}Cl- and/or ^{13}C-PCDD/PCDF.
2. Extract liquid samples with petroleum ether by shaking in bottle for 1 hr.
3. Extract solid samples in a Soxhlet apparatus with benzene.
4. Wash extracts successively with KOH, water and H_2SO_4, and discard washings.
5. Pass extracts through a combination macrocolumn of silica, 33% NaOH-silica, 44% H_2SO_4-silica, and silica, eluting with hexane.
6. Pass eluate through a Woelm basic alumina minicolumn, eluting with 3% CH_2Cl_2-in-hexane and 50% CH_2Cl_2 in hexane, collecting appropriate fractions.

Efficacy of Extraction and Cleanup Methodology for PCDD/PCDF

As already mentioned, the efficacy of extraction and cleanup methodology such as that described above for determining PCDD/PCDF is generally gauged by the extent of recovery of the initially added labeled internal standards. This has been less frequently applied for PCDF determinations because only one labeled PCDF has been available (^{37}C1$_4$-2,3,7,8-tetra-CDF). Collaborative testing of some of these methods has been accomplished through the EPA Dioxin Implementation

Plan, in which Wright State University, the University of Nebraska, EPA (Research Triangle Park, North Carolina) and Dow Chemical Co. laboratories were key participants. Few of the results of these studies have yet been published, however. FDA also conducted a limited assessment of several of these methods for fish analyses [34]. Most of the methods cited seem to be relatively effective, insofar as they have been tested, when accomplished by skilled and experienced analysts.

One type of sample has received particular attention in terms of evaluating the efficiencies of extracting PCDD/PCDF using various procedures: particulates (such as fly ash). Lustenhouwer et al. [35] reported the results of extracting aliquots of the same fly ash sample using six different extraction methods that have been used by various investigators, who have reported data on determinations of PCDD/PCDF in the literature. These results are shown in Table 13.2. Is is clear from these results that Soxhlet extraction with toluene (following acid treatment) is the most effective of the procedures evaluated. Subsequently, more detailed

Table 13.2. Extraction Efficiencies of CP, PCDD and PCDF from Particulate Matter[a] [35]

	Analytical Results					
	CP		*PCDD*		*PCDF*	
Method	*ppb*	*%*	*ppb*	*%*	*ppb*	*%*
Toluene, acid-treated, Soxhlet extraction[b]	1364	100	1308	100	1407	100
CH_2Cl_2 shaking treatment[c]	240	18	124	1	23.5	2
Acetone/hexane, ultrasonic treatment[d]	94	7	No	1	No	1
CH_2Cl_2, Soxhlet extraction[e]	434	32	81.8	6	155	11
$CHCl_3$, Soxhlet extraction[e]	592	44	168	13	260	20
Toluene extraction of solution of fly ash in 48% HF[f]	446	33	113	9	230	16

[a]Samples of fly ash from the same collection were used throughout this investigation.

[b]Fly ash (25 g) was stirred in 200 mL of 1 *N* hydrochloric acid. After centrifugation, the residual fly ash is washed with deionized water over a Buchner funnel and dried at room temperature. The fly ash is extracted for 35 hr using a Soxhlet apparatus.

[c]Fly ash (10 g) is shaken manually in CH_2Cl_2. This is repeated four times. The extracts are combined for cleanup and analysis.

[d]Fly ash (10 g) is treated in an ultrasonic bath for 5 min in acetone/hexane. This is repeated four times. The extracts are combined for cleanup and analysis.

[e]Fly ash (25 g) is extracted for 35 hr using a Soxhlet apparatus.

[f]Fly ash (5 g) is dissolved in a 48% HF solution. The solution is then extracted with toluene. Some particulate matter remains.

Table 13.3. Influence of Time on Extraction Efficiency for PCDD and PCDF[a] [36]

	First 24 hr	*Second 24 hr*	*Third 24 hr*
PCDD			
Tetra-	97.0	2.3	0.7
Penta-	97.4	2.0	0.6
Hexa-	98.4	1.3	0.3
Hepta-	99.3	0.7	0.1
Octa-	99.5	0.5	0
Average	98	1.6	0.3
PCDF			
Tetra-	97.7	1.7	0.8
Penta-	98.4	1.3	0.4
Hexa-	98.6	1.1	0.3
Hepta-	99.4	0.6	0.1
Octa-	99.5	0.5	0
Average	98.5	1.2	0.3

[a]Sample A, pretreatment with 1 *N* HCl, Soxhlet extraction with toluene. Values given in % of PCDD or PCDF extractable after 72 hr.

studies from the same laboratory, in which still other procedures were evaluated, led to the conclusion that Soxhlet extraction with benzene or toluene is about equally effective and far superior to simpler extraction methods [6]. In the latter study, the effect of the period of extraction was also evaluated, with the results shown in Table 13.3. Obviously, a 24-hr extraction period (the minimum evaluated) is sufficient to remove the bulk of the PCDD/PCDF constituents. Cutie has recently reported results of similar studies of the extraction of 2,3,7,8-tetra-CDD from fly ash, soil and activated carbon [37].

Extraction and Cleanup Procedures for Determining PCDD/PCDF in Electrical Equipment PCB Fluids

Because of several accidents involving electrical equipment in the past few years, in which PCB formulations were released into the environment, the need has developed to determine PCDD/PCDF in the fluids used in these devices. Comparatively few measurements of this type have been reported, and the few reports that have appeared have provided little detail on analytical procedures. Apparently, Vos et al. [38] were the first to identify PCDF in European commercial PCB. Subsequently, a Japanese PCB formulation was found to be contaminated with PCDF [39]. PCDF, including 2,3,7,8-tetra-CDF, were later reported to be present in U.S. PCB [40]. In Japan, Morita et al. [41] found PCDF in 16 different PCB formulations, as well as in yusho oil, which was the source of a major Japanese poisoning episode [41]. The major PCDF contaminant in yusho oil was shown to be the 2,3,7,8-tetra-CDF [42]. Several other reports have been issued more recently, confirming the presence of various PCDF in PCB formulations [23,43,46]. In addition, Chittim et al. [47] concluded that CDF were present in PCB electrical fluids sampled from transformers and that the concentrations of these increased with time in service of the transformer.

The extraction and cleanup procedures applied in several studies mentioned above were generally similar to those described earlier in this section for determinations of PCDD/PCDF in other samples. The most detailed account of procedures applicable to PCB electrical fluids [23] also discusses at some length the problems with these analyses that originate from polychlorinated diphenyl ethers, which are also present in some fluids of this type. Some of the latter compounds can give rise to mass spectral fragment ions that coincide with those monitored as indicators of PCDF. It is important therefore, that chlorinated diphenyl ethers be removed from the extracts by the separation before gas chromatography/mass spectrometry (GC/MS) analysis. Albro and Parker have reported that adequate separation can be achieved by careful attention to the elution procedures used in the liquid chromatographic fractionation sequence [23].

Instrumental Methods Applied for Detection and/or Quantification of PCDD/PCDF in in Sample Extracts

Various instrumental methods have been applied for detection and/or quantification of PCDD/PCDF in sample extracts prepared by the procedures described above (Table 13.4). Also shown in Table 13.4 are the types of analytical applications (that is, screening or qualitative detection, quantitative analysis, isomer determinations and so on) for which these various methods have been applied and citations of examples of reports of such applications that have appeared in the literature. As indicated in Table 13.4, instrumental techniques such as ultraviolet (UV) spectroscopy, thin-layer chromatography and gas chromatography are generally useful only for qualitative screening of sample extracts for these compounds. It is generally accepted that the only technique which has both the required sensitivity and specificity for reliable detection and quantification of PCDD/PCDF is GC/MS.

Mass spectrometric analysis for PCDD/PCDF is based on the presumption that these compounds exhibit unique mass spectral fragmentation patterns. Electron impact mass spectra of PCDD and PCDF exhibit intense molecular ions (M^+) with the appropriate ion clustering caused by the two chlorine isotopes. PCDD mainly fragment to M^+-COCl and M^+-COCl-Cl_2 ions. Smaller intensities of various other ions appear in the spectra of both types of compounds. In most instances, monitoring the molecular ions of the PCDD and PCDF will easily distinguish these compounds from other chlorinated aromatic hydrocarbons that may be present in the sample extracts. However, some PCB, as well as DDE, can fragment to yield ion masses that may interfere with detection of certain PCDDs,

Table 13.4. Instrumental Methods Applied to Detect PCDD/PCDF in Sample Extracts

Method	*Application*	*Highly Specific*	*References*
UV Spectroscopy	Tetra-CDD screening	No	48
Thin-Layer Chromatography	Tetra-CDD screening in fats and oils	No	49
Electron-Capture GC	Tetra-CDD and other screening in herbicides and CP	No	26,50–52
Probe MS	Tetra-CDD quantitative analysis in milk, fish and biological tissue	No	53,54
GC/MS	Screening and quantitative analysis	Yes	
Packed-Column			
GC/LRMS-SIM	Tetra-CDD in CP		55
	Tetra-CDD in penta-CP		56,57
	Tetra-CDD in Herbicide Orange formulations		58
	Tetra-CDD in surface wipes and metal surfaces		59
GC/HRMS-SIM	Tetra-, hexa-, hepta- and octa-CDD in beef adipose & liver tissues		60
	Tetra-CDD in chemical wastes, aqueous effluents and soils		
	Tetra-CDD in biological tissue and plant materials		61
	Tetra-CDD in beef adipose tissue		62

Table 13.4. Instrumental Methods Applied to Detect PCDD/PCDF in Sample Extracts *(continued)*

Method	*Application*	*Highly Specific*	*References*
Capillary-Column			
GC/HRMS-SIM	Tetra-CDD in human milk, beef liver, fish, water and sediment		24
	Tetra-CDD in particulates from combustion sources		63
GC/LRMS-SIM	Separation of tetra- and hexa-CDD isomers; PCDD in fly ash		2,64–66
	PCDF in oils, PCB, fly ash, pyrolysis products, human tissue		67–70
	PCDD and PCDF in incinerator effluents		2
	PCDD and PCDF in combustion effluents		20
Liquid-Chromatography/ Packed Column GC/MS-SIM	Tetra-, hexa-, hepta- and octa-CDD in particulates; isomer separation and identification	Yes	2,16
GC/NICI MS[a]	PCDD in animal tissue and wood	?	71
GC/NI MS[b]	PCDF in fish	?	72
GC/API MS[c]	Identification of tetra-CDD isomers in standard mixture	?	73

[a]GC/NICI MS = gas chromatography/negative ion chemical ionization mass spectrometry.
[b]GC/NI MS = gas chromatography/negative ion mass spectrometry.
[c]GC/API MS = gas chromatography/atmospheric-pressure ionization mass spectrometry.

unless rather high mass spectral resolution is employed. Also, the identification of PCDF may be complicated by the presence of chlorinated diphenyl ethers, since, as mentioned earlier, the latter compounds yield intense M^+Cl_2 ions that have the same exact mass and number of chlorine atoms as the corresponding PCDF [23]. In such cases, successful analysis depends on the ability of the pre-separation procedures and/or the gas chromatographic fractionation to separate such interfering compounds from PCDD/PCDF before the latter enter the mass spectrometer. While it is desirable in GC/MS analyses to obtain complete mass spectral scans to identify the analytes of interest, this is not feasible if one needs to achieve very low detection limits in the analyses, for example, a part-per-trillion level sensitivity (in which case a few picograms of the analyte are being detected). In the latter case, which is the more common requirement in analyzing environmental samples, typically only a few ions in the mass spectra of the PCDD/PCDF of interest are monitored, using a technique termed selected-ion monitoring (SIM). This is discussed below.

Earlier analyses of PCDD were aimed primarily at determining low levels of tetra-CDD and used less sophisticated gas chromatographic techniques than those presently available. Indeed, the first attempt to apply mass spectrometry for quantitative determination of tetra-CDD, which was the high-resolution mass spectrometric technique reported by Baughman and Meselson [53], did not use prior gas chromatographic separation at all. Rather, the sample extract was introduced via a direct inlet probe. These investigators modified a high-resolution mass spectrometer to permit repetitive high-resolution scanning over a narrow mass range (typically 0.3 amu) at a rapid rate (4 scans/sec). Mass spectrometric signals were aquired using a signal-averaging computer, which resulted in sufficient sensitivity to detect as little as 1 pg of injected tetra-CDD. This technique was adequate (because of the relatively high mass spectral resolution achieved) to distinguish the molecular ion of tetra-CDD from ions arising from the sample matrix (mostly biological samples), but it was soon realized that the tetra-CDD ions could not be distinguished from those originating from certain PCB, if the latter were present in the sample extract in large concentration compared to the tetra-CDD. Since PCB are often present at the part-per-million level in environmental samples where detection of tetra-CDD at the part-per-trillion level is desired, the quantities of PCB in sample extracts prepared using the cleanup procedures available at the time were frequently too large to permit accurate measurements of tetra-CDD using the Baughman–Meselson method. Another limitation of the latter method was its inability to separate tetra-CDD isomers, and the mass spectrometric response observed with this method was obviously attributable to any tetra-CDD isomers in the sample, not just the 2,3,7,8-tetra-CDD that was of major interest.

During the time period when the Baughman-Meselson method was being applied, several investigators reported the use of coupled gas chromatography/low resolution mass spectrometry (GC/LRMS) for determinations of tetra-CDD in chlorinated phenols and 2,4,5-T herbicide formulations [54-59]. These studies used open-tubular packed gas chromatographic columns that were capable of separating tetra-CDD from other PCDD of different chlorine content (for example, penta-CDD), but had little capacity for separating discrete tetra-CDD isomers. The extent to which other chlorinated species were resolved was not entirely clear, but this was not generally a problem for the sample matrices mentioned. The mass spectrometers used for these analyses were operated in the SIM mode, and typically 1–4 ions were monitored. The detection limits achieved in these determinations were in the range of 0.001–0.1 ppm. These methods derived their specificity from the cleanup methods used, and by requiring that the analyte being detected exhibit the appropriate GC retention time, a mass spectral response at the appropriate ion masses corresponding to tetra-CDD (m/z 320 and 322), and the correct ratio of the chlorine isotope peaks.

As a result of concerns over the possible accumulations of 2,3,7,8-tetra-CDD in the food chain, as a consequence of the use of 2,4,5-T (an herbicide contaminated with tetra-CDD) on rangeland on which cattle grazed, EPA initiated the Dioxin Implementation Plan (DIP) in 1975. This was a program intended to assess quantitatively the levels of 2,3,7,8-tetra-CDD in beef tissues and related environmental samples. This program prompted the development of combined packed-column GC/high-resolution mass spectrometric (GC/HRMS) techniques for determining tetra-CDD at four different laboratories [31,60-62]: Wright State University, the University of Nebraska, EPA at Research Triangle Park, North Carolina, and the Dow Chemical Company. These laboratories were the participants in DIP. The mass spectrometric methods used in these labs were similar to those applied by Baughman and Meselson, but were superior in that high-resolution scanning was generally accomplished over a narrower mass range and at a more rapid scan rate. This improved both the resolution capability and the sensitivity of the technique. Wright State's adaptation of this procedure permitted as many as four ions to be monitored.

A further enhancement of the capabilities of the GC/HRMS technique for analyses of PCDD/PCDF, which was just described, entailed the incorporation of capillary GC columns (wall-coated open-tubular columns). This has been accomplished at the EPA laboratory [24] and at Wright State [63]. The use of capillary GC columns provided the capability to separate discrete PCDD/PCDF isomers; until recently, however, few of these have been available in pure form. Figure 13.2 shows the separation of a mixture of eleven tetra-CDD isomers obtained with this technique at Wright State. The use of capillary column GC/HRMS also further improved the ability to obtain accurate quantitative data on the concentrations of PCDD/PCDF present at very low levels in environmental samples. This capability was extensively tested in the course of DIP by analysis of spiked samples prepared by an EPA laboratory. Typical results obtained by Harless [24] in the analysis of spiked fish samples using this method are shown in Table 13.5. Obviously, the capillary GC/HRMS technique provides the ultimate specificity for analyses of PCDD/PCDF. The criteria applied in identifying tetra-CDD in such analyses have been summarized by Harless [24].

Investigations of the use of capillary GC columns for the separation of isomers of the CDDs were undertaken as early as 1975 by Buser [64], who prepared

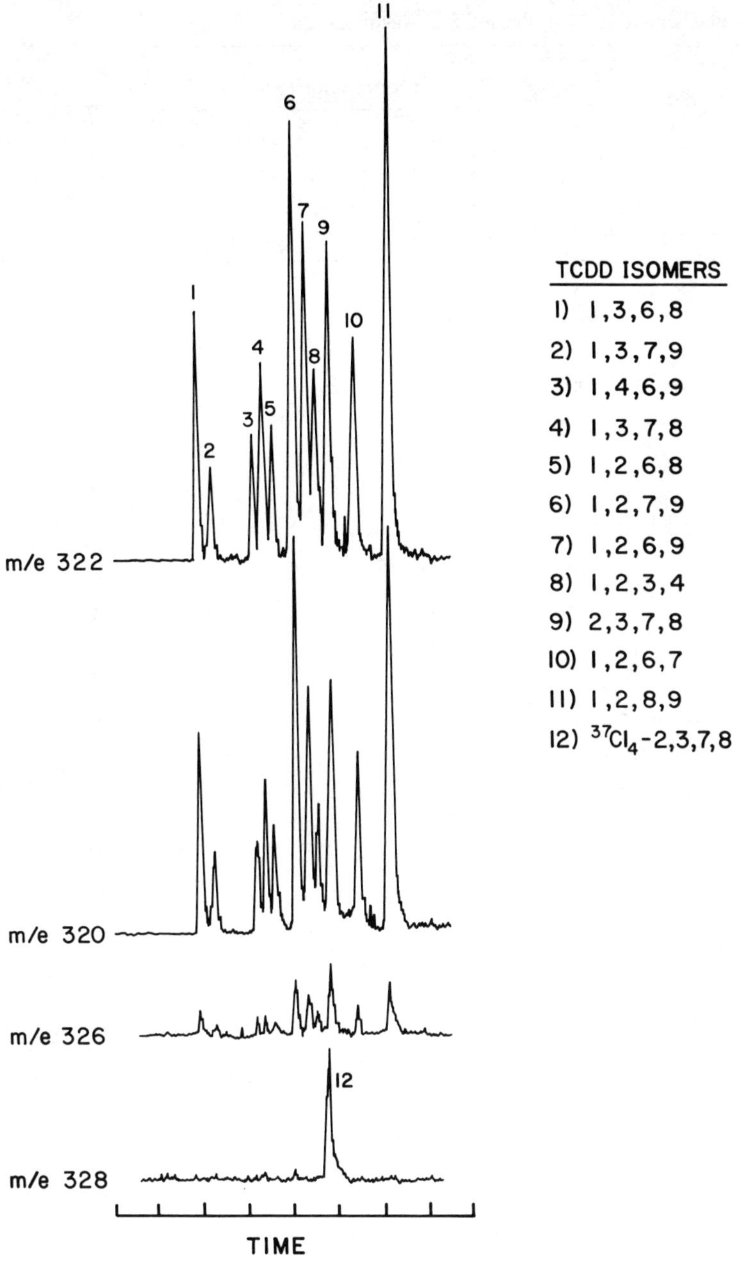

Figure 13.2. Mass Chromatogram of Tetra-CDD Isomer Standard Mixture Obtained with Capillary-Column GC(50-m OV-101, WCOT Column)/HRMS (Resolution of 10,000) at Wright State.

Table 13.5. Typical Analytical Results for Quality Assurance Samples Generated during Analysis of Fish for 2,3,7,8-Tetra-CDD Residues [24]

		Experimental Results			Tetra-CDD Fortification Level	
Sample Weight (g)	*Sample ID*	*^{37}Cl-Tetra-CDD Percent Recovery*[a]	*Tetra-CDD Detection Limit (ppt)*[b]	*Tetra-CDD Detected (ppt)*[b]	*pg*	*ppt*
5	Ocean perch	62	2	20	110	22
		52	4	34	185	35
		82	3	ND[c]	0	0
	Method blank	100	3	ND	0	0
	Beef liver	54	1	19	70	14
		100	2	ND	0	0
		78	2	ND	0	0
	Ocean perch	92	2	19	55	11
		97	5	45	240	48
10	Ocean perch	100+	1	8	130	13
		100+	4	43	600	60
	Lake trout	100+	7	ND	0	0
		100+	3	ND	0	0
	Ocean perch	100+	4	76	1250	125
	Method blank	67	1	ND	0	0
	Ocean perch	93	3	56	650	65
		84	4	73	620	62

[a]Each sample was fortified with 5 or 10 ng of ^{37}Cl-tetra-CDD.

[b]Corrected for % recovery losses. ^{37}Cl-tetra-CDD mean % recovery = 86%; tetra-CDD mean % accuracy = ±15%.

[c]ND = not detected.

eight hexa-CDD isomers and demonstrated that these could be separated on a column having about 35,000 theoretical plates using a temperature program that increased the column temperature by 2°C/min. Somewhat later, Buser [65] demonstrated separation of nine tetra-CDD isomers that he had available at that time. More recently, all twenty-two tetra-CDD isomers have been prepared by Buser [2]; Buser and Rappe [66] reported the results of attempting to separate these isomers on three different capillary columns. These results are shown graphically in Figure 13.3. Still more recently, other capillary columns have been tested for separating the 22 tetra-CDD isomers in a few other laboratories, including that at Wright State. At Wright State, a new hybrid column coated with two stationary phases, OV-17 and Silar-10C, has been developed for this purpose. As shown graphically in Figure 13.4, this column is superior to any of the previously

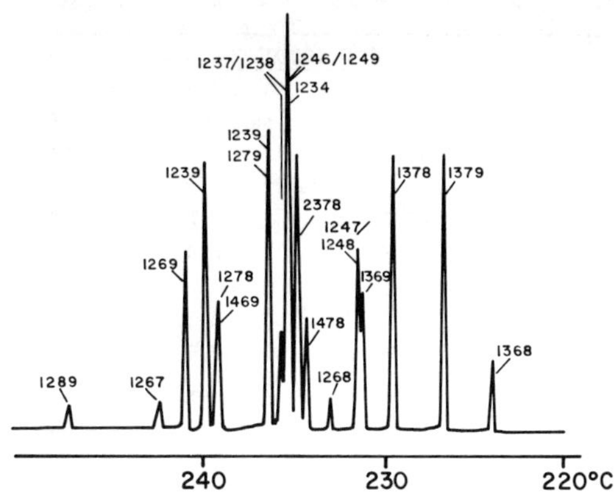

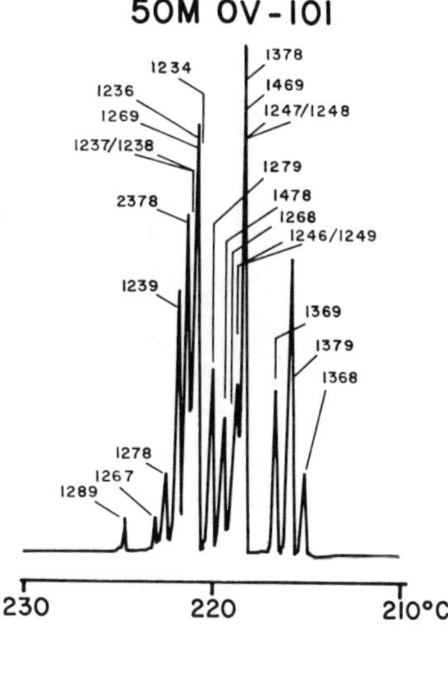

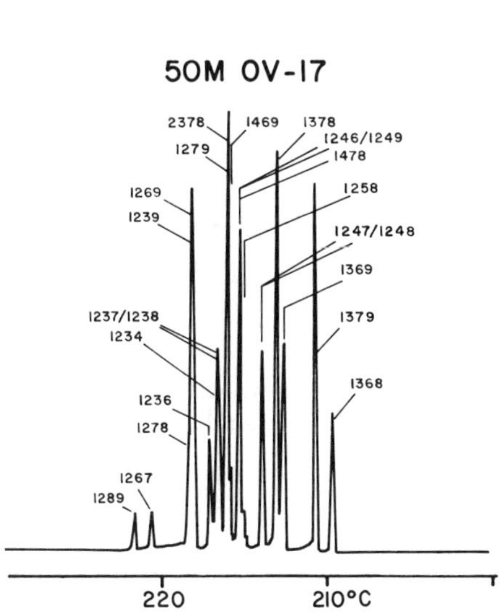

Figure 13.3. Mass Fragmentograms (m/z 320) of a Composite Pyrolysate Sample Showing Elution of All 22 Tetra-CDD Isomers on HRGC Columns: 55-m Silar-10C (left); 50-m OV-17 (bottom); 50-m OV-101 (right) 2,66].

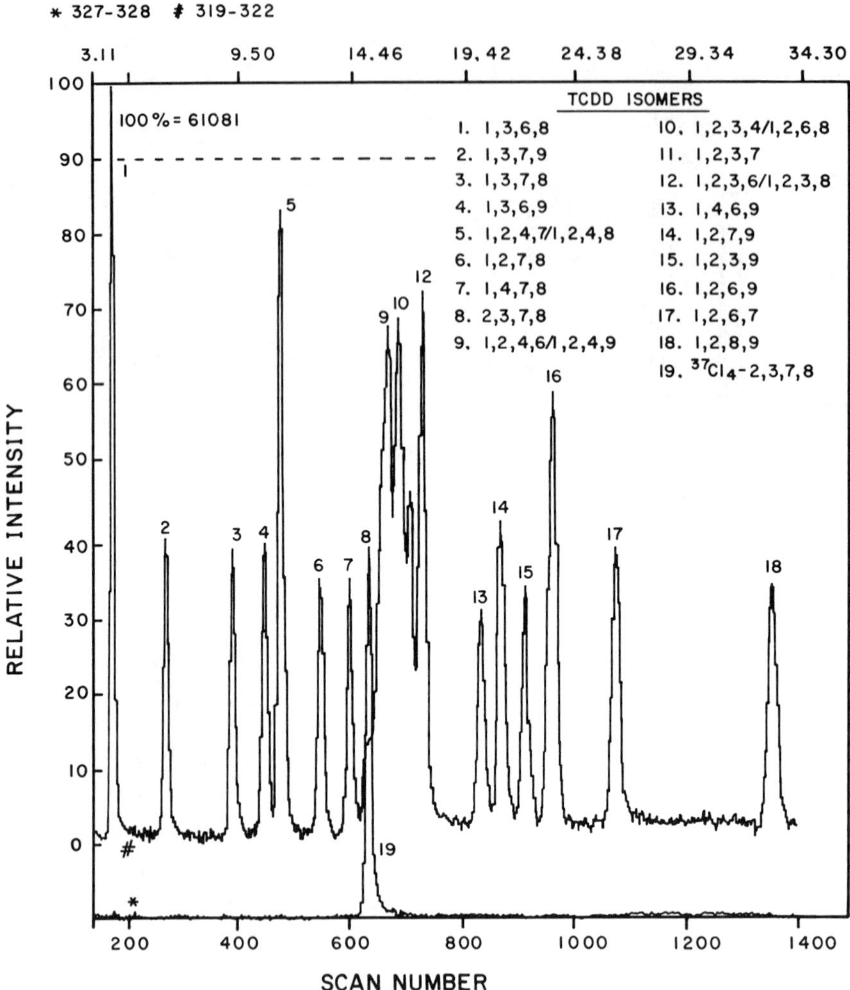

Figure 13.4. SIM Mass Chromatogram Obtained for Calibration Mixture Containing the 22 Tetra-CDD Isomers. Column: 50-m Hybrid-Phase WCOT Capillary.

tested capillary columns for separating tetra-CDD isomers. A summary of the separations of tetra-CDD isomers that can be achieved with the capillary GC columns that have been tested at various laboratories is given in Table 13.6. By using an appropriate set of capillary columns from those listed and accomplishing multiple analyses, it would now be possible to resolve all but two pairs of the tetra-CDD isomers.

Table 13.6. Tetra-CDD Isomers Resolved on Various Capillary GC Columns

Tetra-CDD Isomer	*50-m OV-17 (Glass; 0.37 mm i.d.)*	*50-m OV-101 (Glass; 0.40 mm i.d.)*	*55-m Silar-10C (Glass; 0.25 mm i.d.*	*60m- SP-2330 (Glass; 0.25 mm i.d.)*	*60-m SP-2340 (Fused Silica, 0.25 mm i.d.)*	*50-m Silov (WSU) (Glass; 0.25 mm i.d.)*
1,2,6,7-	X[a]	X	X	X	X	X
1,2,6,8-			X	X	X	
1,2,6,9-			X	X	X	X
1,2,7,8-		X				X
1,2,7,9-		X				X
1,2,8,9-	X	X	X	X	X	X
1,3,6,8-	X	X	X	X	X	X
1,3,6,9-	X	X	X			X
1,3,7,8-	X		X	X	X	X
1,3,7,9-	X	X	X	X	X	X
1,4,6,9-						X
1,4,7,8-		X	X	X	X	X
2,3,7,8-			X	X	X	X
1,2,3,6-	X					
1,2,3,7-						X
1,2,3,8-						X
1,2,3,9-		X	X	X	X	X
1,2,4,6-						
1,2,4,7-						
1,2,4,8-						
1,2,4,9-						
1,2,3,4-						
Total Number of Isomers Uniquely Resolved	7	9	11	10	10	14

[a]X indicates isomer uniquely resolved.

Capillary column GC/LRMS is being used increasingly at several laboratories to provide data on a wide range of PCDD/PCDF isomers in various types of samples [2,64-70], including PCB electrical fluids. Wright State scientists have recently developed and applied a capillary column GC/LRMS SIM procedure [20,38] that permits analyses of the entire range of PCDD/PCDF by chlorinated class, including the tetra- through octachlorinated species. Table 13.7 shows the GC/MS-SIM program used for such analyses, which can be accomplished in a single injection. Since all of the isomers of these compounds are obviously not yet available, the extent to which discrete isomers of these are resolved by the capillary columns used can not yet be assessed. However, it is already clear from the isomeric PCDD/PCDF standards available that many of these isomers are resolved. In any case, this technique is highly useful for screening of sample

Table 13.7. Sequence of Operations in GC/MS (MS-25) Analyses of PCDD and PCDF in Sample Extract

Elapsed Time (min)	*Event*	*GC Column Temperature (°C)*	*Temperature Program Rate (°C/min)*	*Ions Monitored by Mass Spectrometer (m/z)*	*Compounds Monitored*
0.00	Injection, splitless	190			
1.50	Turn on split valve	190			
2.00	Begin temp. program to 220°C	190	5		
5.00	Open column flow to mass spectrometer	202			
6.50	Start PROGRAM 2; sweep = 200 mmp;	220		235.980	Di-CDF
	time on each mass = 0.12 sec			237.977	Di-CDF
				251.974	Di-CDD
				253.972	Di-CDD
10.50	Stop PROGRAM 2	220			
11.00	Start PROGRAM 3; sweep = 200 ppm;	220		269.941	Tri-CDF
	time on each mass = 0.15 sec			271.938	Tri-CDF
				285.935	Tri-CDD
				287.932	Tri-CDD
15.00	Stop PROGRAM 3	220			
15.50	Start PROGRAM 4; sweep = 100 ppm;	220		258.930	Tetra-CDD
	time on each mass = 0.15 sec			303.902	Tetra-CDF
				305.899	Tetra-CDF
				319.897	Tetra-CDD
				321.894	Tetra-CDD
				327.885	$^{37}Cl_4$-Tetra-CDD
23.50	Stop PROGRAM 4	220			
24.00	Start PROGRAM 5; sweep = 750 ppm;	220		337.863	Penta-CDF
	time on each mass = 0.2 sec			339.860	Penta-CDF
				353.858	Penta-CDD
				355.855	Penta-CDD

25.00	Begin temp. program to 230°C	220	5		
34.50	Stop PROGRAM 5				
35.00	Start PROGRAM 6; sweep = 150 ppm; time on each mass = 0.25 sec	230		373.821	Hexa-CDF
				375.818	Hexa-CDF
				389.816	Hexa-CDD
				391.813	Hexa-CDD
50.00	Begin temp. program to 235°C		5		
53.50	Stop PROGRAM 6	235			
54.00	Start PROGRAM 7; sweep =750 ppm; time on each mass = 0.35 sec	235		407.782	Hepta-CDF
				409.779	Hepta-CDF
				423,777	Hepta-CDD
				425.774	Hepta-CDD
				431.765	$^{37}Cl_4$-Hepta-CDD
70.00	Stop PROGRAM 7				
85.00	Start PROGRAM 8; sweep = 750 ppm; time on each mass = 0.50 sec	235		441,732	Octa-CDF
				443.740	Octa-CDF
				457.738	Octa-CDD
				459.735	Octa-CDD
				471.717	$^{37}C1_4$-Octa-CDD
95.00	Stop PROGRAM 8	235			
130.00	Return to initial temp.				

extracts for the various PCDD/PCDF by chlorinated class. Data of this type recently obtained by Wright State for flue gases from a municipal incinerator are presented by Taylor et al. [33].

Dow Chemical scientists have recently achieved considerable success in the separation of CDD isomers by HPLC to achieve fractionation of discrete isomeric compounds [2,16]. This technique appears to be particularly powerful when used in conjunction with GC/MS techniques, and obviously does not necessitate the use of capillary GC columns.

Several newer mass spectrometric techniques have also been investigated for the analysis of PCDD/PCDF. Among these are negative-ion mass spectrometry [72], negative-ion chemical ionization mass spectrometry [71], and atmospheric-pressure ionization mass spectrometry [73]. Preliminary results obtained thus far suggest that all of these offer considerable promise for enhanced specificity in the analyses of these compounds.

Other Screening Methods for Detection of PCDD/PCDF

Several screening methods for PCDD/PCDF depend on biological responses to these compounds. Among such methods are radioimmunoassay, described by McKinney et al. [74], for which quantitative capability has been claimed, an aryl hydrocarbon hydroxylase induction assay described by Bradlaw and Casterline [75], and, most recently, a cytosol receptor assay, reported by Hutzinger et al. [76]. While assays such as these can be highly useful for rapid screening of samples thought to contain toxic PCDD/PCDF, it is not expected that they will be applicable for rigorous quantitative analysis of discrete PCDD/PCDF isomers.

AREAS OF RESEARCH NEEDING MORE EMPHASIS TO ADVANCE THE CAPABILITIES FOR PCDD/PCDF ANALYSES

From the foregoing discussion, it is apparent that several areas of research require additional emphasis to further advance the capabilities for analyses of PCDD/PCDF. Such areas include:

1. synthesis and isolation of additional pure PCDD/PCDF isomer standards for use in calibration and methods development;
2. synthesis of additional compounds that are chemically similar to PCDD/PCDF and that may represent interferences in PCDD/PCDF analyses [diphenyl ethers (DPE), hydroxylated and Methoxy-DPE, and PCB) for methodology evaluation;

3. development of higher-resolution GC and/or LC columns for isolation of PCDD/PCDF isomers; and
4. research on more specific detection methods, possibly including positive and negative ion chemical ionization MS, atmospheric-pressure ionization MS and MS/MS.

REFERENCES

1. Esposito, M.P., T.O. Tiernan and F.E. Dryden. "Dioxins," EPA 600/2-80-197, U.S. EPA, Cincinnati, OH (1980).
2. Hutzinger, O., R.W. Frei, E. Meriam and F. Pocchiari, Eds. *Chlorinated Dioxins and Related Compounds* (New York: Pergamon Press, 1982).
3. Pohland, A.E., and G.C. Yang. *J. Agric. Food Chem.* 20:1093 (1972).
4. Narisada, M. *Yakugaku Zasshi* 79:183 (1959).
5. Aniline, O. *Adv. Chem. Ser.* 120:126 (1973).
6. Grey, A.P., S.P. Cepa, I.J. Solomon and O. Aniline. *J. Org. Chem.* 41:2435 (1976).
7. Nestrick, T.J., L.L. Lamparski and R.H. Stehl. *Anal. Chem.* 51:2273 (1979).
8. Lamparski, L.L., and T.J. Nestrick. *Chemosphere* 10:3 (1981).
9. Lamparski, L.L., and T.J. Nestrick. *Anal. Chem.* 54:402 (1982).
10. Norstrom, A., K. Anderson and C. Rappe. *Chemosphere* 5:21 (1976).
11. Buser, H.R. *J. Chromatog.* 129:303 (1976).
12. Buser, H.R., H.-P. Bosshardt and C. Rappe. *Chemosphere* 7:109 (1978).
13. Choudhry, G.F., G. Sundstrom, R.W.M. van der Wielen and H. Hutzinger. *Chemosphere* 6:327 (1977).
14. Lindahl, R., C. Rappe and H.R. Buser. *Chemosphere* 9:351 (1980).
15. Garå, A., K. Anderson, C.-A. Nilsson and A. Norstrom. *Chemosphere* 10:365 (1981).
16. Lamparski, L.L., and T.J. Nestrick. *Anal. Chem.* 52:2045 (1980).
17. Moore, J.A. Paper presented at the IARC Meeting on PCDDs and PCDFs, Lyon, France, (1978).
18. Masuda, Y., and H. Kuroki. In: *Chlorinated Dioxins and Related Compounds,* O. Hutzinger et al., Eds. (New York: Pergamon Press, 1982), p. 561.
19. U.S. Veterans Administration. "Review of Literature on Herbicides, Including Herbicides and Associated Dioxins, Vol. 1" (1981), p. 283.
20. Tiernan, T.O., M.L. Taylor, J.G. Solch, G.F. VanNess, J.H. Garrett and M.D. Porter. In: *Detoxication of Hazardous Wastes,* J. Exner, Ed. (Ann Arbor, Ann Arbor Science Publishers, 1982), p. 143.
21. American Chemical Society, Subcommittee on Environmental Analytical Chemistry. W.B. Crummett, Chairman. *Anal. Chem.* 52:2242 (1980).
22. National Research Council of Canada. "Polychlorinated Dibenzo-*p*-dioxins. Limitations to the Current Analytical Techniques," Report ISSN 0316-0114, NRCC No. 18576.
23. Albro, P.W., and C.E. Parker. *J. Chromatog.* 197:155 (1980).
24. Harless, R.L., E.O. Oswald, M.K. Wilkinson, A.E. Dupuy, D.O. McDaniel and H. Tai. *Anal. Chem.* 52:1239 (1980).
25. Firestone, D.J. *J. Agric. Food Chem.* 25:1274 (1977).

26. Firestone, D.J., M. Clower, A.P. Borsetti, R.H. Teske and P.E. Long. *J. Agric. Food Chem.* 27:1171 (1979).
27. Huckings, J.N., D.L. Stalling and W.A. Smith. *J. Assoc. Off. Anal. Chem.* 61:32 (1978).
28. Lamparski, L.L., T.J. Nestrick and R.H. Stehl. *Anal. Chem.* 51:1453 (1979).
29. Langhorst, M.L. and L.A. Shadoff. *Anal. Chem.* 52:2037 (1980).
30. Tiernan, T.O., M.L. Taylor, S.D. Erk, J.G. Solch, G.F. VanNess and J. Dryden. "Dioxins, Vol. II, Analytical Method for Chemical Wastes," EPA 600/2-80-57, U.S. EPA, Cincinnati, OH (1980).
31. Lamparski, L.L., and T.J. Nestrick. *Anal. Chem.* 52:2045 (1980).
32. VanNess, G.F., J.G. Solch, M.L. Taylor and T.O. Tiernan. *Chemosphere* 9:553 (1980).
33. Taylor, M.L., T.O. Tiernan, J.H. Garrett, G.F. VanNess and J.G. Solch. "Assessment of Incineration Processes as Sources of Supertoxic Chlorinated Hydrocarbons: Concentrations of Polychlorinated Dibenzo-*p*-dioxins/Dibenzofurans and Possible Precursor Compounds in Incinerator Effluents," Chapter 8, this volume.
34. Brumley, W.C., J.A.G. Roach, J.A. Sphon, P.A. Dreifuss, D. Andzejewski, R.A. Neimann and D. Firestone. *J. Agric. Food Chem.* 29:1040 (1981).
35. Lustenhouwer, J.W.A., K. Olie and O. Hutzinger. *Chemosphere* 9:501 (1980).
36. Kooke, R.M.M., J.W.A. Lustenhouwer, K. Olie and O. Hutzinger. *Anal. Chem.* 53:461 (1981).
37. Cutie, S.S. *Anal. Chim. Acta* 123:25 (1981).
38. Vos, J.G., J.H. Koeman, H.L. Van der Moas, M.C. ten Noever-de Braum and R.H. de Vos. *Food Cosmet. Toxicol.* 8:625 (1970).
39. Roach, J.A.G., and I.H. Pomerants. *Bull. Environ. Contam. Toxicol.* 12:338 (1974).
40. Bowes, G.W., M.J. Mulvihill, B.R.T. Simoneit, A.L. Burlingane and R.W. Risebrough. *Nature* 256:305 (1975).
41. Morita, M., J. Nakagawa, K. Akiyama, S. Minura and N. Isono. *Bull. Environ. Contam. Toxicol.* 18:67 (1977).
42. Rappe, C., A. Garå, H.R. Buser and H.-P. Bosshardt. *Chemosphere* 6:231 (1977).
43. Nagayana, J., M. Kuratsune and Y. Masada. *Bull. Environ. Contam. Toxicol.* 15:9 (1976).
44. Albro, P.W. and C.E. Parker. *J. Chromatog.* 169:161 (1979).
45. Rappe, C. and H.R. Buser. In: *Halogenated Biphenyls, Terphenyls, Nappthalenes, Dibenzodioxins and Related Products,* R.D. Kimbrough, Ed. (Amsterdam: Elsevier Press, 1980).
46. Moseley, C.L., C.L. Geraci and J. Burg. *J. Am. Ind. Hyg. Assoc.* 43:170 (1982).
47. Chittim, B.G., B.S. Clegg, S.N. Safe and O. Hutzinger. Draft Report, Prepared for Environmental Protection Service, Dept. of Fisheries and Environment (Canada) by Wellington Science Associates, Inc., Contract No. 05578-0067.
48. Pohland, E.E., and G.C. Yang. *J. Agric. Food Chem.* 20:1093 (1972).
49. Williams, D.T., and B.J. Blanchfield. *J. Assoc. Off. Anal. Chem.* 55:93 (1972).
50. Firestone, D., J. Ress, N.L. Brown, R.P. Barron and J.N. Damico. *J. Assoc. Off. Anal. Chem.* 55:85 (1972).
51. Woolson, E.A., R.F. Thomas and P.D. Ensor. *J. Agric. Food Chem.* 20:351 (1972).
52. Hutzinger, O., S. Safe and V. Zitko. *Int. J. Environ. Anal. Chem.* 2:95 (1972).
53. Baughman, R.W., and M. Meselson. *Adv. Chem.* 120:92 (1973).
54. Baughman, R.W., and M. Meselson. *Environ. Health Persp.* 5:27 (1973).
55. Crummett, W.B., and R.H. Stehl. *Environ. Health Persp.* 5:15 (1973).

56. Buser, H.R. *J. Chromatog.* 107:295 (1975).
57. Blaser, W.W., R.A. Bredeweg, L.A. Shadoff and R.H. Stehl. *Anal. Chem.* 48:984 (1976).
58. Hughes, B.M., D.C. Fee, M.L. Taylor and T.O. Tiernan. U.S. Air Force, Air Force Systems Command, Report No. ARL TR 75-0110 (1975).
59. Erk, S.D., M.L. Taylor and T.O. Tiernan. *Chemosphere* 1:7 (1979).
60. Taylor, M.L., R.L.C. Wu, C.D. Miller and T.O. Tiernan. Paper presented at the 24th Annual Meeting, American Society for Mass Spectrometry, San Diego, CA. (1976).
61. Shadoff, L.A., and R.A. Hummel. *Biomed. Mass Spectrum et.* 5:7 (1978).
62. Gross, M.L., T. Sun, P.A. Lyon, S.F. Wojinski and D.R. Hilker. *Anal. Chem.* 53: 1902 (1981).
63. Solch, J.G., T.O. Tiernan, G.F. VanNess, J.H. Garrett and M.L. Taylor. Paper presented at 29th Annual Meeting, American Society for Mass Spectrometry, Minneapolis, MN (1981).
64. Buser, H.R. *J. Chromatog.* 114:95 (1975).
65. Buser, H.R. *Anal. Chem.* 49:918 (1977).
66. Buser, H.R., and C. Rappe. *Anal. Chem.* 52:2257 (1980).
67. Buser, H.R., C. Rappe and A. Gara. *Chemosphere* 5:439 (1978).
68. Buser, H.R., H.-P. Bosshardt, C. Rappe and R. Lindahl. *Chemosphere* 5:419 (1978).
69. Rappe, C., H.R. Buser, H. Kuroki and Y. Masuda. *Chemosphere* 4:259 (1979).
70. Kuroki, H. and Y. Masuda, *Chemosphere* 10:771 (1978).
71. Hass, J.R., M.D. Friesen, D.J. Harvan and C.E. Parker. *Anal. Chem.* 50:1474 (1978).
72. Smith, L.M., J.L. Johnson, D.L. Stalling, J.D. Petty and G.R. Dubay. Paper presented at the 29th Annual Meeting for the American Society for Mass Spectrometry, Minneapolis, MN (1981).
73. Mitchum, R.K., W.A. Korfmacher and G.F. Moler. Paper presented at the 29th Annual Meeting for the American Society for Mass Spectrometry, Minneapolis, MN (1981).
74. McKinney, J., P. Albro, M. Luster, B. Corbett, J. Schroeder and L. Lawson, In: *Chlorinated Dioxins and Related Compounds,* O. Hutzinger et al., Eds. (New York: Pergamon Press, 1982), p. 67.
75. Bradlaw, J., and J.L. Casterline. *J. Assoc. Off. Anal. Chem.* 62:904 (1979).
76. Hutzinger, O., K. Olie, J.W.A. Lustenhouwer, A.B. Okey, S. Bandiera and S. Safe. *Chemosphere* 10:19 (1981).

14

New Gas Chromatographic Detector for Tetrachlorodibenzo-*p*-dioxin and Tetrachlorodibenzofuran Analyses

Lawrence H. Keith, L.D. Ogle, A.E. Jones, R.C. Hanisch and R.C. Hall

A need exists for a rapid and inexpensive method to analyze for polychlorinated dibenzo-*p*-dioxins (PCDD) and dibenzofurans (PCDF) at trace levels in complex samples. Present methodology involves identification and quantification using gas chromatography/mass spectrometry (GC/MS) [1]. However, samples may not always contain PCDD and PCDF at levels sufficient to obtain a positive analysis. In these cases, significant amounts of time and money are wasted because no data are obtained. If samples could be screened quickly and cheaply for the presence of these analytes, samples that may contain dioxins or dibenzofurans could be separated for GC/MS analyses.

GC using an electron capture detector (ECD) offers the sensitivity necessary for such a screening procedure, but it does not have the specificity needed to eliminate false positives. The ECD responds to any molecule that has an affinity for capturing electrons in an ionzed field. This includes molecules containing halogens (e.g., pesticides), oxygen (e.g., phthalate esters, ethers and ketones) and other hetero atoms such as sulfur and nitrogen.

A new prototype detector, a thermally modulated electron-capture detector (TMECD), offers greater specificity than an ECD, while at the same time maintaining the very high sensitivity of an ECD to molecules that have a high electron-capturing affinity. PCDD and PCDF are among the types of compounds that respond to a TMECD with high sensitivity and better specificity than an ECD.

The TMECD consists of two ECD with a temperature-controlled catalytic pyrolysis unit between them. Different classes of organic compounds react to the pyrolysis conditions in different ways, i.e., some compounds provide an enhanced signal after pyrolysis; others, a diminshed signal; and others, no change in the magnitude of the signal. The ratio of the peak area from the detector after the reactor to the peak area of the detector before the reactor is a property characteristic of each compound. This peak area ratio can be used in conjunction with its retention time to increase the confidence level of the identity of a given compound.

TMECD CHARACTERISTICS

A simplified block diagram of the TMECD system is shown in Figure 14.1. The peak areas produced by both detectors must be very reproducible if peak area ratios are to be used as properties characteristic of each compound or class of compounds. Table 14.1 presents the average coefficient of variation (CV) of peak areas for seven trihalomethanes and haloacetonitriles (contaminants in many drinking waters). The precision of the peak areas produced is very good (CV = 10%) for all of the compounds except bromoform.

Another requirement for this technique to be generally useful is that the peak area ratios from the two detectors be either independent of the amount of the analyte or be varied with the amount as a straight-line function. Table 14.2 presents the average peak height ratios for three tetrachlorodibenzo-*p*-dioxin (tetra-CDD) isomers and one tetrachlorodibenzofuran (tetra-CDF) isomer. Although a wide range of amounts of these compounds was used, the standard deviations were relatively unaffected and the reproducibility remains good.

In all of the compounds discussed so far (Tables 14.1 and 14.2), the chromatographic peak emerging from the catalytic reactor was diminished in height and peak area (detector 2) from its counterpart signal from detector 1. This is not always the case. At least three experimental parameters can be varied when using a TMECD:

1. reactor temperature,
2. catalytic metal tube composition, and
3. carrier gas.

All three of these parameters can affect the change in the signal size of a given compound in the second detector and, hence, the ratio of signal 2 to signal 1.

Examples of these variations are presented in Tables 14.3 through 14.5. Table 14.3 shows the effects of varying the above three parameters on lindane, a typical chlorinated hydrocarbon pesticide. Some general observations can be made:

1. Increasing the reactor temperature increases the differential between the detector 1 and 2 peak sizes, i.e., decreases the signal in detector 2.
2. The reactivity of the carrier gases is $He/H_2 > N_2 > Ar/CH_4$.
3. The nickel tube is more reactive than the gold tube.

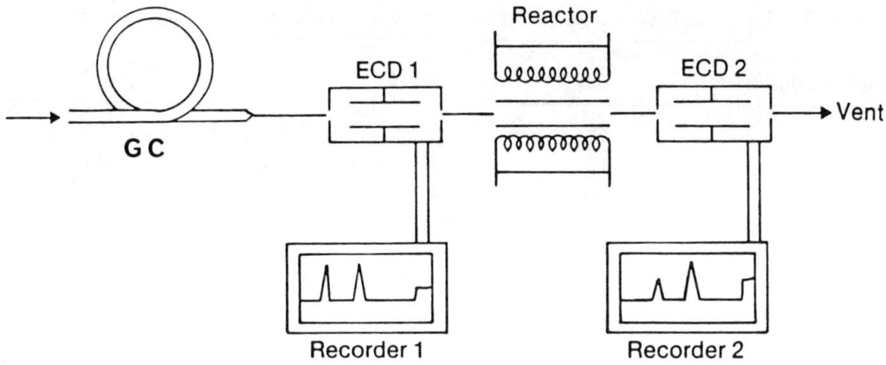

Figure 14.1. Diagram of a TMECD System.

Table 14.1. Reproducibility of TMECD Peak Areas for Trihalomethanes and Haloacetonitriles[a]

	CV[b] (%) of Peak Areas	
Compound	*Detector 1*	*Detector 2*
$CHC1_3$	2.5	0.4
$CHC1_2CN$	5.2	2.8
$CC1_3CN$	3.8	6.2
$CHBrC1_2$	2.7	6.6
$CHBr_2C1$	2.8	8.6
$CHBr_2CN$	4.4	8.6
$CHBr_3$	2.3	11.3

[a]Eight replicate analyses on squalene column; 3 ng of each compound; reactor temperature = 650°C.

[b]$CV = SD\sqrt{x} \times 100\%$.

Table 14.2. Reproducibility of Peak Height Ratios for Tetra-CDD and Tetra-CDF Standards

Compound	*Peak Ratio*	*SD*	*No. of Analyses*	*Ratio Range*	*Picogram Range*
Tetra-CDD					
1,3,6,8-	0.86	0.05	11	0.73–0.91	24–1000
1,2,3,4-	0.775	0.005	6	0.77–0.78	5–1000
2,3,7,8-	0.89	0.04	8	0.81–0.98	50–1080
2,3,7,8-tetra-CDF	0.87	0.02	5	0.83–0.90	10–1070

Table 14.3. Detector 2/Detector 1 Response Ratios for Lindane under Varying Conditions

Reactor Conditions	*350°C*	*600°C*	*800°C*	*900°C*
Gold Tube				
Ar/CH_4	0.93	0.93	0.11	0.04
N_2	0.93	0.91	0.03	0.02
He/H_2	0.14	<0.01	<0.01	0
Nickel Tube				
Ar/CH_4	0.84	0.82	0.01	0
N_2	0.85	0.84	0.35	0

Table 14.4. Detector 2/Detector 1 Response Ratios for Diethyl Phthalate under Varying Conditions

Reactor Conditions	*350°C*	*600°C*	*800°C*	*900°C*
Gold Tube				
Ar/CH_4	1.54	1.58	3.48	8.09
N_2	2.09	2.09	5.78	14.86
He/H_2	3.76	7.08	12.86	20.29
Nickel Tube				
Ar/CH_4	2.29	2.67	2.56	0.31
N_2	2.05	2.46	10.00	0.12

Table 14.5. Detector 2/Detector 1 Response Ratios for Nitrobenzene under Varying Conditions

Reactor Conditions	*350°C*	*600°C*	*800°C*	*900°C*
Gold Tube				
Ar/CH_4	1.00	0.84	0.84	0.90
N_2	0.78	0.84	0.88	0.97
He/H_2	0.08	0	0	0
Nickel Tube				
Ar/CH_4	1.64	1.20	0	0
N_2	1.49	1.49	0	0

The reverse of the signal characteristics is exhibited with the data for diethyl phthalate (Table 14.4). In all but two cases, the signal from the detector 2 is larger than that from the detector 1, so all but two of the ratios are >1.0. The probable cause of the phenomenon is that the products of the catalytic decomposition in these cases have a greater electron-capturing affinity than the parent compounds.

In the most enhanced example, the signal from detector 2 is 20 times that of the original signal of this compound from detector 1. However, the same trends in activity changes from varying reactor temperature, carrier gas and reactor tube composition are generally observed.

The third example, nitrobenzene, is presented in Table 14.5. The size of the signal in detector 2 relative to that from detector 1 was reversed simply by changing the catalytic reactor tube. The gold tube produced peaks that were smaller and the nickel tube produced peaks that were larger than they originally were. A ratio of zero indicates that the signal from detector 2 disappeared.

At this point, we have not had the opportunity to connect the effluent from the TMECD to a mass spectrometer, so the nature of the decomposition products remains unknown. However, their identification is mandatory to a basic understanding of what is happening within the reactor. Until the mechanism of these reactions is known, this new detector's use can only be explored in an empirical way.

Clearly, it is possible to vary the detector parameters to significantly enhance the selectivity of the TMECD to a given analyte, or class of analytes, while at the same time repressing or changing the selectivity of the TMECD to common interfering compounds present in the same samples. Examples of this utility are presented later.

Another very useful technique that can be employed with the TMECD is to operate both detectors with a differential amplifier. Using detector 1 as the reference and subtracting the signal from detector 2 produces a differential gas chromatogram (DGC). As will be seen later, DGC are even more useful than conventional analog gas chromatograms in visually interpreting the chromatographic data produced by this detector system.

It was observed with the tetra-CDD and tetra-CDF standards that the differential chromatograms exhibit better linearity over the wide range of analyte injected (5–1000 pg) than the chromatograms from either of the two ECD separately (Figures 14.2 and 14.3). These linearity curves seem to be fairly typical of the compounds with which we have experimented to date.

ANALYSIS OF TISSUE SAMPLES

Tissues are one of the most challenging sample types because of matrix interferences caused by fats, lipids, sugars and many other naturally occurring organic compounds. The difficulty of analyzing these samples is greatly magnified when the organic analytes are at lower than part-per-million levels.

Generally, the first stage of an analysis involves a series of "cleanup" steps to try to separate the analytes from the naturally occurring compounds in the tissues. Also, at trace levels, the analytes must be concentrated so that they can be detected and measured by some instrumental method; in the case of PCDD and PCDF, this usually means by GC/MS. Generally, the more of these "cleanup" and concentration steps that are involved, the greater will be the time and

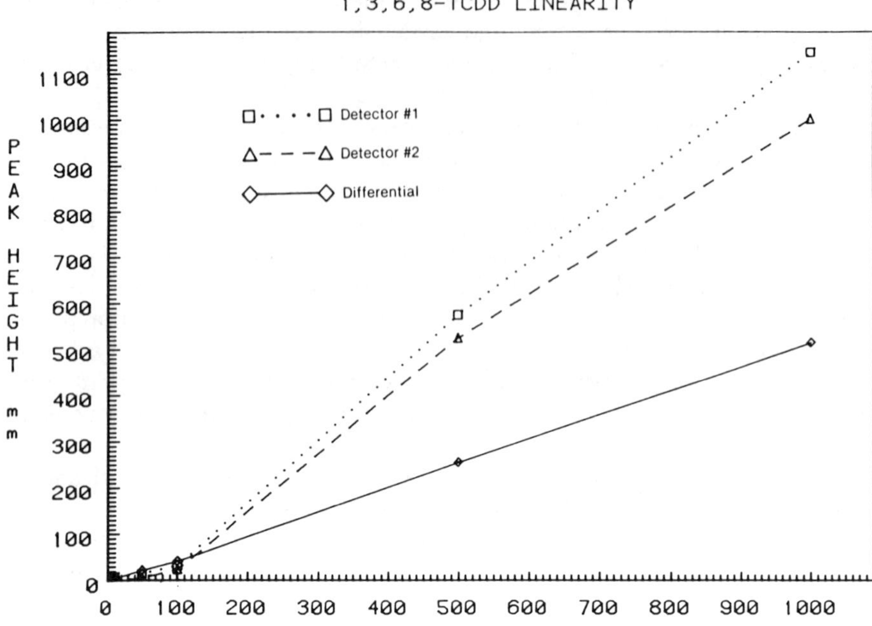

Figure 14.2. Linearity of 1,3,6,8-tetra-CDD Using GC/TMECD.

cost for analyses. Additionally, there is a greater chance for contamination of the sample and/or a loss of the analytes.

One example of current cleanup methodology for tetra-CDD in tissue samples [2] is diagrammed in Figure 14.4. This method is widely accepted, but, as can be seen, it involves a rather lengthy (and thus costly) cleanup procedure before GC/MS analysis. It is particularly frustrating and quite expensive to go through all of this cleanup and then find no detectable levels of tetra-CDD or tetra-CDF in the samples.

Burse [3] is investigating the use of GC/ECD as a screening method for analysis of tetra-CDD in tissues. All of the column chromatography cleanup steps except the final one [reverse-phase high-performance liquid chromatography (HPLC)] are employed before GC/ECD analysis. In an effort to determine whether even more cleanup steps could be eliminated by using GC/TMECD, we participated in a brief study to evaluate this new technique with samples of bovine fat that had been spiked with 7.12 ng 1,3,6,8-tetra-CDD/g fat (7.12 ppb in vitro before saponification). The spiked samples were carried only through the first basic alumina column chromatography step (Figure 14.4). The samples were received as 3.0-mL extracts in hexane and were not concentrated or manipulated

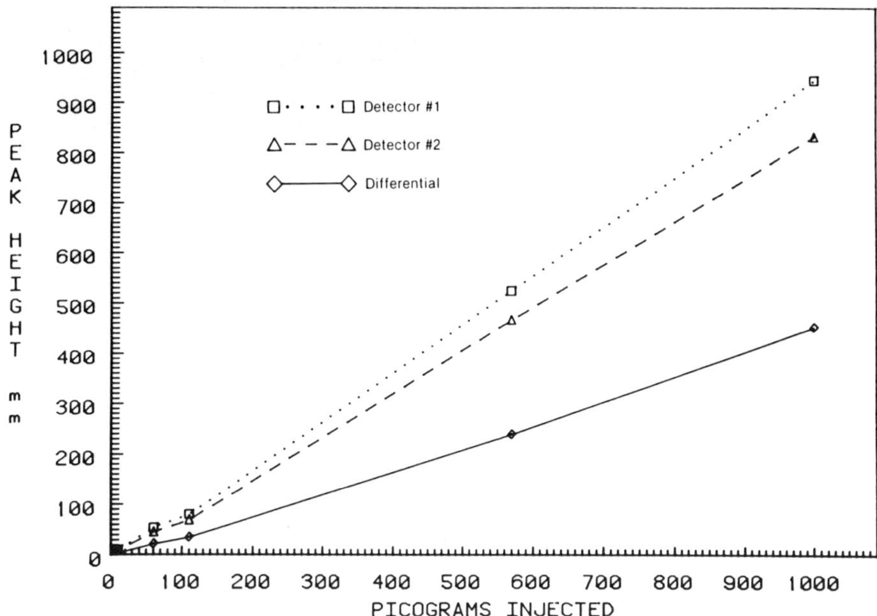

Figure 14.3. Linearity of 2,3,7,8-tetra-CDF Using GC/TMECD.

further. If 100% recovery of the 7.12 ng is assumed, then the concentration in the extracts was 2.33 ng/mL of 1,3,6,8-tetra-CDD. A 10- μL aliquot (23 pg of tetra-CDD) of each sample was injected into a Radian 110B computer-controlled gas chromatograph equipped with the prototype TMECD and analyzed according to the conditions listed in Table 14.6.

The bovine fat blank extract (unspiked) chromatograms are shown in Figure 14.5. The arrows indicate the retention time where 1,3,6,8-tetra-CDD would elute under these chromatographic conditions. There is a small peak at that retention time (corresponding to about 2 ppb) in the chromatogram from ECD 1 (the normal ECD chromatogram). However, it is barely visible in the chromatogram from ECD 2 or in the differential chromatogram.

Several features are noteworthy in Figure 14.5. First, the peak just to the left of the arrow is larger from ECD 1 than it is from ECD 2. This is typical of a halogenated compound (although this has not been confirmed, nor has the peak been identified), and it stands out clearly from all of the other peaks in the differential chromatogram. All of the other impurity compounds in the fat extract appear as negative peaks in the differential chromatogram because they produced larger peaks from ECD 2 than from ECD 1.

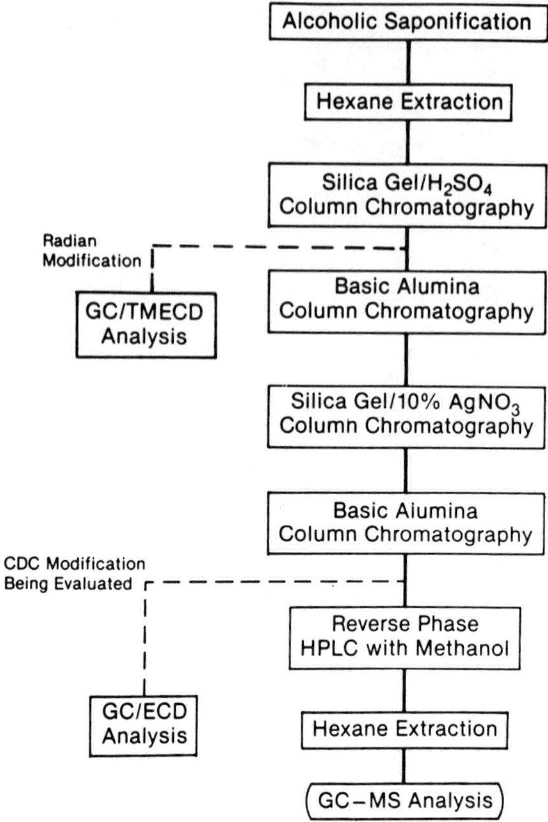

Figure 14.4. Current Methodology for Cleanup and Analysis of tetra-CDD in Tissue Samples Showing Radian and CDC Modifications Being Investigated.

Table 14.6. Analytical Parameters for GC/TMECD Analysis of CDC Bovine Fat Samples for 1,3,6,8-Tetra-CDD

Gas Chromatograph	
Instrument	Radian 110B
Column	5 ft × 2 mm i.d. glass-lined stainless steel packed with 3% 2250 DB on 100/120 mesh Supelcoport
Carrier Gas	95% Argon-5% methane at 30 mL/min
Amount Injected	10 μL
Attenuation	1 × 2
Inlet Temperature	250°C
Temperature Program	100°C with 2-min initial hold followed by programmed rate of 5.3°C/min to 265° with a final hold of 15 min

Table 14.6. Analytical Parameters for GC/TMECD Analysis of CDC Bovine Fat Samples for 1,3,6,8-Tetra-CDD (*continued*)

Detector	
Prototype	TMECD
Temperature	340°C both ECD
Reactor	Gold tube ~7.5 in. long by 1/16 in. o.d.
Reactor Temperature	950°C

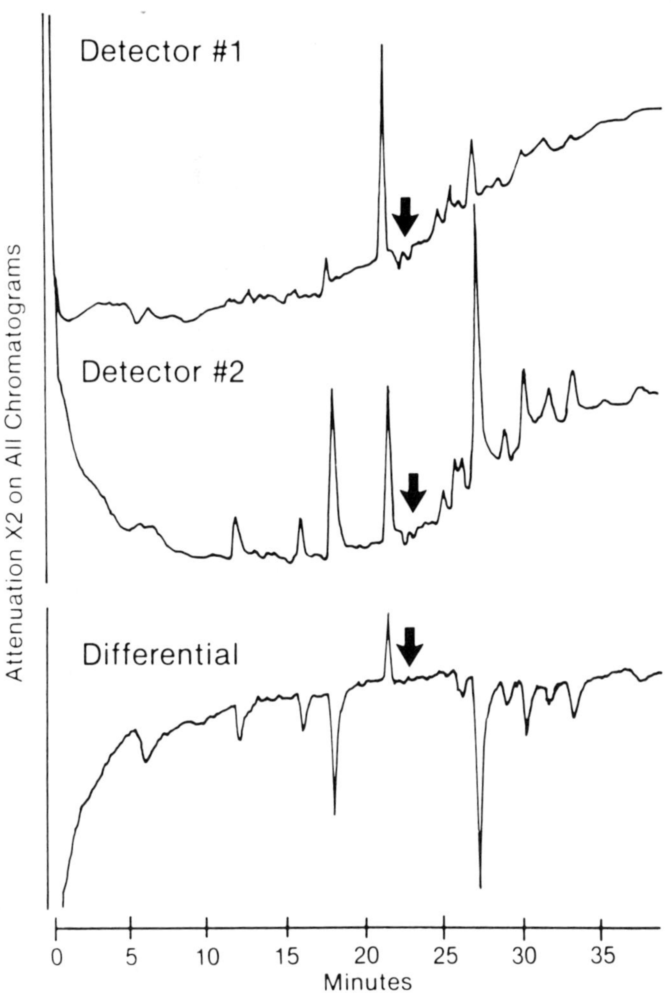

Figure 14.5. Unspiked Bovine Fat Extract Analyzed by GC/TMECD.

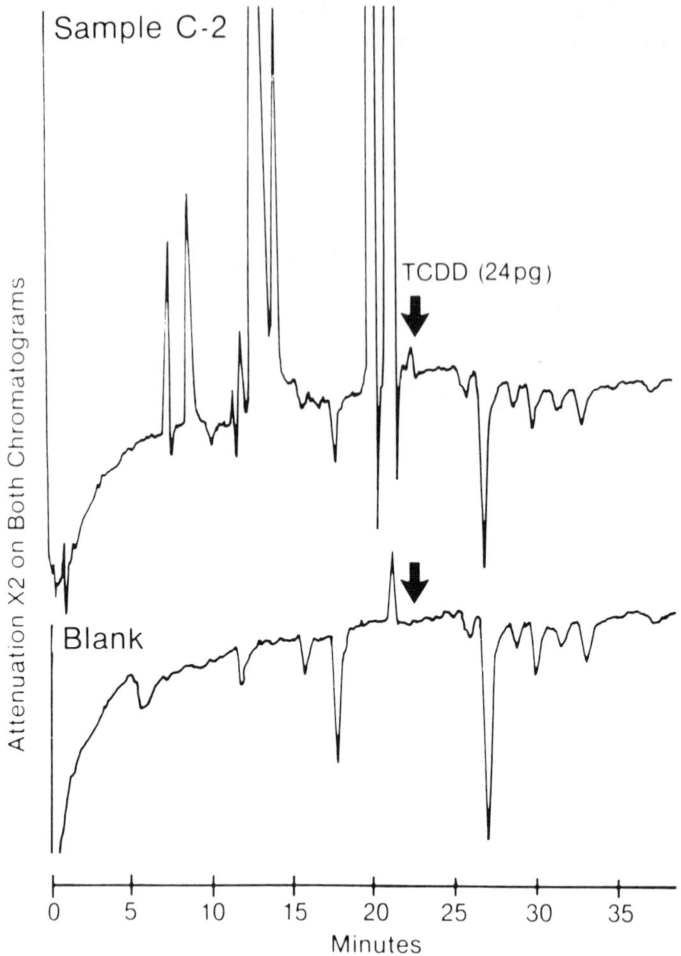

Figure 14.6. Differential GC/TMECD Chromatograms of Bovine Fat Extract Spiked with 7.12 ppb 1,3,6,8-Tetra-CDD and Six Chlorinated Pesticides at 500–2200 ppb.

Figure 14.6 shows only the differential chromatogram of the sample spiked with 7.12 ppb of 1,3,6,8-tetra-CDD plus six chlorinated hydrocarbon pesticides at 70–300 times the tetra-CDD concentration (Sample 2 in Table 14.7). The lower portion of the figure shows the bovine fat blank extract analyzed under the same conditions for reference. A notable feature of the dioxin peak is its characteristic shape. The ratio of the peak above the baseline to that portion of it below the baseline is roughly 2:1. This very characteristic peak shape was observed with all three of the tetra-CDD isomers (1,3,6,8-, 1,2,3,4- and 2,3,7,8-tetra-CDD) and the one dibenzofuran isomer (2,3,7,8-tetra-CDF) studied. This peak shape stands

Table 14.7. Summary of CDC Bovine Fat Analyses for 1,3,6,8-Tetra-CDD

Sample	*Times Analyzed*	*Detector*	*Average Conc. (ppb)*	*Standard Deviation*	*True Conc. (ppb)*
1. 1.00 g Fat with Tetra-CDD	3	1	7.8	2.9	7.12
	3	2	6.3	2.2	7.12
	3	1–2	10.4	3.5	7.12
2. 1.01 g Fat with Tetra-CDD and 1.2 ppm HCB; 0.5 ppm γ-Lindane; 2.2 ppm p,p'-DDT; 1.1 ppm o,p'-DDT; 0.5 ppm o,p'-DDE and 1.5 ppm trans-Nonachlor	3	1	7.1	1.4	7.12
	3	2	5.1	1.1	7.12
	3	1–2	10.3	1.0	7.12
3. 1.01 g Fat with Tetra-CDD and 9.2 ppm p,p'-DDE	3	1	NA[b]	NA	7.12
	3	2	NA	NA	7.12
	3	1–2	8.9	1.0	7.12
4. Duplicate 1.01 g Fat with Tetra-CDD and 9.2 ppm p,p'-DDE	3	1	NA	NA	7.12
	3	2	NA	NA	7.12
	3	1–2	10.0	1.7	7.12

[a]All values are corrected by subtraction of peak heights from an unspiked fat sample analyzed the same day as the samples. Average background contamination was 1.0 ppb ± 1.1 and ranged from 0 to 2.5 ppb.

[b]NA = not analyzed because PCDD peak was a shoulder on the trailing side of the DDE peak.

out clearly and is easy to spot down to the 5 to 10 pg level of detection. We have not observed this phenomenon with pesticides or other halogenated compounds we have examined. A similar (but reversed) peak shape was observed with benzo[a]pyrene (B[a]P); this is discussed later in this chapter.

A total of five fat extracts spiked with 7.12 ppb of 1,3,6,8-tetra-CDD plus the blank were analyzed in triplicate. The results of four of these samples are presented in Table 14.7. In all cases, the concentration was determined using a concentration curve constructed from repetitive analyses of 1,3,6,8-tetra-CDD standards at various dilutions. The fifth sample was spiked with pesticides, and Aroclor 1254 and 1260. Under the conditions employed for these experiments, the polychlorinated biphenyl (PCB) peaks caused too much interference for the dioxin peak to be analyzed.

It is noteworthy that the dioxin could not be quantified using detectors 1 or 2 in the sample containing p,p'-DDE at 1300 times the concentration of the 1,3,6,8-tetra-CDD. The dioxin was observed as an unresolved shoulder at the base of the tailing edge of the p,p'-DDE peak. However, a distinct but partially resolved dioxin peak was observed under these same chromatographic conditions in the differential chromatogram. Duplicate samples were analyzed in triplicate to verify this phenomenon.

Table 14.8. Bovine Fat Matrix Effect on 1,3,6,8-TCDD Peak Ratios

	Tetra-CDD Injection Range (pg)	*Average Ratio*	*Standard Deviation*	*Ratio Range*
11 Standards	22–1000	0.86	0.05	0.73–0.91
6 Samples	24	0.73	0.05	0.67–0.81

It is also important to stress that the conditions under which these samples were analyzed were not optimized. Experiments have not been conducted using other carrier gases or other metal reaction tubes. Only the temperature of the reactor was varied between 800 and 950°C, and the higher temperature produced better differential signals from the 1,3,6,8-tetra-CDD standard.

A comparison of the average peak ratio of 1,3,6,8-tetra-CDD standards to the same ratios in the samples indicates a probable matrix effect (Table 14.8). At this point, it is too early to speculate on the cause of this effect, but it does seem to be real and is reflected by the overall lower average concentration of the dioxin determined with detector 2 (5.7 ± 1.7 ppb) as compared to an average of 7.5 ± 2.1 ppb with detector 1. The overall average determined in the differential mode was 9.9 ± 1.8 ppb.

ANALYSIS OF TETRA-CDD AND POLYNUCLEAR AROMATIC HYDROCARBONS (PAH) STANDARDS

PCDD and, especially, PCDF may be found in conjunction with incineration samples. However, PAH also may be present in these samples at much greater concentrations. Since all of these molecules are planar, they are sometimes difficult to separate. Therefore, a mixture of 1,2,3,4-tetra-CDD at 5 ng/mL with three randomly chosen representative PAH (naphthalene, fluorene, and B[a]P) at 50,000 ng/mL each was analyzed by GC/TMECD. The chromatograms from detectors 1 and 2 are shown in Figure 14.7. The TCDD peak is visibly smaller in the chromatogram from detector 2, while the fluorene and B[a]P peaks are larger. Naphthalene did not give a significant signal. There are also two phthalate ester peaks in the standard as impurities [di-*n*-butyl phthalate (DPB) and di-(2-ethylhexyl) phthalate (2-ETHP)].

The differential chromatogram of this same standard is shown in Figure 14.8. The tetra-CDD peak is clearly visible as a positive peak while the PAH, the two phthalate esters and small amounts of other unknown impurities are plotted as negative peaks. The odd shape of the B[a]P peak is reminiscent of the tetra-CDD and tetra-CDF peak shapes, but differs in that one-third of it is above the

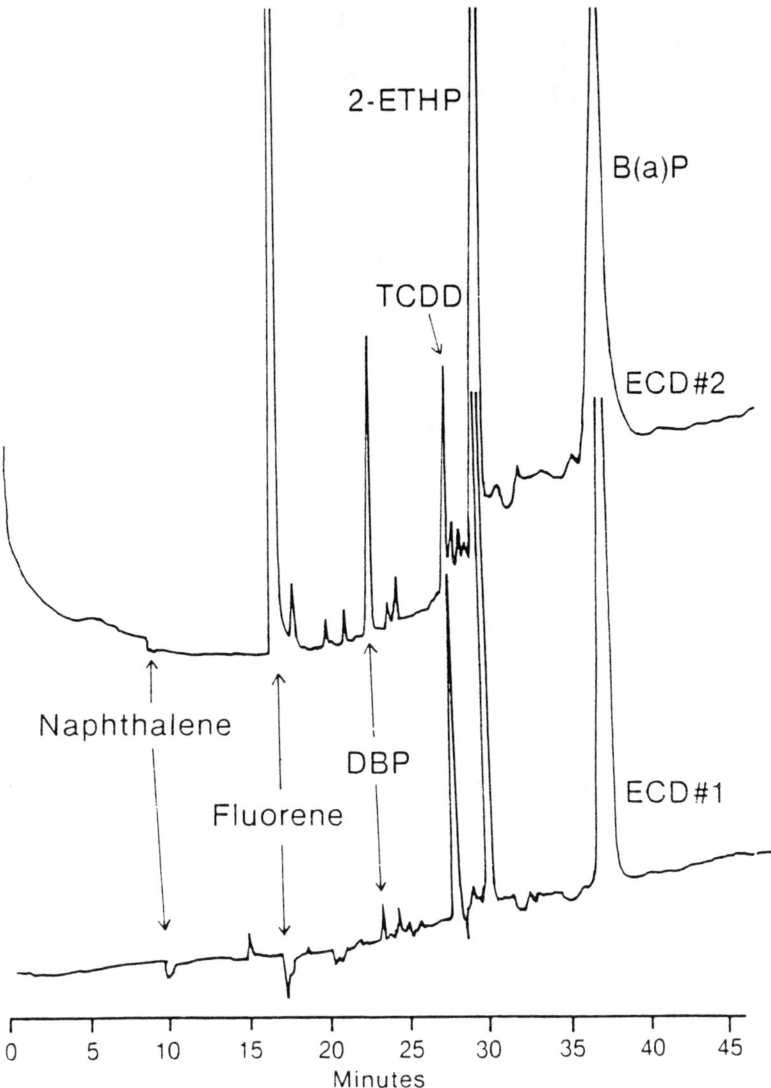

Figure 14.7. GC/TMECD Analysis of 50 g of 1,2,3,4-Tetra-CDD with 500,000 pg each of Naphthaline, Fluorene and B[a]P. DBP and 2-ETHP are Impurities in the Benzene-Hexane Solvent Mixture Chromatographed on a 5-ft 3% SP-2250 DB Column Using Ar/CH_4 Carrier Gas and a Gold Reaction Tube at 950°C.

baseline and two-thirds of it is below the baseline. This ratio is the reverse of that observed with the tetra-CDD and tetra-CDF signals. Figure 14.9 shows the combined TMECD chromatograms of the hexane-benzene solvent blank used to prepare the tetra-CDD/PAH standard solution.

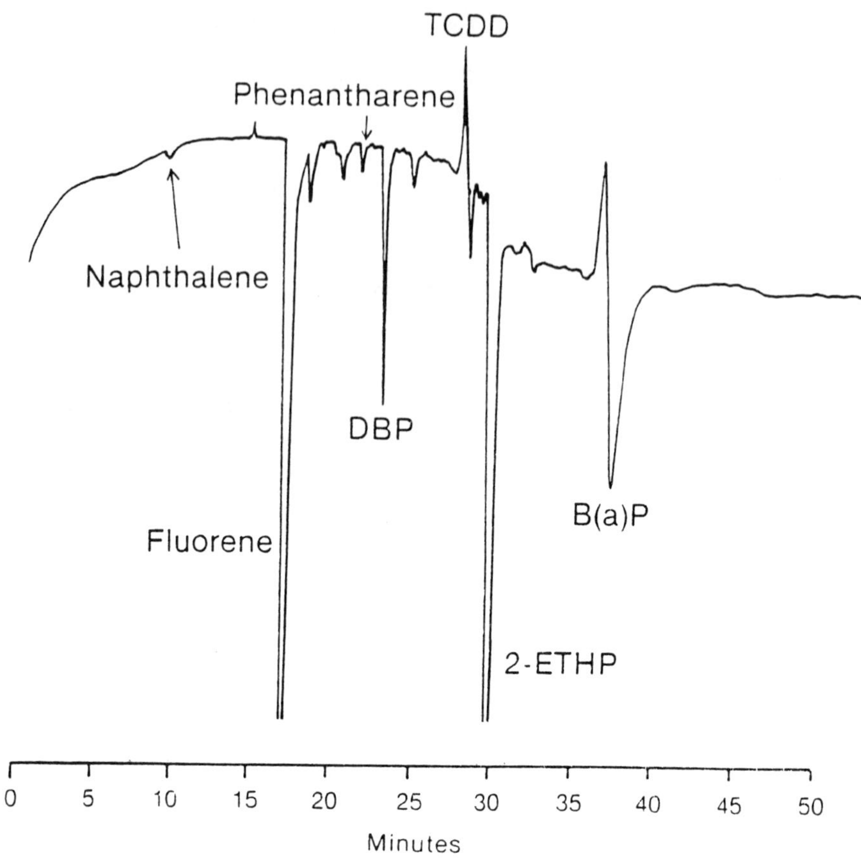

Figure 14.8. Differential GC/TMECD Analysis of 50 pg of 1,2,3,4-Tetra-CDD with 500,000 pg Each of Naphthalene, Fluorene and B[a]P. Analytical Conditions are the Same as in Figure 14.7.

ANALYSIS OF PESTICIDE STANDARDS

Figures 14.10 and 14.11 illustrate one of the advantages that can be gained from TMECD analysis of complex mixtures. Chromatogram A of Figure 14.10 is a mixture of 0.5 ng of each of several priority pollutant pesticides; chromatogram B shows 5 ng of Aroclor 1254 mixture. Chromatogram C is a mixture of the solutions analyzed in A and B. The response of all three chromatograms was generated from detector 1 of the TMECD system and therefore represents the usual GC/ECD chromatographic patterns observed with these mixtures. From Figure 14.10C it is seen that the identification, and certainly the quantification, of the pesticides is made difficult or impossible by the coeluting peaks of the Aroclor 1254. The current procedure for solving this problem is to make a column chro-

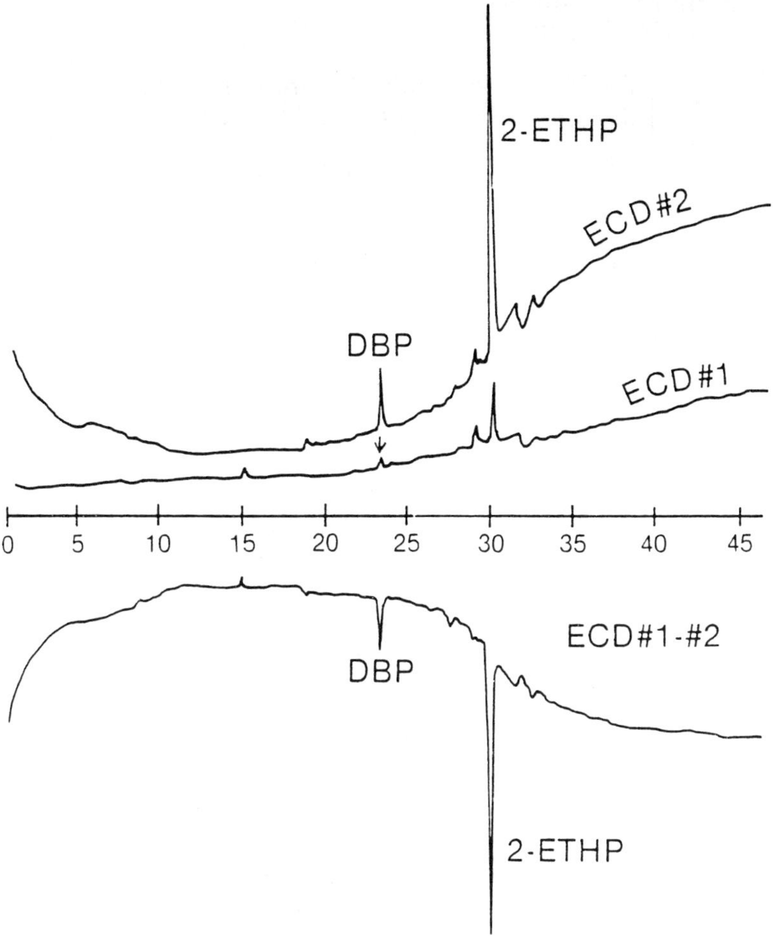

Figure 14.9. GC/TMECD Analysis of the Hexane-Benzene Solvent Used to Prepare the Tetra-CDD-PAH Standards Chromatographed in Figures 14.7 and 14.8. Top Chromatograms are From Detectors 1 and 2. Bottom Shows the Differential GC/TMECD Chromatogram.

matographic separation of the mixture using alumina or silica gel, collecting and concentrating several fractions of the eluate and then reanalyzing them by gas chromatography.

However, use of the TMECD detector system eliminates the need for prior separation by column chromatography—the Aroclor 1254 signals are simply removed electronically. Figure 14.11A shows the same mixture of priority pollutant pesticides after they have passed through a gold reaction tube at 850°C and through detector 2. The signals generated in all three chromatograms of Figure

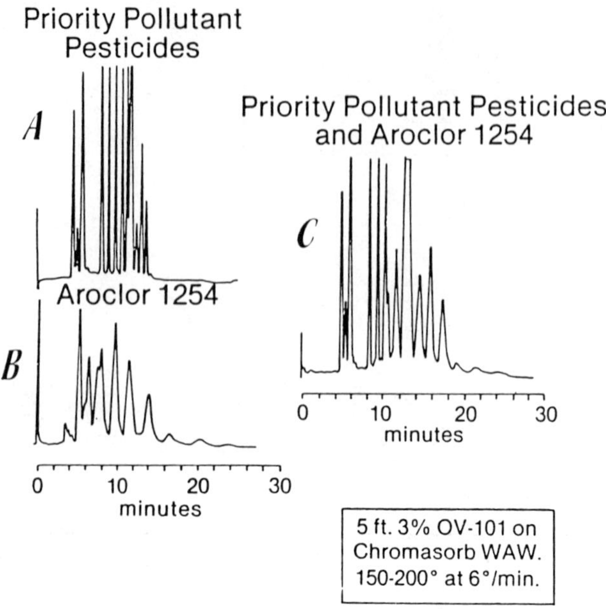

Figure 14.10. Chromatograms Using ECD 1. (A) Mixture of 0.5 ng each of Priority Pollutant Pesticides (Aldrin, Dieldrin, 4,4′-DDT, 4,4′-DDD, 4,4′-DDE, Lindane, Heptachlor, Heptachlor Epoxide, α-Endosulfan, β-Endosulfan, Endosulfan Sulfate, α-BHC, β-BHC and δ-BHC); (B) 5.0 ng of Arochlor 1254 Mixture of PCB, and (C) a 1:1 Mixture of Solutions A and B.

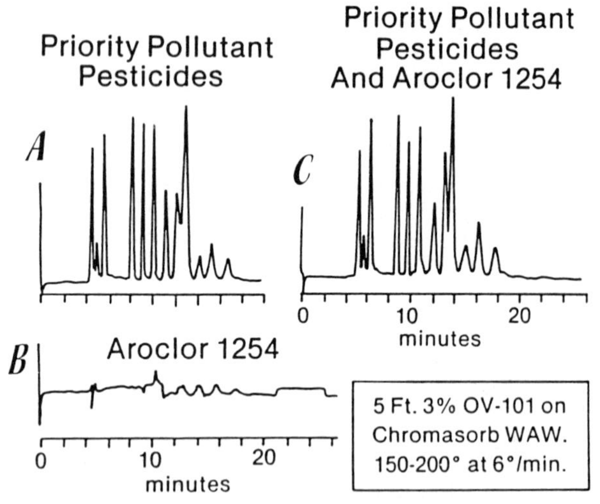

Figure 14.11. Differential GC/TMECD Chromatograms of the Same Solutions as in Figure 14.10.

14.11 are differential amplifier outputs. Since all of the priority pollutant pesticide signals are diminished after the compounds pass through the catalytic pyrolysis unit, the differential between ECD 1 and 2 is positive. Under the same reaction conditions, Aroclor 1254 signals are unchanged and the differential output from the two detectors (chromatogram 14.11B) is zero; only a few impurity peaks are detectable. Chromatogram 14.11C is that of the mixture of pesticides and Aroclor 1254. Since the PCB signals are effectively zeroed out of the mixture, chromatogram 14.11C is identical in appearance to 14.11A. The effect is the same as that achieved by labor-intensive column chromatographic separations followed by reanalysis with a conventional ECD, but the cost in time and money is less than half.

Figure 14.12A illustrates another troublesome mixture—selected chlorinated hydrocarbon pesticides, PCB (Aroclor 1016) and four phthalate esters (dimethyl phthalate, diethyl phthalate, DBP and 2-ETHP). Chromatogram 14.12B shows this same mixture after passing through a gold reaction tube at 850°C and output from detector 2. Four of the peaks in 14.12B are distinctly enhanced over their counterparts in 14.12A. These are the four phthalate esters. All of the pesticide peaks in 14.12B are smaller than their counterparts in 14.12A. Although not easily discerned, the Aroclor 1016 peaks are unchanged in size in chromatogram 14.12B.

Chromatogram 14.12C is the DGC. Since the PCB signals were unchanged by the pyrolytic conditions chosen for this experiment, they are absent from the

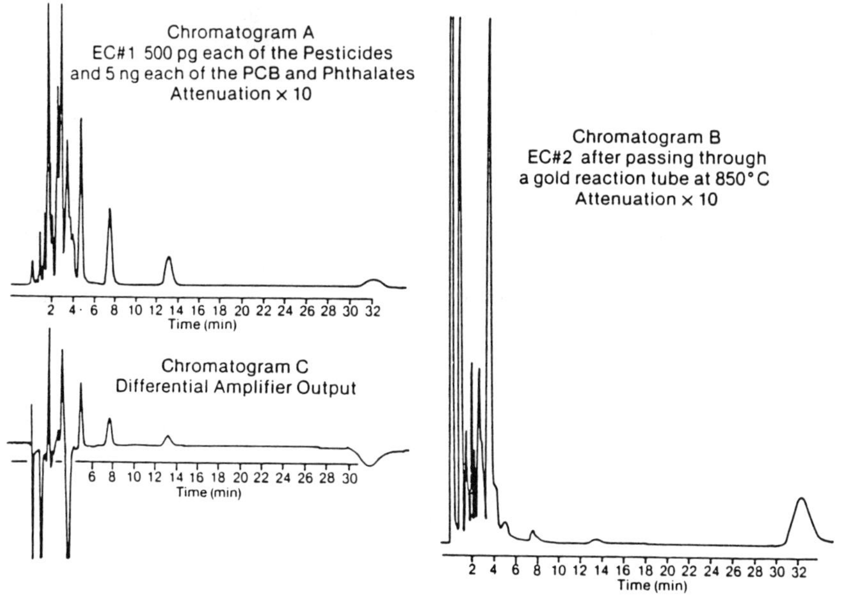

Figure 14.12. Chromatograms of Five Chlorinated Hydrycarbon Pesticides (Lindane, Heptachlor, Heptachlor Expoxide, 4,4′-DDE and 4,4′-DDT) and Aroclor 1016.

DGC. However, the pesticide peaks were smaller from detector 2, so the products of differential amplification are positive. Conversely, the four phthalate esters were enhanced by pyrolysis, so negative peaks resulted. The Radian 11OB GC data system, of course, is capable of handling both positive and negative peak area integrations.

SUMMARY

GC/TMECD appears to offer a unique new method for screening for PCDD and PCDF at trace levels in complex samples. An unusual peak shape in differential chromatograms makes PCDD and PCDF easily visible. This visibility is enhanced even more by the fact that common impurities with these analytes (phthalate esters and PAH) produce negative peaks in the differential chromatograms while the analytes of interest produce positive peaks. This differentiation using the detector itself can be very useful when selectively applied to difficult samples.

In addition to greater selectivity gained by use of a differential amplifier, the peak ratios from the two detectors, in conjunction with a GC retention time, can increase the confidence level of an identification. However, extremely "dirty" samples may exhibit a matrix effect than can change the peak ratios somewhat.

A great benefit of the TMECD is its high sensitivity to most pollutants of interest. In the present example, tetra-CDD analyses were conducted on tissue extracts in which the tetra-CDD concentrations were only 2.33 ng/mL. The lower limit of detection from these samples appears to be about 2 pg (which corresponds to about 1 ppb from these extracts). However, if the 3.0-mL extracts had been concentrated to 0.03 mL, the detection level would have been 10 ppt. If 10 g of tissue had been used and the extract was concentrated to 0.03 mL, the theoretical detection level would have been 1 ppt. This is probably the lowest practical detection level to be expected using tissue samples. However, extraction of 100 g of fly ash and other solid materials is feasible so that a lower detection level of 1 ppt should easily be possible from these types of samples. Extracts of 1-L water samples should result in lower than part-per-trillion detection levels. The lower quantification level for the tissue samples in this work was five times the lower detection level (about 5 ppb); presumably, this would approximately follow through on the above calculations.

Thus, it appears that GC/TMECD has much to offer as a new screening method for tetra-CDD and tetra-CDF in many types of samples, e.g., tissues, fly ash and other incineration residues, soil, water, and transformer oils. It has the potential to save large sums of money by removing negative samples from expensive GC/MS analyses. It also has the potential to simplify existing rigorous and labor-intensive cleanup schemes and leave those only for samples requiring GC/MS analyses.

Examples with mixtures of various pesticides, PCB and phthalate esters have shown that the detector itself can be used to perform some unique electronic separations and thereby also save significant amounts of time and money.

However, this work has raised at least as many questions as it has answered. Further work needs to be pursued in order to:

1. determine the reaction products responsible for the signals detected after the reactor;
2. determine if there are molecular positional effects on the response factors of various isomers;
3. demonstrate the method using capillary columns;
4. define the limits of matrix effects on peak ratios;
5. optimize selectivity and sensitivity for tetra-CDD, tetra-CDF and other selected analytes; and
6. combine GC/TMECD with two-dimensional post-detector peak recycling (PDPR).

With respect to the last point, combining a selective detector, such as TMECD, with two-dimensional PDPR GC [4] should achieve the ultimate in sample cleanup simplicity and compound identification reliability using GC. Retention times on two different columns plus the enhanced selectivity from peak ratios and a differential gas chromatogram all from a single analysis will increase the confidence level of compound identifications and quantifications, leading GC analyses to a new level of sophistication and reliability.

REFERENCES

1. EPA. "Method 613: 2,3,7,8-Tetrachlorodibenzo-*p*-dioxin," *Federal Register* 44(233): 69526–69531 (1979).
2. Lamparski, L.L., T.J. Nestrick and R.H. Stehl. *Anal. Chem.* 51(9):1453–1458 (1979).
3. Burse, V., Center for Disease Control. Personal communication (July 1982).
4. Keith, L.H., R.C. Hall, R.C. Hanisch, R.G. Landolt and J.E. Henderson. "New Methods for Gas Chromatographic Analysis of Water Pollutants," in *Water Chlorination: Environmental Impact and Health Effects, Vol. 4, Book 1,* R.L. Jolley et al., Eds. (Ann Arbor, MI: Ann Arbor Science Publishers, 1983), pp. 563–582.

15

Parameters for Identification and Confirmation in Trace Analyses of Polychlorinated Dibenzo-*p*-dioxins and Dibenzofurans

C. Rappe, S. Marklund, M. Nygren and A. Garå

Because of the extreme toxicity of some of the polychlorinated dibenzo-*p*-dioxins (PCDD) and dibenzofurans (PCDF), very sensitive, selective and specific analytical techniques are required. Detection levels in biological and environmental samples should be orders of magnitude below the usual detection limits obtained in pesticide analysis. A sensitivity of 1 pg might be required to detect the most toxic isomers, such as 2,3,7,8-tetra-CDD, in a 1-g sample. Any analysis at such low levels is complicated by the presence of a multitude of other, possibly interfering compounds, and also by the large number of PCDD (75 isomers) and PCDF (135 isomers). The different isomers of PCDD and PCDF may vary greatly in their biological and toxicological properties. Consequently, the analytical method should allow the separation, identification, quantification and confirmation of specific isomers.

In recent years many analytical methods were developed for analysis of PCDD and PCDF, especially for the most toxic 2,3,7,8-tetra-CDD in environmental samples. The most specific of these methods use high-resolution gas chromatography/mass spectrometry (HRGC/MS). Prerequisites for best analyses are representative sampling, good storage, efficient extraction and sample purification (containment) followed by good separation, ultrasensitive detection and quantification, and most desirable confirmation.

This chapter discusses the use of various capillary columns for the separation of critical isomers and also a few confirmation methods.

ISOMER SEPARATION ON HRGC

The identification of specific PCDD and PCDF isomers is primarily based on separation and retention time measurements using glass capillary and/or fused-silica columns of different polarity. Of special interest is the analysis of the highly toxic isomers, e.g. 2,3,7,8-tetra-CDD, 1,2,3,7,8-penta-CDD and 2,3,7,8-tetra-CDF (figure 15.1). For unambiguous identification of these isomers it is necessary to have access to all isomers within a specific group: all 22 tetra-CDD and all 38 tetra-CDF. Synthesis and separation of the 22 tetra-CDD have been reported earlier, and synthetic programs aiming at synthesis of the 38 tetra-CDF, 14 penta-CDD and 28 penta-CDF are in progress in our laboratory.

For the separation of the TCDD isomers we have used a Silar-10C column [1]. On this column we can separate the highly toxic 2,3,7,8-tetra-CDD from the other 21 isomers; the two isomers eluting most closely are the 1,2,3,4- and the 1,4,7,8- isomers. Figure 15.2 shows the separation of these three isomers.

Concerning the 38 tetra-CDF isomers, it has been found that the 2,3,7,8-isomer coelutes with 2,3,4,8-tetra-CDF on the Silar-10C column. In this case the SP-2330 column by Supelco can separate these two isomers (Figure 15.3).

In the case of the penta-CDF, it is a problem to separate the 1,2,3,7,8- and 1,2,3,4,8- isomers, which coelute on both the Silar-10C and the SP-2330 columns. However, in this case a 50-m OV-17 column can be used [4].

ISOMER-SPECIFIC MS RESPONSE AND FRAGMENTATION

In 1978 it was reported that PCDD show differences in the lower mass spectra range due to differences in the chlorine substitution [3]. Two chlorine atoms on one ring yields a m/z 74 fragment, three chlorines, a m/z 108 fragment; and four chlorines, a m/z 142 fragment. Among the 22 tetra-CDD, we have thirteen 2:2 isomers, eight 3:1 isomers and only one 4:0 isomer [3].

We have now investigated the mass spectral behavior of a mixture of the 2:2 tetra-CDD isomers. Figure 15.4 illustrates the separation pattern of molecular ions of the isomers using a Silar-10C column and the MS instrument working in the electron impact (EI) mode. Using this column all the isomers were separated, except the 1,4,6,9- and 1,2,7,8- isomers. In this investigation we found the 1,3,7,8- and 2,3,7,8- isomers to be the dominating peaks.

When the same mixture is analyzed on the same instrument working in the negative chemical ionization (NCI) mode, the pattern is quite different (see Figure 15.4), indicating a difference in the response ratios (NCI/EI) for the different tetra-CDD isomers. However, as seen in Table 15.1, for all these isomers the response was much higher in EI than in NCI (NCI/EI $> > 1$). The lowest response ratio was observed for the 2,3,7,8- isomer.

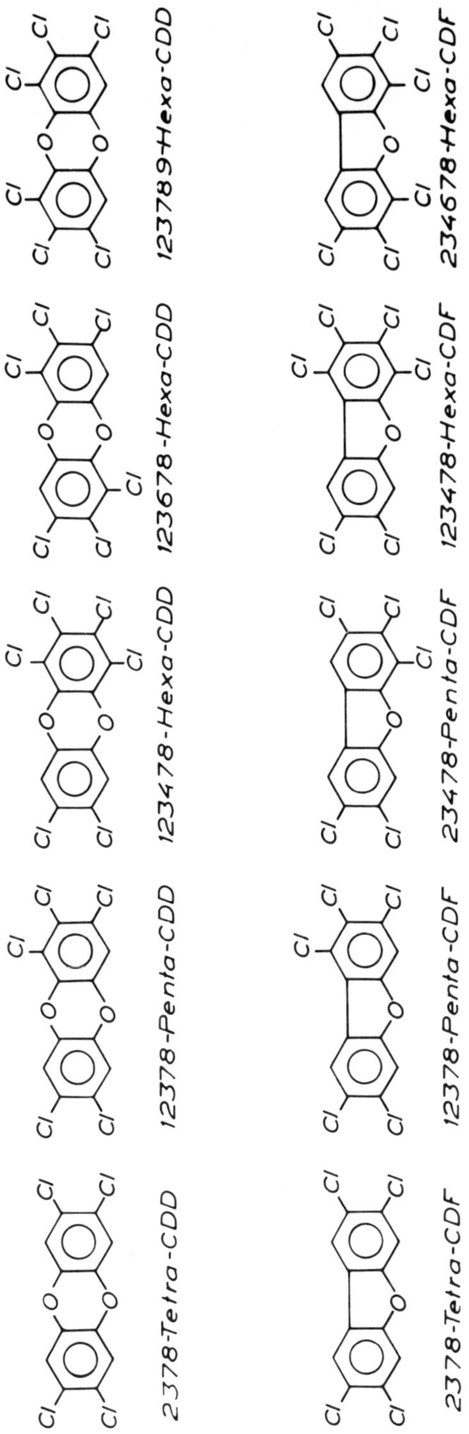

2378-Tetra-CDD

12378-Penta-CDD

123478-Hexa-CDD

123678-Hexa-CDD

123789-Hexa-CDD

2378-Tetra-CDF

12378-Penta-CDF

23478-Penta-CDF

123478-Hexa-CDF

234678-Hexa-CDF

Figure 15.1. The Most Toxic PCDD and PCDF Isomers.

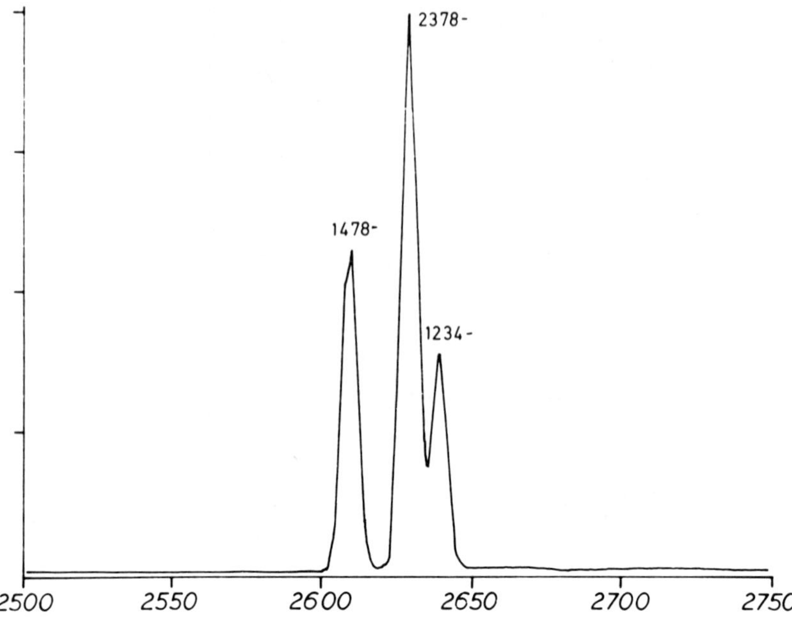

Figure 15.2. Separation of 1,4,7,8-, 2,3,7,8- and 1,2,3,4-Tetra-CDD Using a 55-m Silar-10C Column.

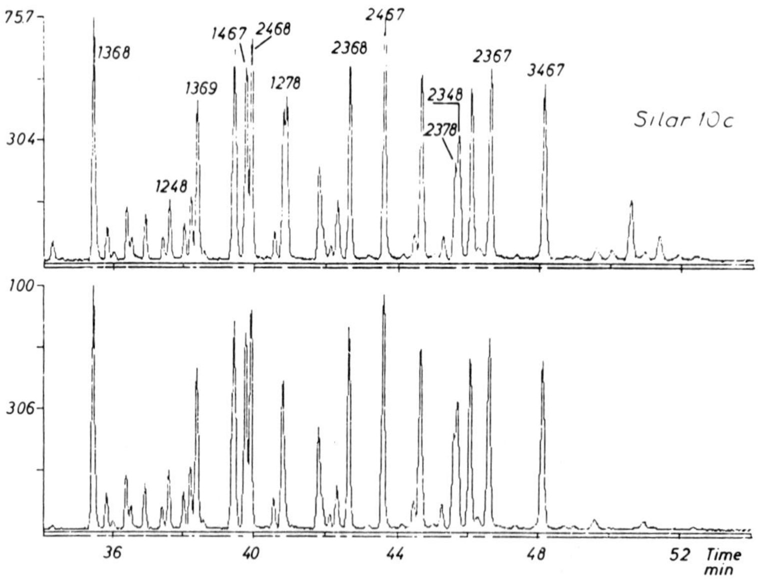

Figure 15.3. Analyses of Tetra-CDF in a Flue Gas Condensate Using 55-m Silar-10C and 60-m Supelco SP-2330 Columns.

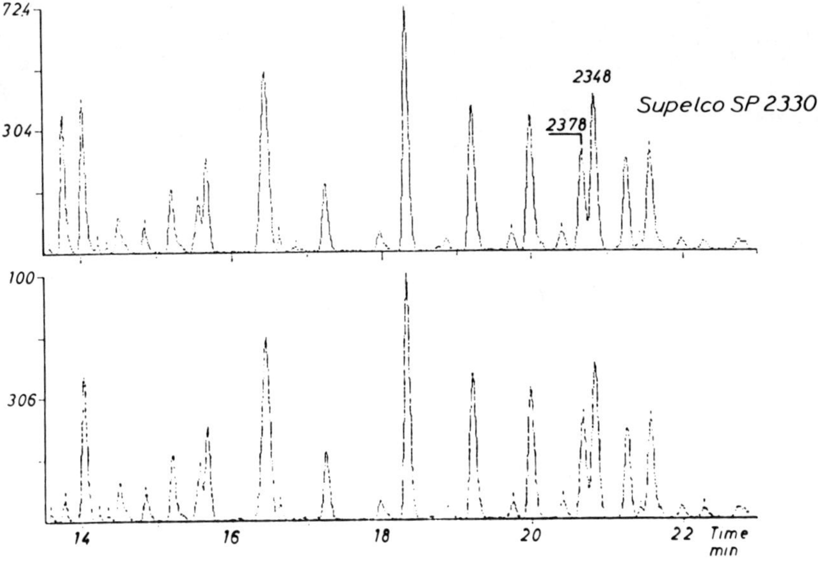

Figure 15.3. *(continued)* Analyses of Tetra-CDF in a Flue Gas Condensate Using 55-m Silar-10C and 60-m Supelco SP-2330 Columns.

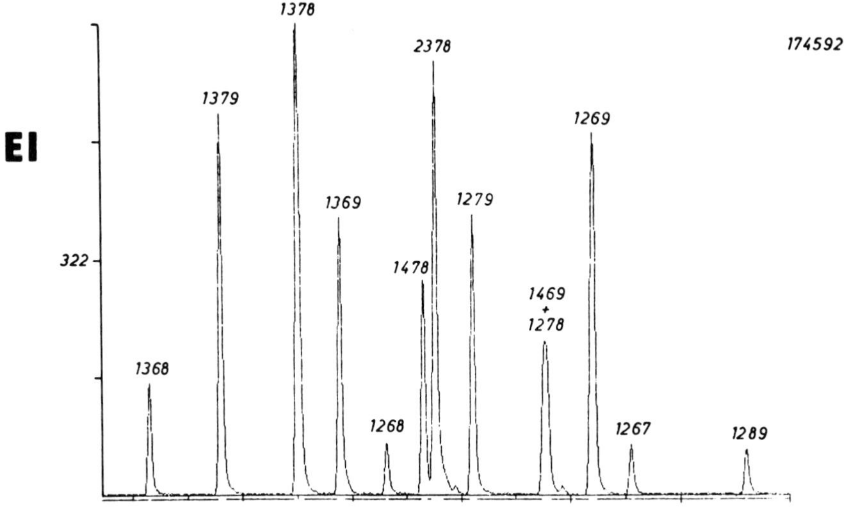

Figure 15.4. ECI and NCI Analyses of 13 2:2 Tetra-CDD Isomers Using a 55-m Silar-10C Column.

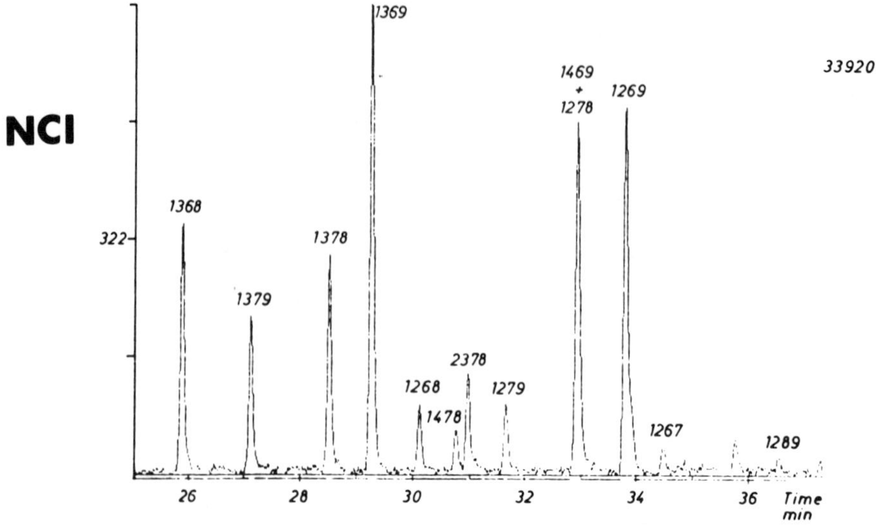

Figure 15.4. *(continued)* ECI and NCI Analysis of 13 2:2 Tetra-CDD Isomers Using a 55-m Silar-10C Column.

Table 15.1. Response Ratios and Fragmentation Ratios (NCI) for Eleven 2:2 Tetra-CDD Isomers

Isomer	*NCI/EI*	*M/M-C1*	*M/M-C1$_2$*
1,3,6,8-	0.48	0.40	0.70
1,3,7,9-	0.090	0.076	0.24
1,3,7,8-	0.089	0.28	0.27
1,3,6,9-	0.32	0.28	0.61
1,2,6,8-	0.30	0.37	0.98
1,4,7,9-	0.047	0.086	0.23
2,3,7,8-	0.049	1.64	0.28
1,2,7,9-	0.053	0.060	0.18
1,2,6,9-	0.23	0.50	0.69
1,2,6,7-	0.12	0.44	0.23
1,2,8,9	0.066	0.11	0.17

Figure 15.5 shows the isomeric separation of the M, M-C1 and M-C1$_2$ fragments for the same tetra-CDD isomers using the NCI technique. We can observe large differences between the isomers. We have calculated the fragmentation ratio M/M-C1 and M/M-C1$_2$, and these values are also given in Table 15.1.

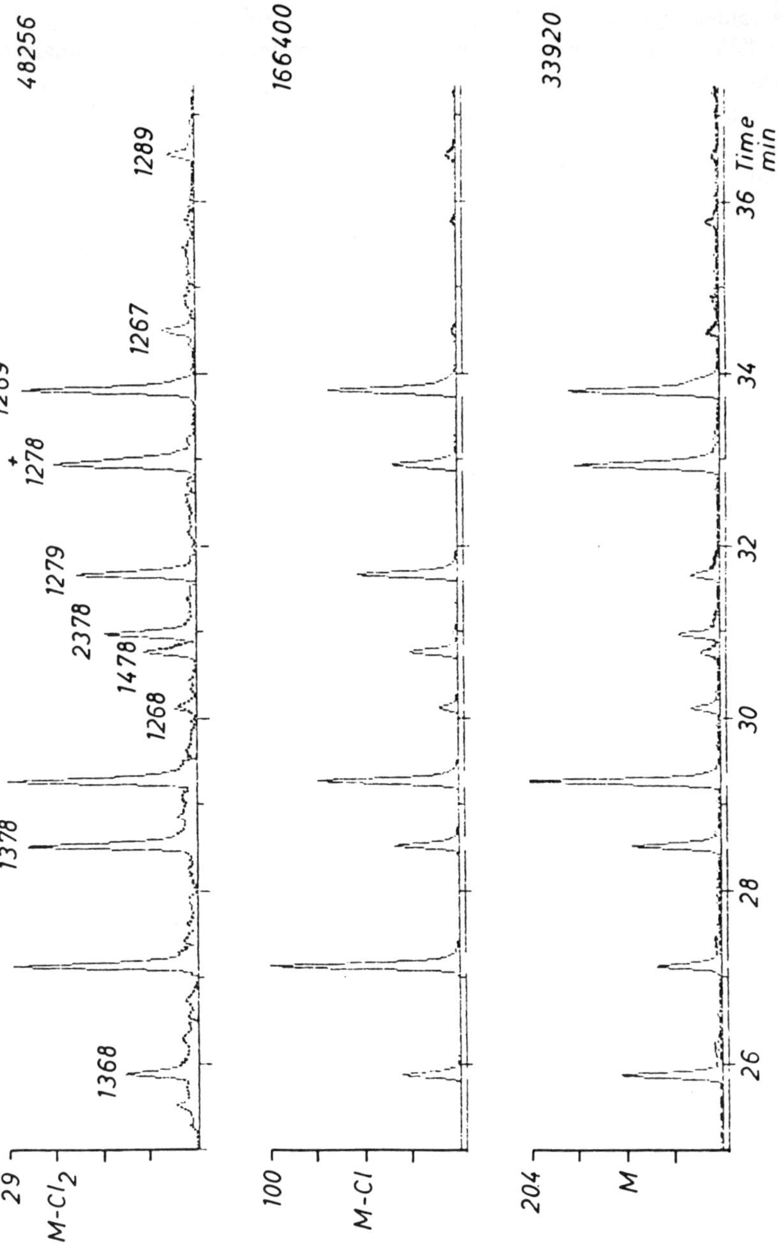

Figure 15.5. M, M-C1 and M-$C1_2$ traces for 13 2:2 Tetra-CDD Isomers Using a 55-m Silar-10C Column.

Of special interest is the observation that the M-Cl fragment is much stronger than the M-ion (M/M-Cl = 0.2), and that for 2,3,7,8-tetra-CDD and M-Cl fragment is very small, but the M-Cl_2 is quite good (see Figure 15.5).

The values reported in Table 15.1 show remarkable differences among the 2:2 tetra-CDD isomers. These isomer-specific parameters are useful supplements to the high-resolution gas chromatography (HRGC) separation for the confirmation of unknown tetra-CDD isomers in trace analyses of environmental and occupational samples.

In our laboratory we have also access to about 30 pure PCDF standards. Carefully weighed standard solutions were prepared, and 100 pg of each isomer could be injected in the GC/MS, the MS operating in the EI or NCI mode.

Figure 15.6 shows the separation of 13 tetra-CDF isomers using a 60-m SP-2330 column, with the MS operating in the EI and NCI mode. We injected 100 pg of each isomer. In the EI analysis we found the difference in peak heights between the isomers to be about a factor of two. 1,3,6,8-Tetra-CDF was found to have the highest response; and the 1,2,6,7- and 2,3,6,7- isomers, the lowest (Table 15.2). In the NCI analyses (Figure 15.6) we found the difference in response between the isomers to be much larger, a 15-fold difference was observed between the 2,3,6,8- and 1,2,7,9- isomers (see Table 15.2).

We have also calculated the response ratios (NCI/EI) for the 13 tetra-CDF isomers, and the values are included in Table 15.2. In contrast to the tetra-CDD

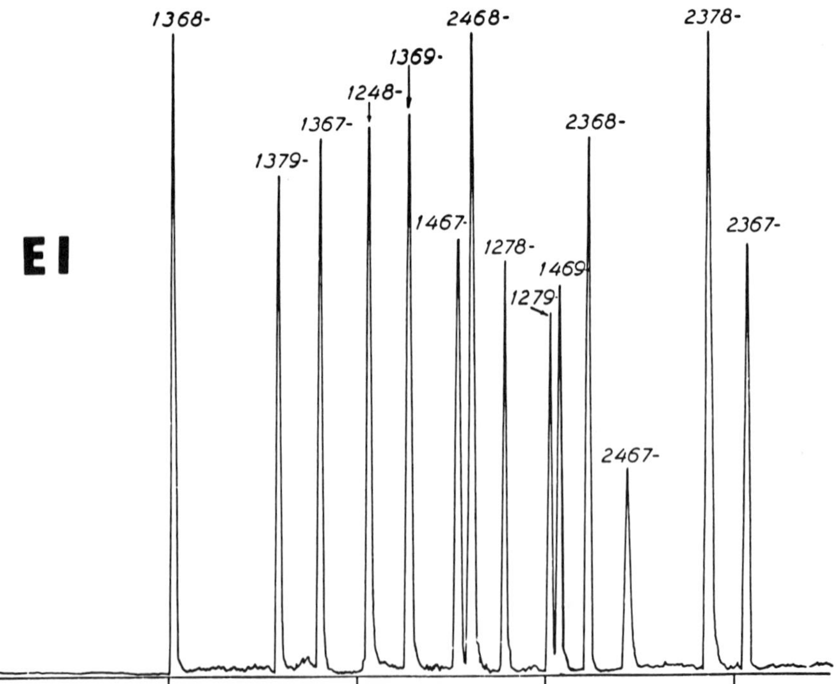

Figure 15.6. Separation and Response of 13 Tetra-CDF (100 pg each) Using a 60-m Supelco SP-2330 Column.

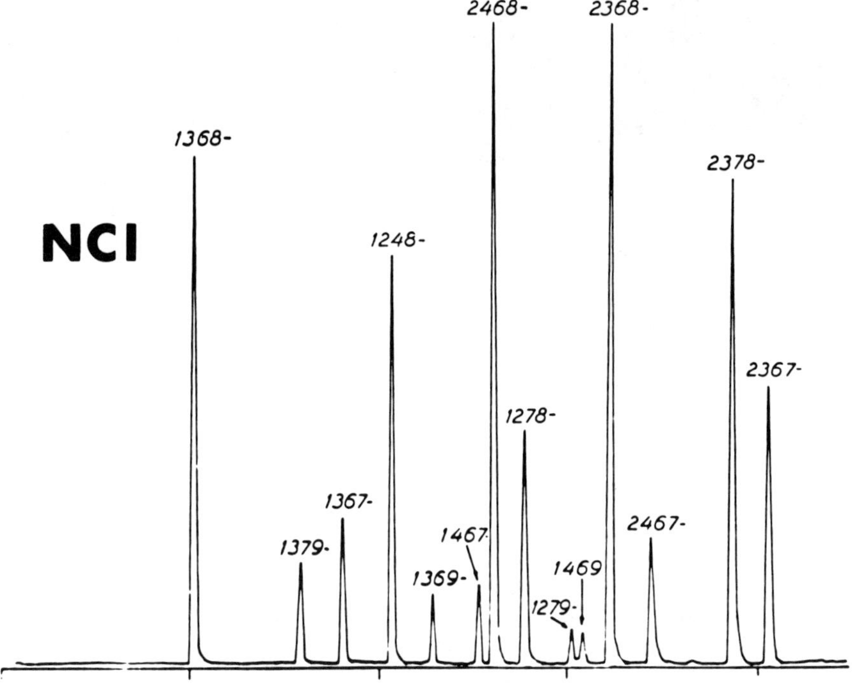

Figure 15.6. *(continued)* Separation and Response of 13 Tetra-CDF (100 pg each) Using a 60-m Supelco SP-2330 Column.

Table 15.2. Peak Areas and Response Ratio for 100 pg of 13 Tetra-CDF (Supelco SP-2330)

	Peak Area		
Isomer	*EI*	*NCI*	*NCI/EI*
1,3,6,8-	32,989	188,198	5.7
1,3,7,9-	28,786	38,128	1.3
1,3,6,7-	30,420	50,404	1.7
1,2,4,8-	23,872	164,101	6.9
1,2,7,8-	25,896	109,960	4.2
1,2,6,7-	19,592	42,855	2.2
2,3,6,8-	25,483	281,546	11.0
2,3,7,8-	24,698	231,206	9.4
2,3,6,7-	19,617	97,476	5.0
1,2,3,7-	22,987	120,795	5.3
1,2,7,9-	22,400	19,168	0.86
1,3,7,9-	27,296	44,457	1.6
1,2,3,8-	25,408	187,012	7.4

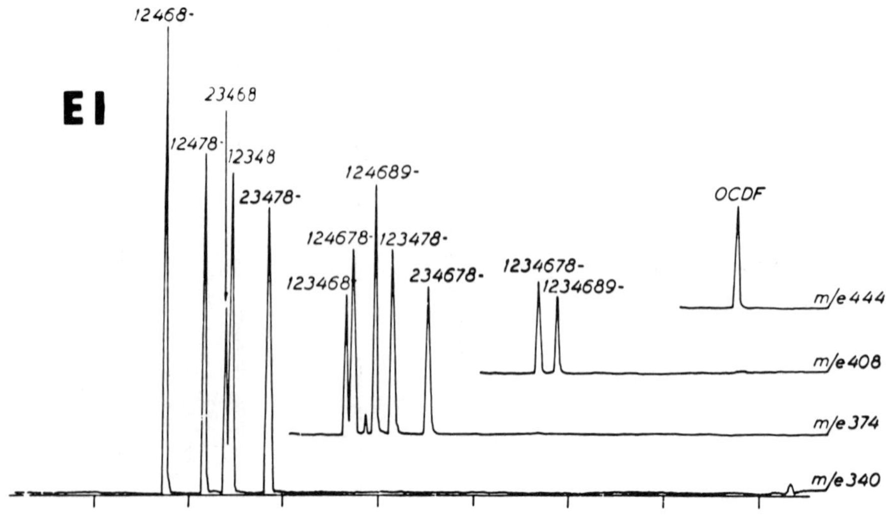

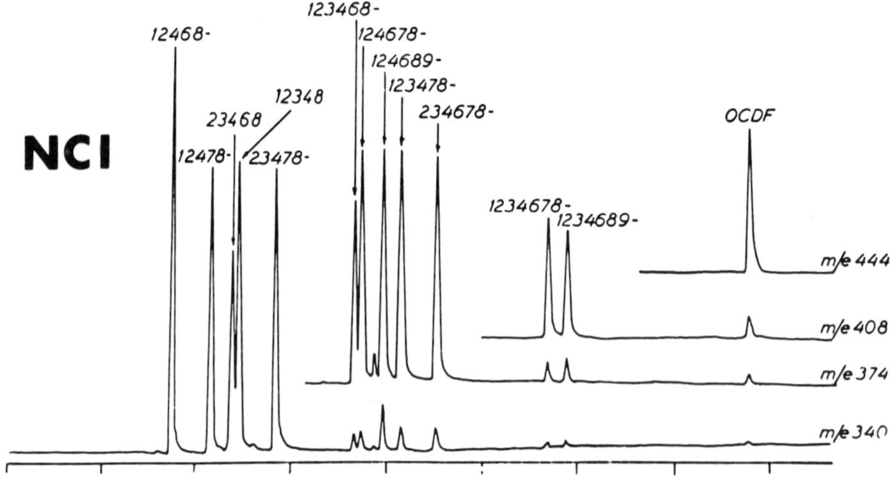

Figure 15.7. Separation and Response of Five Penta-, Five Hexa-, Two Hepta-and Octa-CDF (100 pg each) Using a 25-m SE-54 Column.

(Table 15.1), we found that the tetra-CDF in general have a higher response in NCI than in EI (NCI/EI > 1), the 1,2,7,9- isomer being the only exception among the isomers studied.

Figure 15.7 shows the separation of five penta-, five hexa-, two hepta- and octa-CDF using a 25-m SE-54 column, with the MS operating in the EI and NCI mode. Here we have also injected 100 pg of each isomer. The response factors and

Table 15.3. Peak Areas and Response Ratios for 100 pg of Five Penta-CDF, Five Hexa-CDF, Two Hepta-CDF and Octa-CDF (SE-54)

	Peak Area		
Isomer	*EI*	*NCI*	*NCI/EI*
Penta-CDF			
1,2,4,6,8-	28,820	785,988	27
1,2,4,7,8-	19,150	524,249	27
2,3,4,6,8-	11,533	380,954	33
1,2,3,4,8-	22,671	577,840	26
2,3,4,7,8-	20,040	540,560	27
Hexa-CDF			
1,2,3,4,6,8-	10,499	375,129	36
1,2,4,6,7,8-	13,159	453,573	35
1,2,4,6,8,9-	15,093	342,121	23
1,2,3,4,7,8-	9,814	329,306	34
2,3,4,6,7,8-	15,154	435,710	29
Hepta-CDF			
1,2,3,4,6,7,8-	8,532	345,299	40
1,2,3,4,6,8,9-	8,317	265,231	32
Octa-CDF	16,822	302,123	18

response ratios are given in Table 15.3. The differences between the isomers were much smaller here than for the tetra-CDF (see also Table 15.2).

From our work with pure standards and standard mixtures we can draw the conclusion that EI (or M-C1 in NCI) is the preferred analytical method for tetra-CDD, while NCI is the preferred method for ultratrace analyses of PCDF. In our Finnigan 4021 instrument we have a detection level of less than 0.1 pg for some of the PCDF isomers using methane NCI. Another conclusion is that quantification of individual PCDD and PCDF isomers (especially tetra-CDF with NCI) should be based on individual response factors experimentally found for the specific isomers.

Using the NCI technique and our synthetic standards we have observed a typical $M-19$ fragmentation for hexa-, hepta- and octa-CDF. Figure 15.8 shows the M, $M+2$, $M-19$ and $M-17$ traces for five hexa-CDF standards. The M and $M+2$ ratio is 0.515, in agreement with the theoretical value for a hexachloro compound, while the $M-19$ and $M-17$ ratio is 0.606, indicating a pentachloro fragment (theoretical value 0.617). The M-19 fragment is a $M-Cl+O$ fragment.

Examination of Figure 15.8 shows that the fragmentation ratio $M/M-19$ varies between 0.20 and 0.05 for the different isomers. This fragmentation ratio has been used to confirm the identification of 1,2,3,4,7,8-hexa-CDF at a level of 100 ppt in blood plasma samples from Taiwanese Yusho patients. The ratio found in this sample was 0.19, the value for the synthetic standard was found to be 0.20 [4].

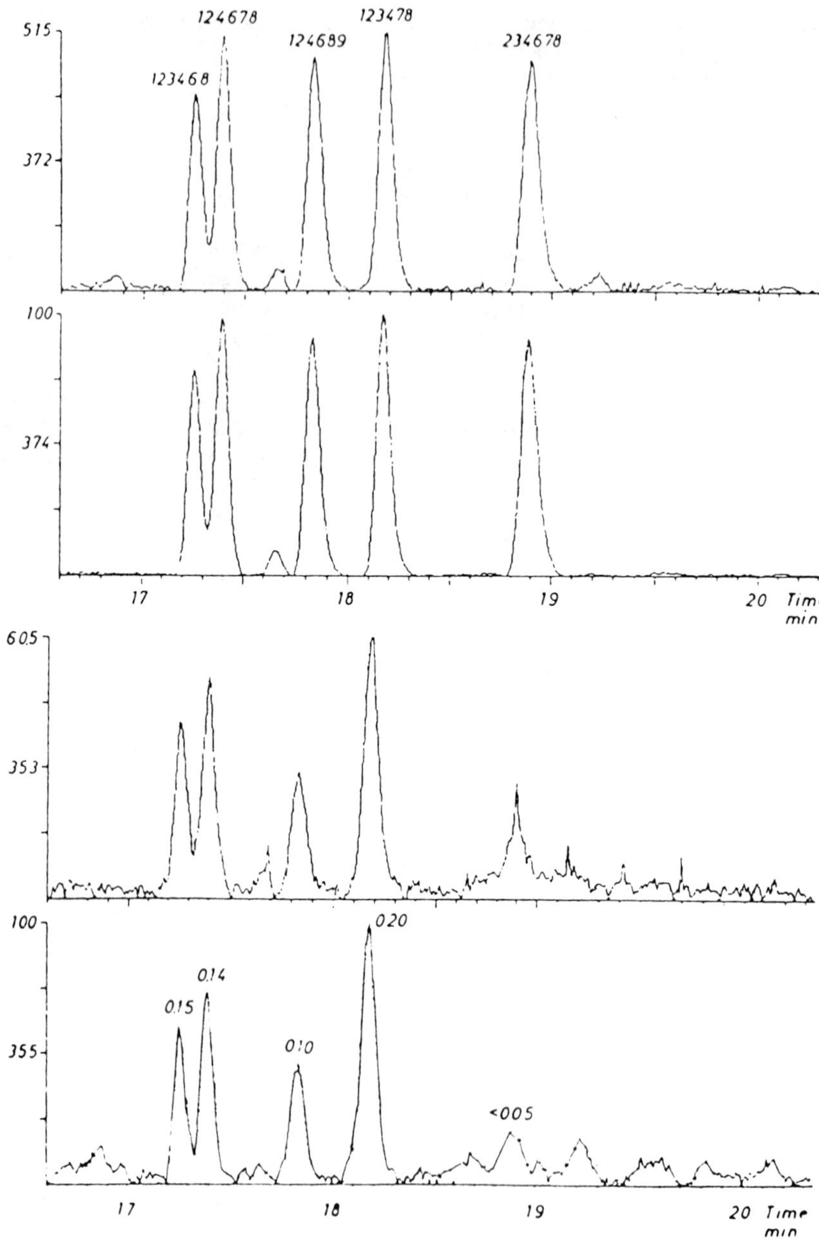

Figure 15.8. M. M+2, M-19 and M-17 Curve for Five Hexa-CDF Isomers Using a 25-m SE-54 Column.

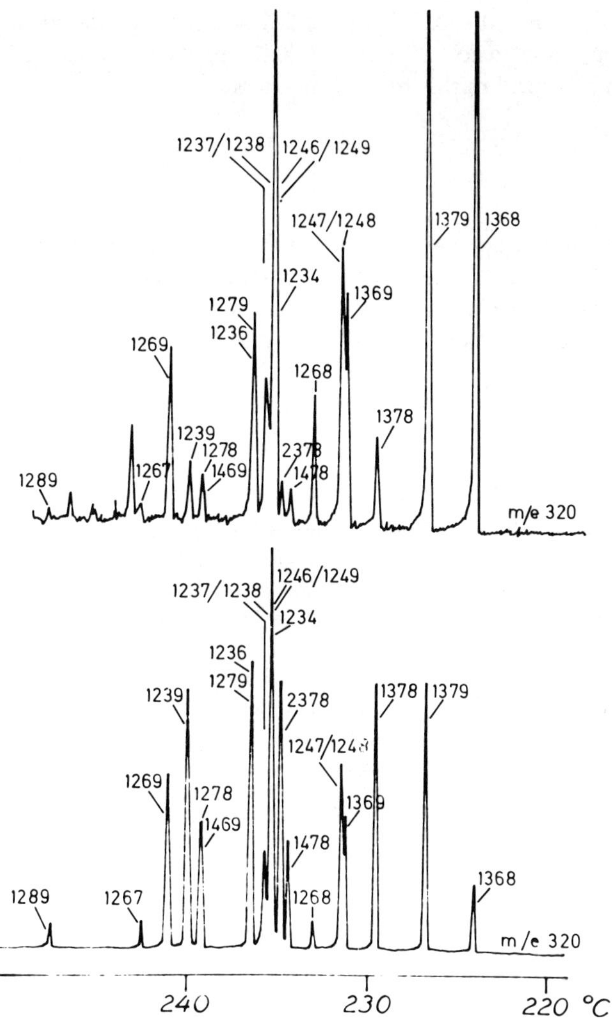

Figure 15.9. Fragmentogram of a Fly Ash Sample (top) and a Synthetic Mixture of 22 TCDD (bottom).

CLUSTER ANALYSES

In the case when a sample contains a multitude of PCDD and PCDF isomers, e.g, fly ash, flue gas condensate and other incineration emissions, positive identification of PCDD and PCDF could be confirmed by cluster analyses. Figure 15.9 shows the fragmentogram of tetra-CDD in a fly ash sample and a mixture of the synthetic tetra-CDD isomers. Perfect overlap is observed, and calculation shows the probability of false positive to be extremely small ($p > 10^{-10}$). This probability decreases drastically by an increase in overlapping peaks; 25 tetra-CDF, 10 penta-CDD and 15-penta-CDF, which also can be found in this type of matrices.

SUMMARY

Polar capillary columns such as Silar-10C and SP-2330 can be used for separation of the 22 tetra-CDD and 38 tetra-CDF isomers. PCDD and PCDF can also be separated by isomer-specific response ratios (NCI/NI) or fragmentation ratios. Cluster analyses can be used when the sample contains a multitude of PCDD and PCDF.

REFERENCES

1. Buser, H.R., and C. Rappe. "High-Resolution Gas Chromatography of the 22 Tetrachlorodibenzo-*p*-dioxin Isomers," *Anal. Chem.* 52:2257–2262 (1980).
2. Buser, H.R., H.-P. Bosshardt, C. Rappe and R. Lindahl. "Identification of Polychlorinated Dibenzofuran Isomers in Fly Ash and PCB Pyrolyses," *Chemosphere* 7:419–429 (1978).
3. Buser, H.R., and C. Rappe. "Identification of Substitution Patterns in Polychlorinated Dibenzo-*p*-dioxins (PCDDs) by Mass Spectrometry," *Chemosphere* 7:199–211 (1978).
4. Rappe, C., M. Nygren and G. Gustafsson. "Human Exposure to Polychlorinated Dioxins and Dibenzofurans," Chapter 21, this volume.

16

Validation Study for the Gas Chromatography/Atmospheric Pressure Ionization/Mass Spectrometry Method for Isomer-Specific Determination of 2,3,7,8-Tetrachlorodibenzo-*p*-dioxin

R.K. Mitchum, W.A. Korfmacher and G.F. Moler

Over the past several years, there has been an increasing interest in the determination of 2,3,7,8-tetrachlorodibenzo-*p*-dioxin (2,3,7,8-tetra-CDD); in response, various laboratories have developed methods for the determination of 2,3,7,8-tetra-CDD at the parts-per-trillion level [1]. Recently, emphasis has shifted to the need for an isomer-specific method for 2,3,7,8-tetra-CDD, and a few laboratories [2–6], including our own [2], have responded to this requirement.

Buser and Rappe [3] described an isomer-specific method for 2,3,7,8-tetra-CDD using high-resolution gas chromatography (GC) combined with low-resolution electron impact mass spectrometry (MS). The isomer-specific method for 2,3,7,8-tetra-CDD reported by Lamparski and Nestrick [6] is based on high-performance liquid chromatography (HPLC) combined with low-resolution GC and low-resolution electron impact MS.

An alternative method of mass spectral ionization using negative chemical ionization (NCI) MS has been employed in the determination of 2,3,7,8-tetra-CDD [2,7–9]. Mitchum et al. [2] demonstrated that the combination of high-resolution GC and negative ion atmospheric pressure ionization (NIAPI) MS (in which an oxygen-rich plasma is produced) will also provide a method for isomer-specific determination of 2,3,7,8-tetra-CDD. In this report, the use of GC/NIAPI/MS methodology for 2,3,7,8-tetra-CDD determination is summarized and the preliminary results of a major validation study of this analytical method are presented.

EXPERIMENTAL

Apparatus

The combined GC/NIAPI/MS apparatus has been described in detail in previous reports [2,8]. Briefly, this system consisted of a 50-m SP-2100 (Hewlett-Packard, No. 19091-60050) 0.1-mm i.d. fused-silica capillary GC column interfaced directly into the API source. The GC injector was a moving needle injector (Chrompack Model No. 9001). The fused-silica column was attached directly to the moving needle injector via a silicone rubber adhesive sealant [10].

The gas chromatograph (modified Hewlett-Packard 5710) was operated isothermally at 220°C. The GC carrier gas flowrate was 0.5 cm^3/min (the inlet head pressure was set at 50 psi). The GC effluent flowed directly into the API source (Figure 16.1) and was combined with makeup gas to give a total flowrate of 18 cm^3/min into the API source. The makeup gas consisted of 0.1% oxygen in ultrahigh-purity nitrogen.

The mass spectrometer and data system have been previously described [8]. For the present experiments, the selected ion monitoring data were obtained using a window of $\pm$0.125 amu about each ion centroid. For most of the analyses, the ions monitored were m/z 176, 178 and 182, with integration times of 0.2, 0.2 and 0.1 sec, respectively.

Reagents

The uniformly ring-labeled ^{13}C-2,3,7,8-tetra-CDD (Midwest Research Institute, Contract No. 222–76–2037[C]) was found to be 97.8 atom % ^{13}C with an isotopic distribution of 78.4% $^{13}C_{12}$, 17.4% $^{13}C_{11}$ and 4.2% $^{13}C_{10}$. The unlabeled 2,3,7,8-tetra-CDD was obtained from ECO-Control, Inc. (Cambridge, Massachusetts), and was used as received. The uniformly ring-labeled ^{14}C-2,3,7,8-tetra-CDD was obtained from KOR Isotopes and had a specific activity of 107 Ci/mol.

All solvents used were Burdick and Jackson "distilled-in-glass" grade, or equivalent. Other chemicals were ACS grade.

Sample Cleanup Procedure

The method used for the cleanup of fish samples is a modification of the method reported previously [8]. The current sample cleanup method consists of four steps: (1) homogenization and digestion, (2) extraction, (3) silica HPLC, and (4) reverse-phase HPLC. The final separation and quantification were performed by GC/NIAPI/MS.

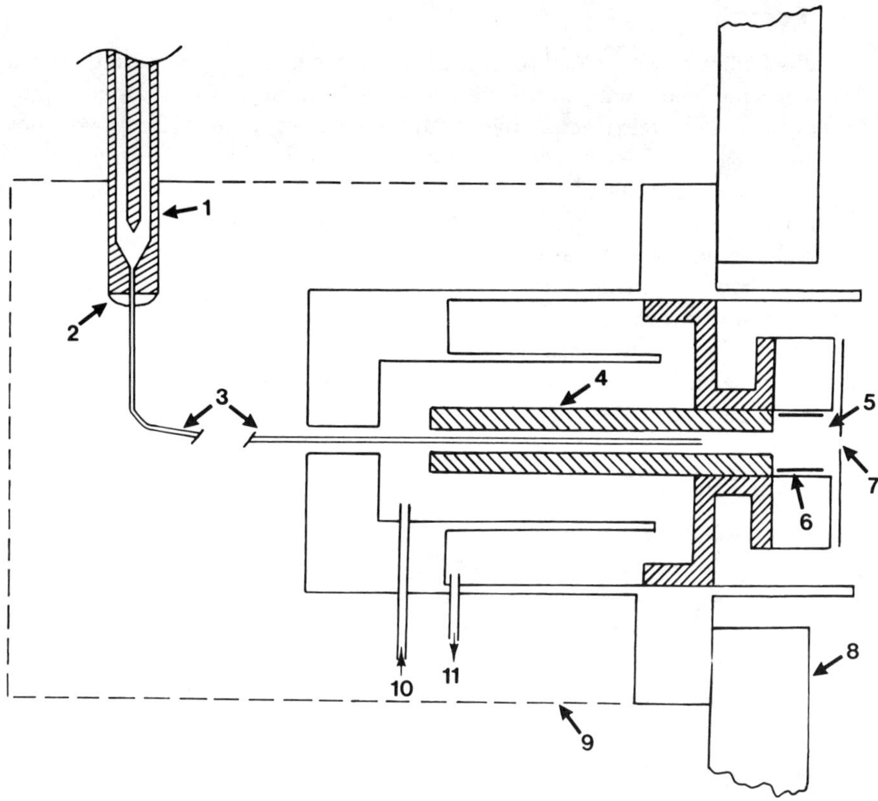

Figure 16.1 GC/NIAPI/MS Inlet Schematic: (1) Moving Needle Injector; (2) Silicone Rubber Adhesive Sealant; (3) 50-m Fused-Silica Capillary Column; (4) Quartz Injector Tube; (5) Source Volume; (6) ^{63}Ni Foil; (7) 50-μm Aperture; (8) Vacuum Housing; (9) GC Oven Wall; (10) Makeup Gas Inlet; (11) Overflow Gas Outlet.

Homogenization and Digestion

A large quantity of fish (200–500 g) was homogenized in a blender and 10-g samples were weighed out into 250-mL Erlenmeyer flasks for analysis. After addition of about 50 mL of deionized water (DW), about 30 mL of 50% aqueous KOH solution was added and the sample was stirred over low heat until most of the tissue was digested. At this point the ^{13}C-2,3,7,8-tetra-CDD spike was added to the sample at a level of 100 ppt. (For 2,3,7,8-tetra-CDD-spiked fish samples, the unlabeled 2,3,7,8-tetra-CDD was also added at this time.) Then, another 10 mL of 50% KOH solution was added and the sample was stirred for an additional 5–10 min while it was allowed to cool.

Extraction

Benzene (100 mL) was added to the flask, and the mixture was stirred for 5 min. The benzene phase was transferred to a 500-mL separatory funnel, and the homogenate was reextracted two additional times with 100-mL portions of benzene. The combined benzene extracts were washed with DW, then acidified DW and again with DW. The benzene extract was then passed through a column consisting of anhydrous Na_2SO_4 (2.54 x 8 cm) on top of a mixed bed of 50% acidic alumina and 50% basic alumina (2.54 x 6 cm).

The benzene extract was then reduced in volume in a 400-mL beaker on a hot plate and under a stream of air in a hood. The concentrated extract was transferred to to 2-mL reaction vial. The extract was then further reduced in volume under a stream of N_2, using toluene to rinse down the sides of the vial. This resulted in a final extract volume of 10 μL with toluene as the solvent.

Silica HPLC

The 10- μL toluene sample was injected into the silica HPLC column and the fraction corresponding to 2,3,7,8-tetra-CDD was collected in a 2-mL vial. For this work, an Altex "Ultrasphere" 25-cm x 4.6-mm i.d. silica HPLC column was used with hexane as the eluant at a flowrate of 2.0 mL/min. 2,3,7,8-Tetra-CDD eluted with the solvent front. The silica HPLC column was cleaned with CH_3CN between fish sample runs.

Reverse-Phase HPLC

The silica HPLC fraction was reduced in volume with purified N_2 using toluene to rinse down the sides of the vial. This resulted in a 10- μL solution with toluene as the solvent. This solution was injected into the reverse-phase HPLC system and the fraction corresponding to the 2,3,7,8-tetra-CDD retention time was collected in a 2-mL vial. The reverse-phase HPLC system consisted of two Altex "Ultrasphere" ODS 25-cm x 4.6-mm i.d. columns connected in series. The flowrate was set at 1.0 mL/min with acetonitrile as the eluant. 2,3,7,8-Tetra-CDD typically eluted at 15.5 min under these conditions. The reverse-phase HPLC system was cleaned with $CHCl_3$ between fish sample runs.

GC/NIAPI/MS

The fraction collected from reverse-phase HPLC was evaporated to 10 μL under purified N_2 while the sides of the vial were washed with toluene. Then 20–100% of the sample was injected into the GC/NIAPI/MS system for analysis.

Sample Quantification

Quantification was performed by obtaining the ratio of the peak areas for (unlabeled) 2,3,7,8-tetra-CDD and ^{13}C-2,3,7,8-tetra-CDD for a sample and comparing this to the standard curve. These quantification steps were performed by using the UNKNOWN program [11,12].

Recovery Experiments

Several recovery experiments were performed using ^{14}C-2,3,7,8-tetra-CDD. In these experiments, a fish sample was spiked with ^{14}C-2,3,7,8-tetra-CDD at a level equivalent to about 200 ppt of 2,3,7,8-tetra-CDD. In separate experiments, ^{14}C-2,3,7,8-tetra-CDD–spiked fish samples were extracted: two times, and then a third time; three times, a fourth time, and then a fifth time. These five extract sets were each washed and carried through a Na_2SO_4/alumina column step. The sample volume was reduced to 100 μL under a stream of nitrogen with toluene as the final solvent and then transferred to a 20-mL scintillation vial containing 10 mL of scintillation cocktail (Scintisol). These samples were counted using a liquid scintillation counter (Beckman LS7500) and compared to a ^{14}C-2,3,7,8-tetra-CDD standard, which represented a 100% recovery.

In additional experiments, one ^{14}C-2,3,7,8-tetra-CDD–spiked fish sample was carried through the silica HPLC cleanup step, and another was carried through the reverse-phase HPLC cleanup step. These samples were then counted to monitor the losses at each step in the cleanup. In each of these experiments, the sample vial and syringe used for injection into the HPLC were later rinsed several times with 100-μL portions of toluene. These rinses were combined in scintillation vials and counted to determine the amount of ^{14}C-2,3,7,8-tetra-CDD that was "lost" in the concentration and transfer steps.

Other Validation Experiments

A series of four standard solutions was analyzed by the GC/NIAPI/MS system. Each standard solution was injected three times. In a separate study, another standard solution was analyzed by GC/NIAPI/MS sixteen times during a one-month period.

In an interlaboratory study, our laboratory received a series of solutions containing 2,3,7,8-tetra-CDD plus, in selected samples, Aroclor 1254, DDE and other tetra-CDD isomers. These 60-μL solutions represented the amount of 2,3,7,8-tetra-CDD that would be obtained from extracting a 2-g sample with a recovery of 100%. The samples were spiked (by the outside laboratory) in the range of 0–200 ppt with 2,3,7,8-tetra-CDD and at 200 ppt with ^{13}C-2,3,7,8-tetra-CDD. These samples were analyzed directly by GC/NIAPI/MS.

In another interlaboratory study, our laboratory received four fish extracts that had already undergone a sample cleanup process. These samples were analyzed by GC/NIAPI/MS. The results were compared with those obtained by eight other laboratories that participated in a round-robin study of these four fish samples. (One of the eight participating laboratories sent the four fish extracts to our laboratory after cleanup and analysis by their method.)

RESULTS AND DISCUSSION

At atmospheric pressure, in an oxygen-rich plasma, the major negative ions are formed from both electron capture and clustering processes:

$$O_2 + e^- + N_2 \rightarrow O_2^- + N_2 \tag{1}$$

$$O_2^- + O_2 \rightleftharpoons O_4^- \tag{2}$$

$$O_2^- + N_2 \rightleftharpoons O_2N_2^- \tag{3}$$

$$O_2N_2^- + N_2 \rightleftharpoons O_2N_4^- \tag{4}$$

The relative abundance of these ions is a function of both O_2 concentration and ion source temperature. Under the conditions used in this work, 2,3,7,8-tetra-CDD reacts with O_2^- as shown in Figure 16.2 [2,8].

The NIAPI mass spectrum of 2,3,7,8-tetra-CDD in an oxygen-rich plasma is presented in Figure 16.3 (using the system described by Mitchum and Korfmacher [13]). For quantification, 2,3,7,8-tetra-CDD is monitored at m/z 176 and ^{13}C-2,3,7,8-tetra-CDD at m/z 182. In addition, m/z 178 is monitored as a confirmatory ion for 2,3,7,8-tetra-CDD. Figure 16.4 shows the selected ion chromatograms obtained from injection of a standard solution containing 150 pg of 2,3,7,8-tetra-CDD and 320 pg of ^{13}C-2,3,7,8-tetra-CDD into the GC/NIAPI/MS system.

Figure 16.5 shows the standard curve obtained from a series of standards with a 2,3,7,8-tetra-CDD/^{13}C-2,3,7,8-tetra-CDD mass ratio range of 0.024–4.7. A blank standard (containing only ^{13}C-2,3,7,8-tetra-CDD) was also run. The standard curve parameters were generated by the program STANDARD [11,12]. Figure 16.5 also demonstrates that the linearity of the GC/NIAPI/MS system was very good over this range of ratios. In making this standard curve, no correction was made for the isotropic purity of ^{13}C-2,3,7,8-tetra-CDD because the same ^{13}C-2,3,7,8-tetra-CDD was used for all standards and samples, and so the correction factor was canceled out.

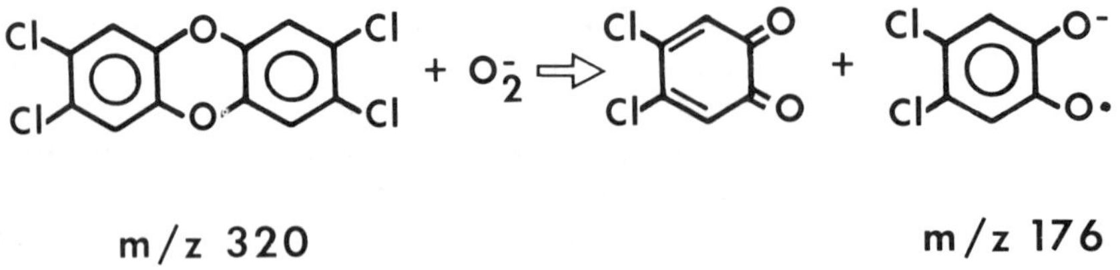

m/z 320 m/z 176

Figure 16.2. Reaction of 2,3,7,8-Tetra-CDD with O_2^-.

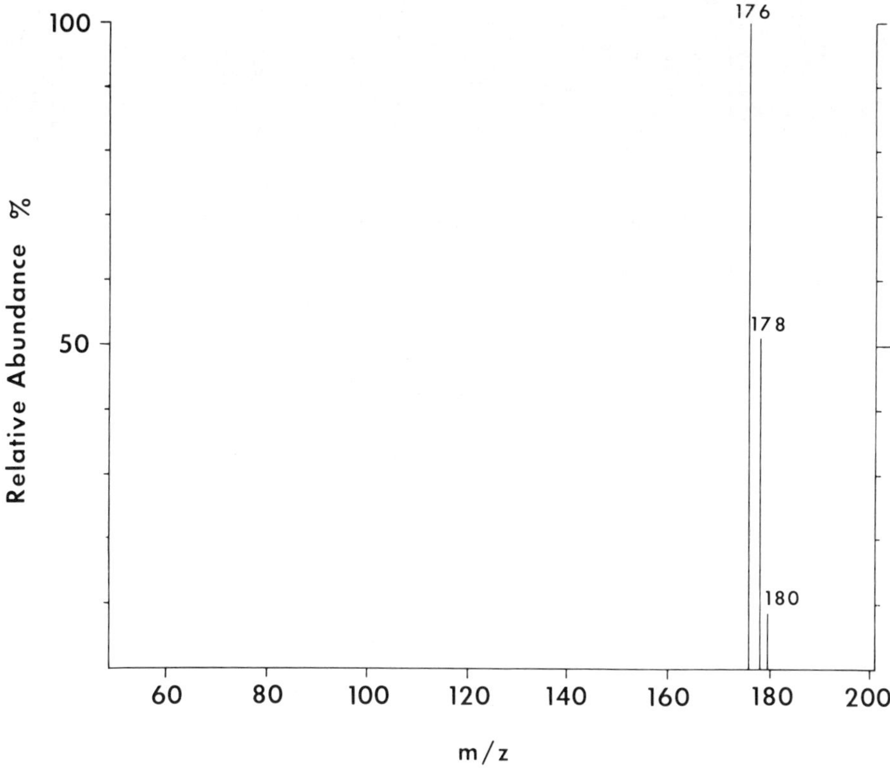

Figure 16.3. NIAPI Mass Spectrum of 2,3,7,8-Tetra-CDD [13].

Mitchum et al. [2] demonstrated that the GC/NIAPI/MS technique is isomer-specific for 2,3,7,8-tetra-CDD. As stated in that report and shown in Figure 16.6 [2], only tetra-CDD isomers with two chlorines in each benzene ring form the m/z 176 ion under the described NIAPI/MS conditions. Thus, only the thirteen 2:2 tetra-CDD are monitored at m/z 176, and 2,3,7,8-tetra-CDD is easily separated (by fused-silica GC) from this smaller group of tetra-CDD.

Validation data for 2,3,7,8-tetra-CDD determination methodologies have been reported [14–17]. The full validation study for the GC/NIAPI/MS method for determination of 2,3,7,8-tetra-CDD is summarized in Table 16.1. This chapter details the results obtained in the first four parts of this study. When the study is completed, a full report will be published. As shown in Table 16.1, this validation study includes both intra- and interlaboratory portions. The goal of this study is to validate both the cleanup and the GC/NIAPI/MS methodology. The first step of the study measures the recovery of the cleanup method. Parts 2–4 of this study test the validity of the GC/NIAPI/MS procedure, and parts 5 and 6 of this study measure the validity of the extraction and cleanup methodology combined with the GC/NIAPI/MS quantification step.

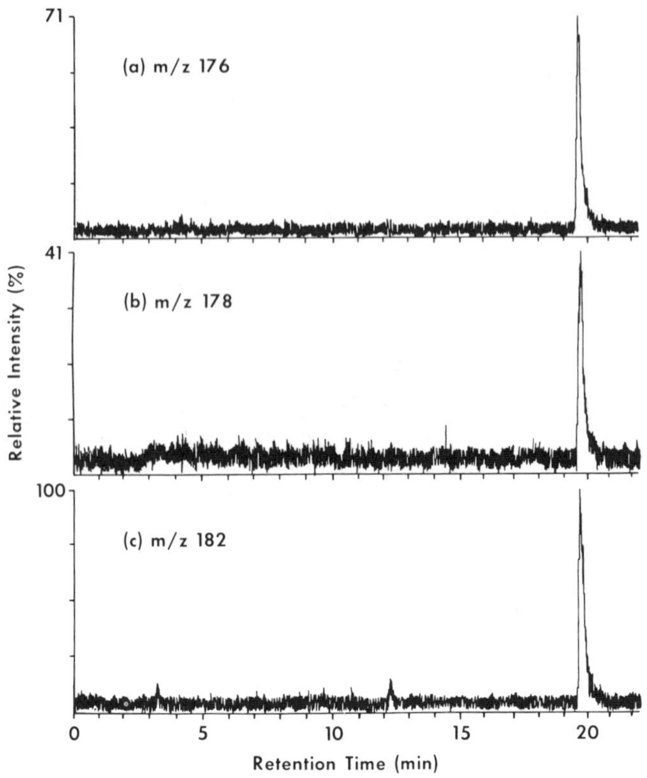

Figure 16.4. GC/NIAPI/MS Response for a Standard Mixture of 0.15 ng of 2,3,7,8-tetra-CDD and 0.32 ng of ^{13}C-2,3,7,8-tetra-CDD: (a) Response at m/z 176; (b) Response at m/z 178; (c) Response at m/z 182.

Figure 16.7 shows the cleanup scheme used in this study for the catfish samples. The recovery at various stages of this cleanup procedure was examined by ^{14}C-2,3,7,8-tetra-CDD. The results of these experiments are summarized in Table 16.2. As shown in Table 16.2, the recovery of a ^{14}C-2,3,7,8-tetra-CDD-spiked catfish sample extracted three times with benzene and carried through the Na_2SO_4/alumina cleanup step was 81%. When the same sample was extracted a fourth time, the gain was an additional 11%. On the other hand, when a catfish sample was extracted only two times and taken through the Na_2SO_4/alumina cleanup step, the recovery was only 67%. Therefore, it is important to extract the fish sample at least three times with benzene.

As a sample is taken further through the cleanup procedure, the recovery decreased. For a spiked catfish sample carried through the silica HPLC step, the

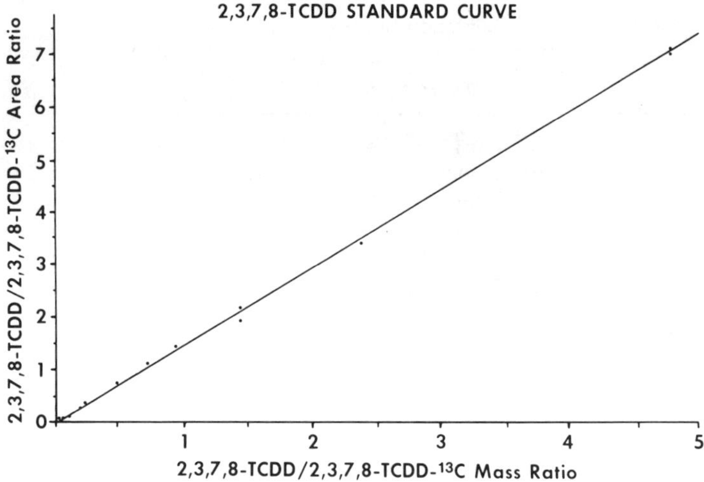

Figure 16.5. Standard Curve (26 Points) Showing (m/z 176)/(m/z 182) Peak Area Ratios vs. (2,3,7,8-tetra-CDD)/(^{13}C-2,3,7,8-tetra-CDD) Mass Ratios of Standards.

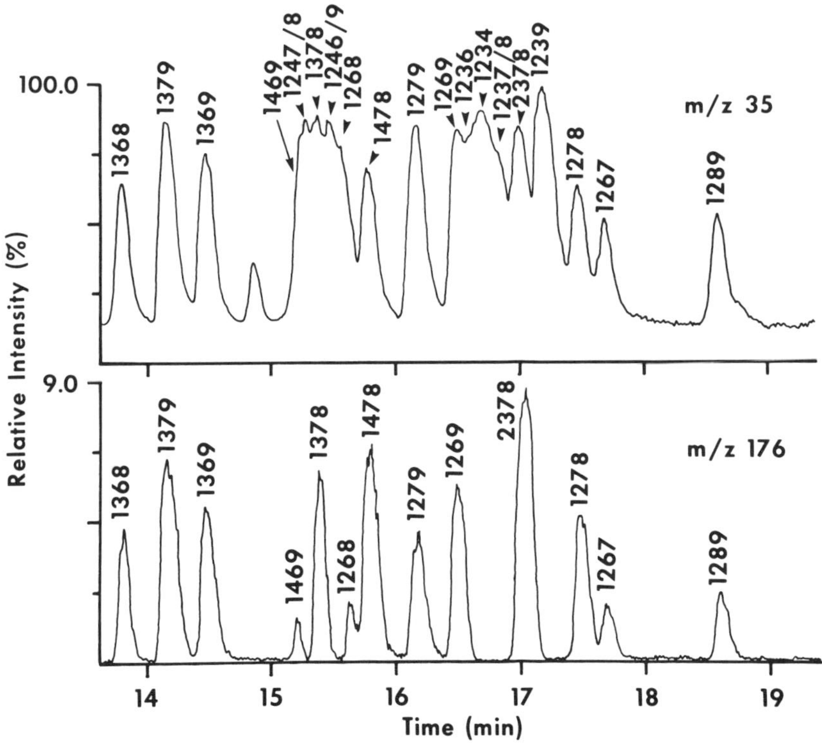

Figure 16.6. GC/NIAPI/MS Response for the 22 Tetra-CDD Monitoring at m/z 35 (Upper Trace) and m/z 176 (Lower Trace) [2].

Table 16.1. GC/NIAPI/MS Validation Study

Part	*Description*	*Type*
1	^{14}C-2,3,7,8-TCDD Recovery Study	Intralaboratory
2	Standard Solutions	Intralaboratory
3	Spiked Solutions	Interlaboratory
4	Fish Extracts	Interlaboratory
5	Spiked Catfish Samples	Intralaboratory
6	Spiked Fish Samples	Interlaboratory

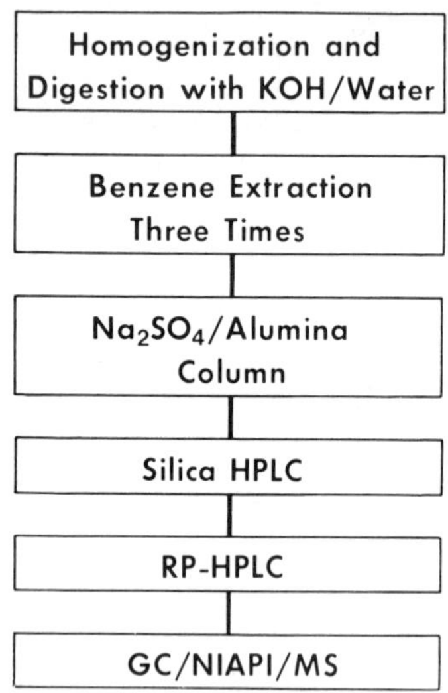

Figure 16.7. Sample Cleanup and Analysis Scheme.

recovery was 57% (see Table 16.2), but the recovery dropped to 36% for a sample carried through the entire cleanup procedure. At least half of the losses after the initial extraction step could be attributed to 2,3,7,8-tetra-CDD remaining in the HPLC injection syringe or the reaction vial in which the sample was concentrated before being injected into the HPLC (see experiments 7 and 9 in Table 16.2).

Table 16.3 lists the results obtained in the first half of the intralaboratory standard solution study. This is the easiest test of the quantification step because it involves merely analyzing standard solutions of 2,3,7,8-tetra-CDD and ^{13}C-2,3,7,8-tetra-CDD, which were generated in our laboratory. The results of this study are in good agreement with the ratios in which the standards were prepared.

Table 16.2. ^{14}C-2,3,7,8-TCDD Recovery Experiments[a]

Exp. No.	*Description*[b]	*% Recovery*
1	Extract catfish B1 two times and cleanup extract to the Si HPLC step[c]	67
2	Extract catfish B1 a third time and cleanup extract to the Si HPLC step[c]	12
3	Extract catfish B2 three times and cleanup extract to the Si HPLC step[c]	81
4	Extract catfish B2 a fourth time and cleanup extract to the Si HPLC step[c]	11
5	Extract catfish B2 a fifth time and cleanup extract to the Si HPLC step[c]	4
6	Extract catfish B3 three times and cleanup extract to the reverse-phase HPLC step[c]	57
7	Rinse the HPLC sampling syringe and the reaction vial used in experiment 6 for the Si HPLC injection step	14
8	Extract catfish B4 three times and cleanup extract to the GC/NIAPI/MS step[c]	36
9	Rinse the HPLC sampling syringe and the reaction vial used in experiment 8 for the reverse-phase HPLC step	10

[a]All catfish samples were spiked with ^{14}C-2,3,7,8-tetra-CDD at a level of 150–200 ppt (the catfish samples ranged from 10 to 13 g).

[b]For additional description, see the Experimental section.

[c]Not including this step.

Table 16.3. Intralaboratory Standard Solution Study

Experiment	*Standard Ratio*	*Result by GC/NIAPI/MS*[a]
1		
A	0.0479	0.0504 ± 0.0087
B	0.0479	0.0470 ± 0.0083
C	0.0479	0.0494 ± 0.0078
Weighted Mean		0.0489 ± 0.0058
2		
A	0.191	0.215 ±0.023
B	0.191	0.198 ± 0.019
C	0.191	0.210 ± 0.021
Weighted Mean		0.207 ± 0.014

Table 16.3. *(continued)* Intralaboratory Standard Solution Study

Experiment	*Standard Ratio*	*Result by GC/NIAPI/MS*[a]
3		
A	0.957	0.956 ± 0.073
B	0.957	0.972 ± 0.080
C	0.957	0.997 ± 0.079
Weighted Mean		0.973 ± 0.058
4		
A	2.39	2.33 ± 0.22
B	2.39	2.38 ± 0.24
C	2.39	2.38 ± 0.24
Weighted Mean		2.36 ± 0.16

[a] ± bands are 95% confidence limits [12].

Table 16.4. GC/NIAPI/MS Stability Study

Day	*Standard Ratio*	*Result*[a]
1	0.479	0.493 ± 0.067
4	0.479	0.477 ± 0.051
5	0.479	0.450 ± 0.069
5	0.479	0.526 ± 0.056
11	0.479	0.503 ± 0.053
13	0.479	0.504 ± 0.051
14	0.479	0.500 ± 0.048
14	0.479	0.498 ± 0.052
15	0.479	0.494 ± 0.049
21	0.479	0.500 ± 0.043
21	0.479	0.471 ± 0.045
26	0.479	0.528 ± 0.045
27	0.479	0.479 ± 0.046
28	0.479	0.464 ± 0.040
28	0.479	0.479 ± 0.044
29	0.479	0.474 ± 0.036
Weighted Mean		0.489 ± 0.012

[a] ± bands are 95% confidence limits [12].

Table 16.4 lists the results of the second half of the standard solution study. This step consisted of analyzing one standard solution sixteen times over a one-month period. This series tested the stability and reproducibility of the GC/NIAPI/MS instrument. As the data show, the day-to-day stability and reproducibility of the quantification system was very good.

In the third part of the validation study, our laboratory received a series of solutions containing various levels of 2,3,7,8-tetra-CDD plus, in selected samples, Aroclor 1254, DDE and other isomers of tetra-CDD. This was a blind study. Table 16.5 presents a representative listing of the results obtained in this study. Considering the added difficulties of an interlaboratory study, these results are good. In only one case (sample 7) was there a large difference between the spiked level of the solution and the level reported by the GC/NIAPI/MS method.

In the fourth part of this study, our laboratory analyzed four fish extracts that had already been carried through a sample cleanup process. Table 16.6 lists the results of our 2,3,7,8-tetra-CDD quantification for these extracts and com-

Table 16.5. Spiked Solution Study[a]

Sample	*2,3,7,8-TCDD Spike Level (ppt)*	*Results by GC/NIAPI/MS Method (ppt)*[b]
1	6	ND (3.0)[c]
2	15	20 ± 5.7
3	150	168 ± 15.1
4	6	ND (2.8)
5	30	28 ± 4.0
6	30	42 ± 5.5
7	150	65 ± 7.3
8	30	41 ± 5.5
9	150	169 ± 13.9
10	75	91 ± 8.6

[a]Results are based on a ^{13}C-2,3,7,8-tetra-CDD spike of 200.0 ppt.
[b]± bands are 95% confidence limits [12].
[c]ND = none detected; the number in parentheses is the detection limit for the analysis.

Table 16.6. Fish Extract Study

Sample	*Results[a,b] by GC/NIAPI/MS Method (ppt)*	*Results[c,d] of Eight Laboratories (ppt)*
1	53.5 ± 4.7	61.2 ± 8.5
2	ND (2.3)	1
3	32.1 ± 3.6	30.4 ± 5.6
4	31.1 ± 3.3	32.3 ± 8.2

[a]Results are based on a ^{13}C-2,3,7,8-tetra-CDD spike of 100.0 ppt.
[b]± bands are 95% confidence limits [12].
[c]Results obtained by a group of eight laboratories [18], not including GC/NIAPI/MS data.
[d]± bands are standard deviations.
[e]ND = none detected; the number in parentheses is the detection limit for the analysis.

pares these to the results obtained by eight other laboratories that participated in the round-robin study [18]. As these data show, our results compared very well with the average values.

The final two parts of the validation study are still in progress. A final report of the validation of this method for the determination of 2,3,7,8-tetra-CDD will be prepared when the study is completed. The results in this initial report show that the GC/NIAPI/MS quantification step works very well. The final report will provide further validation of the quantification step as well as validating the cleanup procedure.

ACKNOWLEDGMENTS

The authors thank R.L. Harless, U.S. Environmental Protection Agency, North Carolina, for providing the 2,3,7,8-tetra-CDD–spiked solutions; J.J. Ryan, Health and Welfare Canada, for the data on the round-robin study; and D.L. Stalling, for providing the samples used in the fish extract study. The authors are also grateful to C. Partridge for the use of his liquid scintillation counter.

REFERENCES

1. Cairns, T., L. Fishbein and R. Mitchum. *Biomed. Mass Spectrom.* 7:484 (1980).
2. Mitchum, R.K., W.A. Korfmacher, G.F. Moler and D.L. Stalling. *Anal. Chem.* 54:719 (1982).
3. Buser, H.R., and C. Rappe. *Anal. Chem.* 52:2257 (1980).
4. Nestrick, T.J., L.L. Lamparski and R.H. Stehl. *Anal. Chem.* 51:2273 (1979).
5. Nestrick, T.J., L.L. Lamparski and D.I. Townsend. *Anal. Chem.* 52:1865 (1980).
6. Lamparski, L.L., and T.J. Nestrick. *Anal. Chem.* 52:2045 (1980).
7. Hass, J.R., M.D. Friesen, D.J. Harvan and C.E. Parker. *Anal. Chem.* 50:1474 (1978).
8. Mitchum, R.K., G.F. Moler and W.A. Korfmacher. *Anal. Chem.* 52:2278 (1980).
9. Mitchum, R.K., J.R. Althaus, W.A. Korfmacher and G.F. Moler. *Adv. Mass Spectrom.* 8:1415 (1980).
10. Mitchum, R.K., W.A. Korfmacher and G.F. Moler. *J. High Resolut. Chromatog. Chromatog. Commun.* 4:180 (1981).
11. Moler, G.F., R.R. Delongchamp and R.K. Mitchum. *Anal. Chem.* 55:842 (1983).
12. Moler, G.F., R.R. Delongchamp, W.A. Korfmacher, B.A. Pearce and R.K. Mitchum. *Anal. Chem.* 55:835 (1983).
13. Mitchum, R.K., and W.A. Korfmacher. *Spectra* 8:12 (1982).
14. Gross, M.L., T. Sun, P.A. Lyon, S.F. Wojinski, D.R. Hilker, A.E. Dupuy, Jr. and R.G. Heath. *Anal. Chem.* 53:1902 (1981).
15. Langhorst, M.L., and L.A. Shadoff. *Anal. Chem.* 52:2037 (1980).
16. Lamparski, L.L., T.J. Nestrick and R.M. Stehl. *Anal. Chem.* 51:1453 (1979).
17. Harless, R.L., E.O. Oswald, M.K. Wilkinson, A.E. Dupuy, Jr., D.D. McDaniel and H. Tai. *Anal. Chem.* 52:1239 (1980).
18. Ryan, J.J. *J. Assoc. Off. Anal. Chem.* (in preparation).

17

Confidence Limits for Isotope Dilution–Gas Chromatographic/ Mass Spectrometric Determination of 2,3,7,8-Tetrachlorodibenzo-*p*-dioxin in Environmental Samples

G.F. Moler, R.R. Delongchamp, R.K. Mitchum, W.A. Korfmacher and B.A. Pearce

2,3,7,8,-Tetrachlorodibenzo-*p*-dioxin (tetra-CDD) is a toxic by-product produced during the manufacture of 2,4,5-trichlorophenol, a precursor used in the synthesis of the herbicides 2,4,5-trichlorophenoxyacetic acid (2,4,5-T) and 2,4,5-trichlorophenoxypropionic acid (Silvex). Due to the extreme toxicity of tetra-CDD and the fact that it has entered the environment and continues to enter the environment (mainly through application of 2,4,5,-T and Silvex), analytical chemists have had to address problems encountered in the determination of tetra-CDD at very low concentrations, generally in the low parts-per-trillion range.

The need for high specificity, in addition to sensitivity, has led to the use of isotope dilution-gas chromatographic/mass spectrometric (ID-GC/MS) methodology. The goal of ever lowering the detection limit for tetra-CDD has led to ID-GC/MS techniques that produce detection limits of <1.0 ppt.

The Mass Spectroscopy Laboratory of the National Center for Toxicological Research (NCTR), has recently been involved with developing methodology [1,2] for determination of tetra-CDD in environmental samples. Quantification is by ID-GC/MS. It was early in our GC/MS methods development project that a problem involving ID-GC/MS standard curves presented itself: conventional linear regression techniques produced inaccurate confidence limits. We soon discovered that other mass spectroscopists were aware of this problem but had resigned themselves to the use of conventional methods because other, more accurate methods had not been presented or verified. Such a resignation on our part was unacceptable because we, as a research branch of a federal regulatory agency, would have had to deal with the legal ramifications of an admission that reported confidence limits for environmental pollutants such as tetra-CDD were inaccurate.

What began as a limited search for a solution to a problem grew into a comprehensive examination [3] and comparison of the accuracies of conventional approaches to deal with the problem. It was found that of three conventional ID-GC/MS methods, none was completely satisfactory, as each had its limitations. The characteristics of ID-GC/MS data were then reexamined. It was determined that since the ordinate of an ID-GC/MS standard curve is the ratio of the areas of two chromatographic peaks, what was needed was a weighted linear regression technique that used weights based on the variance of the area of a single chromatographic peak. A search of the literature failed to uncover any acceptable method to estimate such peak area variances, so we developed such a method [4] for this purpose.

This chapter reviews the results of our research and shows how resulting new quantitative and qualitative analysis schemes have been applied to the problem of analysis of tetra-CDD by ID-GC/MS. Implementation of linear regression procedures is accomplished with the use of FORTRAN IV programs running under a Finnigan INCOS Data System. This software is capable of not only producing accurate confidence limits for concentrations of tetra-CDD found in environmental samples, but also of performing qualitative analysis by analyzing molecular fragment m/z ratios. This software is available from the authors on request.

SOURCES OF VARIANCE

An ID-GC/MS standard curve is the line

$$Y = aX + b \tag{1}$$

where X is the weight ratio:

$$X = \frac{\text{analyte weight}}{\text{isotope diluent weight}} \tag{2}$$

and Y is the area ratio:

$$Y = \frac{\text{analyte GC/MS peak area}}{\text{isotope diluent GC/MS peak area}} \tag{3}$$

After the constants a and b are obtained from linear regression calculations, a portion of an isotope diluent–spiked extract of a sample S is injected into the GC/MS and the resulting sample area ratio Y_s is used to calculate the sample weight ratio X_s:

$$X_s = (Y_s - b)/a \tag{4}$$

If the measured weight of the sample is M and the sample is spiked with a measured volume J of an isotope diluent solution of concentration Q, then the analyte concentration C in the sample is

$$C = JQX_s/M \tag{5}$$

After estimates of the variances of components J, X_s, Q and M are obtained, the variance of C is estimated using the formula for the variance σ_{rs}^2 of a product of two random variables r and s, which have means $\bar{r}$ and $\bar{s}$ respectively,

$$\sigma_{rs}^2 = \sigma_r^2\bar{s}^2 + \sigma_s^2\bar{r}^2 + \sigma_r^2\sigma_s^2 \tag{6}$$

and the power series estimator, $V_{r/s}$, of the variance of a ratio of two random variables,

$$V_{r/s} = (\bar{s}^2\sigma_r^2 + \sigma^2\bar{r}_s^2 + \sigma_r^2\sigma_s^2)/\bar{s}^4 \tag{7}$$

There are two components of the variance $\sigma_{X_s}^2$ of X_s. One is the component from the ID-GC/MS linear regression, σ_{reg}^2, and the other is due to errors made in preparing standards for this regression. σ_{reg}^2 will be derived in the following section on weighted least-squares linear regression. As far as errors in preparing standards are concerned, it was found in the experiments conducted in this study that variances from pipetting, diluting and combining solutions with class A glassware were not significant compared to variances from weighing the standards. If the measured weights of analyte standard and isotope diluent standard are G1 and G2, respectively, and the standard deviations of these weights are S1 and S2, respectively, then the relative errors, E1 = S1/G1 and E2 = S2/G2, will be propagated throughout the concentrations C_{AN_i} and C_{ID_i} of analyte and isotope diluent, respectively, in prepared standard solution i in such a manner that C_{AN_i} and C_{ID_i} will have standard deviations of $C_{AN_i}E1$ and $C_{ID_i}E2$, respectively. Further, if the isotope diluent spiking solution is prepared from the same stock as the standard curve standards, then the standard deviation component in Q due to the isotope diluent will cancel with the component of the standard deviation of X_s due to the isotope diluent. The remaining component of the standard deviation of X_s due to errors in making the standards is $E1X_s$ and the variance of X_s may therefore be written:

$$\sigma_{X_s}^2 = \sigma_{reg}^2 + (E1X_s)^2 \tag{8}$$

Using Equations 5 through 8, and noting that Q can now be considered a constant as its standard deviation has canceled, σ_C^2 may be calculated with:

$$\sigma_C^2 = \frac{Q^2}{M^4} (\sigma_J^2 + J^2)\{X_s^2\sigma_M^2 + (M^2 + \sigma_M^2)[\sigma_{reg}^2 \quad (9)$$

$$+ (E1X_s)^2]\} + X_s^2\sigma_J^2M^2)$$

Confidence limits of C are estimated as

$$\text{Conf. Limits (C)} = C \pm t_{N-2}\sigma_C \quad (10)$$

where t is Student's t (for some specified confidence interval), and N is the number of points in the ID-GC/MS standard curve. Equation 10 is a valid estimator if the coefficients of variation (CV = standard deviation/mean) of M and of J $\leq$ 0.1. If these CV were larger, the probability density function of C would deviate significantly from being symmetric, and one-sided confidence limits would be required. Under these CV restrictions, the minimum weight M_{min} and minimum volume J_{min} required to use Equation 10 as valid approximation are, therefore:

$$M_{min} = \sigma_M/0.1 \quad (11)$$

$$J_{min} = \sigma_J/0.1 \quad (12)$$

In practice, the largest component of σ_C^2 is usually σ_{reg}^2. This component is derived in the following section.

WEIGHTED LEAST-SQUARES LINEAR REGRESSION

Formulas used for computing a weighted least-squares linear regression are reviewed below. Given the ordered pairs, (X_i, Y_i), i = 1 to N, a and b are obtained [5–8] by minimizing $\chi^2(N)$:

$$\chi^2(N) = \sum_{i=1}^{N} (Y_i - aX_i - b)^2/\sigma_{Y_i}^2 \quad (13)$$

where $\sigma_{Y_i}^2$ is the variance of Y_i. Define the weight W_i of the ith point as:

$$W_i = 1/\sigma_{Y_i}^2 \quad (14)$$

The sum of weights W1 is

$$W_1 = \sum_{i=1}^{N} W_i \tag{15}$$

The weighted means, $\overline{X}$ and $\overline{Y}$, are:

$$\overline{X} = \sum_{i=1}^{N} W_i X_i / W1 \tag{16}$$

$$\overline{Y} = \sum_{i=1}^{N} W_i Y_i / W1 \tag{17}$$

Defining SSx, SSy and SSxy as:

$$SSx = \sum_{i=1}^{N} W_i (X_i - \overline{X})^2 \tag{18}$$

$$SSy = \sum_{i=1}^{N} W_i (Y_i - \overline{Y})^2 \tag{19}$$

$$SSxy = \sum_{i=1}^{N} W_i (X_i - \overline{X})(Y_i - \overline{Y}) \tag{20}$$

a and b are given by:

$$a = SSxy/SSx \tag{21}$$

$$b = \overline{Y} - a\overline{X} \tag{22}$$

After a and b are calculated, let Y_s be measured K times. Then if the weight of sample point j is $W_{sj} = 1/\sigma_{sj}^2$, then the sum of weights, W2, the weighted mean $\overline{Y}_s$, and X_s are given by:

$$W2 = \sum_{j=1}^{K} W_{sj} \tag{23}$$

$$\overline{Y}_s = \sum_{j=1}^{K} W_{sj} Y_{sj} / W2 \tag{24}$$

$$X_s = (\bar{Y}_s - b)/a \tag{25}$$

It is usually desirable to construct upper and lower confidence intervals around X_s of equal probability, and when one Student's t value is so chosen for both confidence intervals, the resulting confidence limits around X are given by:

$$\text{Conf. Limits} = \left(\bar{X} + \frac{X_s - \bar{X}}{1 - g}\right) \pm \left(\frac{1}{1 - g}\right)\frac{ts}{a} \sqrt{\left(\frac{1}{1 - g}\right)\left(\frac{1}{W1}\right) + \frac{1}{W2} + \frac{(X_s - \bar{X})^2}{SSx}} \tag{26}$$

where

$$s = \sqrt{\frac{SSy - a^2SSx}{N - 2}} \tag{27}$$

$$g = \frac{t^2s^2}{a^2SSx} \tag{28}$$

However, when g is small (e.g., less than 0.1 [7]), it can be ignored [7,8] without significantly altering the computed confidence limits; when $g \approx 0$ the resulting limits are then symmetrical about X_s:

$$\text{Conf. Limits} = X_s \pm \frac{ts}{a}\sqrt{\frac{1}{W1} + \frac{1}{W2} + \frac{(X_s - \bar{X})^2}{SSx}} \tag{29}$$

σ^2_{reg}, for use in Equation 9, is:

$$\sigma^2_{reg} = \frac{s^2}{a^2}\left[\frac{1}{W1} + \frac{1}{W1} + \frac{(X_s - \bar{X})^2}{SSX}\right] \tag{30}$$

ID-GC/MS LINEAR REGRESSION METHODS

Convenvional methods used to compute ID-GC/MS standard curves include unweighted, Poisson weighted and replicate measurement weighted linear regressions. The methods differ in their estimators of $\sigma^2_{Y_i}$ used in the regression.

The unweighted regression is used under an implicit assumption of uniform variance and can be computed from Equations 13–29 by simply setting $\sigma^2_{Y_i}(i = 1$ to N) equal to 1, which produces an unweighted estimator V_i^{uw} of $\sigma^2_{Y_i}$:

$$V_i^{uw} = 1 \tag{31}$$

In the Poisson regression method, the mass spectrometer is considered to be a simple pulse-counting device, with the counts being Poisson-distributed. In this distribution, the variance of the counts is equal to the counts. The GC/MS peak areas are then treated simply as counts. Let N_i be the analyte GC/MS peak area and let D_i be the isotope diluent GC/MS peak area. Then

$$Y_i = N_i/D_i \tag{32}$$

If allowance is made for constant hardware or computer software gain $G^{1/2}$, the Poisson variance estimator V_{Ni}^P of N_i and the Poisson variance estimator V_{Di}^P 0 will be the gain squared times N_i and D_i, respectively:

$$V_{Ni}^P = GN_i \tag{33}$$

$$V_{Di}^P = GD_i \tag{34}$$

By combining Equations 7, 32, 33 and 34, one obtains the Poisson estimator V_{Yi}^P of σ_{Yi}^2:

$$V_{Yi}^P = (D_iN_i + N_i^2 + GN_i)G/D_i^3 \tag{35}$$

In the third conventional regression procedure, L replicate measurements of Y_i are obtained. The replicate measurement estimator V_{Yi}^{Rm} of σ_{Yi}^2 is:

$$V_{Yi}^{Rm} = \sum_{j=1}^{L} (Y_{i,j} - Y_i)^2/(L-1) \tag{36}$$

Replicate measurements must also be obtained for Y_{si} to estimate σ_{Ysi}^2 with V_{Ysi}^{Rm}.

A new ID-GC/MS linear regression method [3] uses a squared successive differences (SSD) [4] estimator V_A^{SSD} of the variance of the area A of a single chromatography peak. This SSD estimator of area variance for a peak consisting of P intensity values I distributed over K background intensity values B is given by (see Appendix for derivation):

$$V_A^{SSD} = \tfrac{1}{2}\sum_{i=1}^{P-1} (I_i - I_{i+1})^2 - \tfrac{1}{2}\sum_{i=1}^{P-1} (SF_i - SF_{i+1})^2 \tag{37}$$

$$+ \frac{p^2}{K(K-1)} \sum_{i=1}^{K} (B_i - \bar{B})^2$$

where SF is a smoothing function [9] estimator of a theoretical smooth peak about which the chromatography peak is randomly distributed. In the SSD regression procedure, $V_{N_i}^{SDD}$ and $V_{D_i}^{SSD}$ are computed using Equation 37 and then are combined with Equations 7 and 32 to yield the SSD estimator $V_{Y_i}^{SSD}$ of $\sigma^2_{Y_i}$:

$$V_{Y_i}^{SSD} = (D_i^2 V_{N_i}^{SSD} + N_i^2 V_{D_i}^{SSD} + V_{N_i}^{SSD} V_{D_i}^{SSD})/D_i^4 \tag{38}$$

EXPERIMENTAL

Gas Chromatography/Mass Spectrometry

The GC/MS used in this study was described previously [1,2]. It consisted of an atmospheric-pressure ionization/mass spectrometer (API/MS) (Extranuclear Laboratories, Pittsburgh, Pennsylvania) interfaced to a Hewlett-Packard 5710 gas chromatograph. The chromatograph used a 50-m SP-2100 fused-silica capillary column connected to a falling needle injector (Chrompack, Bridgewater, New Jersey). The API/MS was monitored with a Model 2400 INCOS Data System (Finnigan Corp., Sunnyvale, California), which included the MSDS (revision 3) data acquisition software package.

The GC oven was operated isothermally at 220°C. Helium was the GC carrier gas and the linear flowrate was 25 cm/sec. An ion molecule reaction between tetra-CDD and O_2^- produced 4.5-dichloro-1,2-benzoquinone ions [1,2]. These ions were monitored at m/z 176 and 178 for ^{12}C-tetra-CDD and at m/z 182 for ^{13}C-tetra-CDD.

Other Equipment

The balance used to weigh the standards was a Cahn Electrobalance G-2 (Ventron Instruments, Paramount, California). The sample balance was a Model 2610 Dial-O-Gram (Ohaus, Florham Park, New Jersey). The spiking syringe was a 100-μL GC syringe.

Tetra-CDD Standards

Uniformly ring-labeled ^{13}C-tetra-CDD was synthesized under contract [1], and ^{12}C-tetra-CDD was purchased from Eco-Control (Cambridge, Massachusetts) and used as received (reported purity was 98%). ^{12}C-Tetra-CDD (383 μg) and ^{13}C-tetra-CDD (160 μg) were weighed out on the Electrobalance and placed into separate volumetric flasks, each containing 100 mL of toluene. These were the stock solutions from which 12 standards were prepared. The weight ratios (g ^{12}-C-tetra-CDD/g ^{13}C-tetra-CDD) of the 12 standards ranged from 0.0 to 4.79. All dilutions and volumetric transfers were conducted at 20°C.

Estimates of σ_M^2, σ_J^2, E1 and E2

The standard deviations S1 and S2 of the weighings of ^{12}C-tetra-CDD and ^{13}C-tetra-CDD, respectively, were calculated from repetitively weighing 1- and 5-mg class M calibration weights on the electrobalance. Then E1 = S1/383 and E2 = S2/160.

An unweighted linear regression was performed with the sample balance (scale reading vs. weight) using temperature-controlled H_2O for standard weights. A total of 15 points were used and weights ranged from 3.0 to 18.0 g. An unweighted linear regression was performed with the spiking syringe (syringe reading vs. volume) using temperature-controlled H_2O for standard volumes. A total of 20 points were used and volumes ranged from 10 to 100 μL. σ_M^2 and σ_J^2 were then estimated for any specified weight or volume, respectively, from Equation 29.

ID-GC/MS Standard Curves

Each of the 12 tetra-CDD standards was injected three times into the GC/MS, for a total of 36 injections. The ordered pairs (X_i, Y_i), i = 1–36, where X was the standard weight ratio and Y was the ratio of the GC/MS peak area at m/z 176 to the GC/MS peak area at m/z 182, were used to compute least-squares linear regression lines by the four discussed methods. In the SSD method, a nine-point quadratic-cubic smoothing function [9] was used to compute SF for Equation 37.

Relative Contributions to Error (RCE)

After the tetra-CDD SSD ID-GC/MS standard curve was computed, the standard curve point with weight ratio = 4.79 was sent back through the standard curve as 19 "samples," with these simulated samples having sample weights (M) ranging from 0.5 to 10.0 g. Each of these were simulated to have been spiked at 100 ppt, so that the spike volume (J) changed with sample size. Confidence interval widths ($2t_{N-2}\sigma_c$) for each of these samples were calculated.

For each of the 19 simulated samples, RCE of σ_J, σ_M, σ_{reg} and $E1X_S$ were calculated as follows. The relative increase in σ_C due to σ_J was calculated by computing σ_C (Equation 9) normally with σ_J included and then with $\sigma_J = 0$. Then,

$$\text{relative increase } (\sigma_J) = \frac{\sigma_C - \sigma_C(\sigma_J = 0)}{\sigma_C} \qquad (39)$$

Relative increases were calculated for the other three components. Then the normalized RCE of component Z (Z = σ_J, σ_M, σ_{reg} or $E1X_s$) was computed as:

$$\text{RCE(Z)} = \frac{\text{relative increase (Z)}}{\sum \text{relative increases}} \qquad (40)$$

Poisson Hypothesis Experiments

Two experiments testing the Poisson hypothesis (Equations 33 and 34) were conducted with the API/MS operating in the positive ion mode and the GC disconnected from the API source. In the first experiment, using single ion monitoring (SIM) at m/z 36.0, a mass peak width of 0.01 amu and an integration time of 0.2 sec, 100 consecutive counts were recorded and the mean and variance of these 100 counts were computed. This process was repeated, incrementing the m/z by 0.1 amu up to m/z 37.9. The second experiment was conducted similarly to the first, except m/z ranged from 27.0 to 32.8 in increments of 0.2 amu; the scan times were 0.2 sec; and the mass peak widths were 0.2 amu.

Standardized Variable Cumulative Probability Distributions

Simulation experiments were conducted with an IBM 4341 computer using FORTRAN IV and Statistical Analysis System (SAS Institute, Raleigh, North Carolina) software packages. Each experiment consisted of the simulation and statistical analysis of 500 ID-GC/MS standard curves based on the line $Y = X$ $(a = 1, b = 0)$.

The simulated standard weight ratios were 0.0, 0.1, 0.5, 0.75, 1.0 and 1.5. At each standard weight ratio X_i, one or more area ratios $Y_i = N_i/D_i$ were generated as follows. $\bar{D}_i$ was randomly selected from the uniform distribution, $20{,}000 \leq \bar{D}_i \leq 80{,}000$. $\bar{D}_i$ so selected was then the area of a Weibull function [10] with peak shape parameter = 2.0 about which a particular simulated chromatography peak of area D_i was generated. This peak consisted of 240 points randomly normally distributed about the Weibull function with standard deviations (noise levels) equal to 40% of the Weibull function value (Figure 17.1). Every third point of the peak was sampled (to simulate ID-GC/MS peak switching [11]) to yield a net of 80 points per peak. D_i was the sum of 80 intensity values of this random peak. [D_i was not corrected for peak switching (i.e., multiplied by 3) as this factor would simply cancel in all pertinent equations.] Then $\bar{N}_i = X_i\bar{D}_i$, and N_i was the area of a peak that was similarly distributed about a Weibull function of area $\bar{N}_i$.

After all Y for a standard curve were generated, linear regression parameters were calculated. Then, for each of the 500 standard curves in an experiment, a "true" sample weight ratio X_t was randomly selected from either of the uniform distributions, $0 \leq X_t \leq 1.5$ or $0 \leq X_t \leq 0.5$. A sample area ratio, $Y_s = N_s/D_s$, was randomly generated about $Y_t(=X_t)$ by the method described above. Then the estimated sample weight ratio X_E was computed as:

$$X_E = (Y_s - b)/a \tag{41}$$

The standardized variable U for each of the 500 standard curves and X values was:

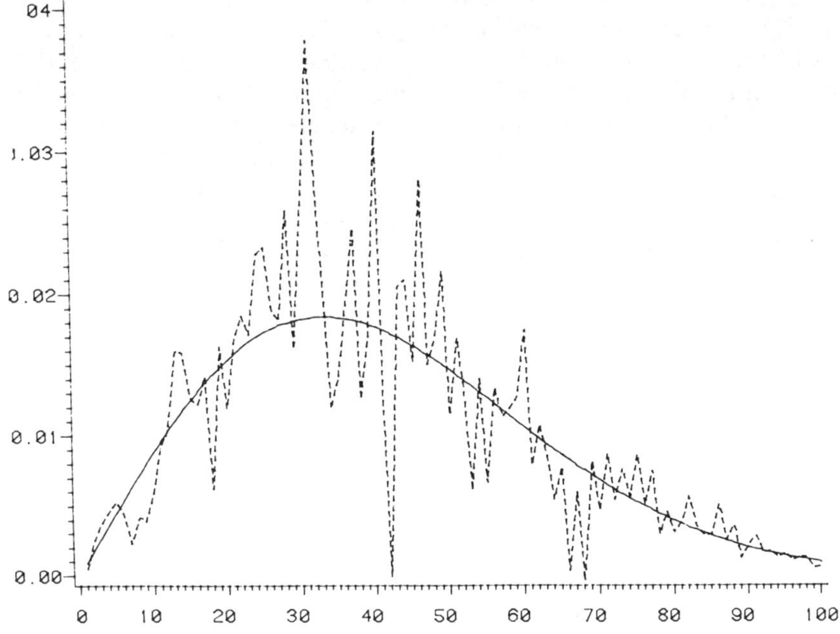

Figure 17.1. Simulated Chromatography Peak (Dashed Line) with a Noise Level Equal to 40% of a Weibull Function (Solid Line) Height. Weibull Peak Shape Parameter = 2.0.

$$U = \frac{X_E - X_t}{\sigma_{reg}} \tag{42}$$

The results of each experiment were expressed as a coplot of the cumulative probability distributions (CPD) of U and Student's t, and a Kolmorgorov-Smirnov (KS) [12] one sample goodness-of-fit statistic. The KS test result was expressed as a probability (p) that the CPD of U could have been a CPD of Student's t. Experiments were conducted using the unweighted, SSD and replicate measurement regression methods.

Implementation

Two tetra-CDD ID-GC/MS quantitative analysis programs were written for and implemented on the INCOS Data System. One program, called "STANDARD," produced an SSD ID-GC/MS weighted linear regression standard curve. The other program, called "SAMPLE," used regression parameters calculated by

STANDARD to calculate the concentration of ^{12}C-tetra-CDD in a sample. After a tetra-CDD standard was injected into the GC/MS, scan boundaries around the tetra-CDD selected ion chromatography peaks (i.e., at m/z 176, 178 and 182) were chosen visually, as were two background regions on either side of the tetra-CDD peaks. Mass spectrometry data system data files for these three regions were submitted to STANDARD, which computed an SSD weighted least-squares linear regression, stored the regression parameters in a disk file and produced a written report. The report included detailed statistical analyses of the GC/MS data produced from every injection. For qualitative analysis purposes (^{12}C-tetra-CDD confirmation), STANDARD also computed the SSD variance estimate $V_{R_i}^{SSD}$ of the ratio $R_i = A_{178_i}/A_{176_i}$. The weight W_{R_i} of this ratio:

$$W_{R_i} = 1/V_{R_i}^{SSD} \tag{43}$$

was used in computing a sum of weights W3,

$$W3 = \sum_{i=1}^{N} W_{R_i} \tag{44}$$

and a weighted mean $\bar{R}$:

$$\bar{R} = \sum_{i=1}^{N} R_i W_{R_i}/W3 \tag{45}$$

Both W3 and $\bar{R}$ were added to the standard curve disk file by STANDARD.

SAMPLE analyzed ID-GC/MS data of samples based on the standard curve parameters stored in the disk file created by STANDARD. SAMPLE first determined if ^{12}C-tetra-CDD chromatography peaks (at m/z 176 and 178) existed. A ^{12}C-tetra-CDD peak was defined to exist if its area was above its threshold area TA, which was defined as three times the SSD estimate of the standard deviation of the area of the peak. TA were computed for peaks at m/z 176 and 178:

$$TA_{176} = 3\sqrt{V_{A176}^{SDD}} \tag{46}$$

$$TA_{178} = 3\sqrt{V_{A178}^{SSD}} \tag{47}$$

For K injections of the sample, the weight $W_{R_{si}}$, the sum of weights W4, and weighted mean $\bar{R}_s$ were computed by SAMPLE as:

$$W_{R_{si}} = 1/V_{R_{si}}^{SSD} \tag{48}$$

$$W4 = \sum_{i=1}^{K} W_{R_{si}} \tag{49}$$

$$\bar{R}_s = \sum_{i=1}^{K} W_{R_{si}} R_{si} / W4 \qquad (50)$$

These were used, along with W3 computed by STANDARD, to calculate 99% confidence limits for $\bar{R}_s$:

$$\text{Conf. Limits } (\bar{R}_s) = \bar{R}_s \pm t_{N-1} \sqrt{1/W3 + 1/W4} \qquad (51)$$

These confidence limits, along with TA_{176} and TA_{178}, were used by SAMPLE to determine if the sample injection passed the following three criteria for confirming ^{12}C-tetra-CDD in the sample:

1. $A_{176} \geq TA_{176}$;
2. $A_{178} \geq TA_{178}$; and
3. $\bar{R}$ was between conf. limits $(\bar{R}_s)$.

SAMPLE then computed 95% confidence limits of C for every injection in which ^{12}C-tetra-CDD was confirmed. SAMPLE also calculated a detection limit, DL, for every injection. DL was defined as the concentration of ^{12}C-tetra-CDD that would have been computed by SAMPLE had the ^{12}C-tetra-CDD chromatography peaks just barely passed the confirmation criteria. Defining the minimum area $A_{176}(min)$ of the m/z 176 peak as either $A_{176}(min) = TA_{176}$ or as $A_{176}(min) = TA_{178}/\bar{R}$, whichever was found to be greater, DL was defined as the net minimum signal above the system blank, i.e., the y-intercept b:

$$DL = \frac{(A_{176}(min)/A_{182} + b) - b}{a} = \frac{A_{176}(min)}{a\,A_{182}} \qquad (52)$$

RESULTS

E1, E2, J_{min} and M_{min}

The standard deviation of the Electrobalance weighings was 4 μg. Since the ^{12}C-tetra-CDD standard weighed 383 μg, E1 = 4/383 = 0.0104. ^{13}C-tetra-CDD weighed 160 μg, so E2 = 4/160 = 0.025. M_{min} was calculated by using Equation 29 of the sample balance's linear regression using $\sigma_M = M_{min}/0.1$ and was 0.95 g. J_{min} was calculated similarly and was 3.1 μL.

ID-GC/MS Linear Regression Parameters

Table 17.1 lists linear regression parameters for the ID-GC/MS standard curves computed by each of the four described methods. Each of the ID-GC/MS regres-

Table 17.1. ID-GC/MS Linear Regression Parameters

Parameter	SSD Regression	Repl. Meas. Regression	Poisson Regression	Unweighted Regression
a	1.379	1.349	1.397	1.420
b	−0.005945	−0.0008127	−0.01546	−0.01997
$\bar{X}$	0.1012	0.1208	0.04926	0.9474
$\bar{Y}$	0.1335	0.1621	0.05335	1.325
SSx	4,744	9,695	68,910	65.10
SSy	9,076	17,690	136,000	131.80
SSxy	6,541	13,080	96,260	92.46
W1	10,900	194,600	5,710,000	36
g	7.36×10^{-4}	3.24×10^{-4}	1.37×10^{-3}	4.93×10^{-4}

sion methods used the same 36 points generated from 12 standards with weight ratios from 0.0 to 4.79, using three injections per standard.

Relative Contributions to Error (RCE)

Figure 17.2a shows the influence of sample size on the 95% confidence interval width produced by the SSD ID-GC/MS linear regression for a simulated sample containing 479 ppt ^{12}C-tetra-CDD. Figure 17.2b shows the corresponding RCE for each of the four contributors to total error (σ_C).

ID-GC/MS Confidence Limits

Table 17.2 lists 95% confidence limits computed by each of the four ID-GC/MS linear regression methods [the confidence limits were computed with Equation 29, as g (Equation 28) was negligible (see Table 17.1) in all cases]. The data were simulated by sending 11 of the standard curve points back through the standard curves as "samples" spiked at 100 ppt. The spikes were defined to have zero variance (i.e., $\sigma^2_m = 0$, $\sigma^2_J = 0$, and $E_1 = 0$) so that variances due solely to the ID-GC/MS standard curves could be compared. The 95% confidence limits for the standards (column 1 of Table 17.2) were calculated using the determined values of E1, E2 and the estimator (Equation 7) of the variance of a ratio.

Poisson Hypothesis Experiments

Figure 17.3a shows a reconstructed positive ion mass spectrum between m/z 36.0 and 37.9. The counts are the means of 100 consecutive measurements. Figure

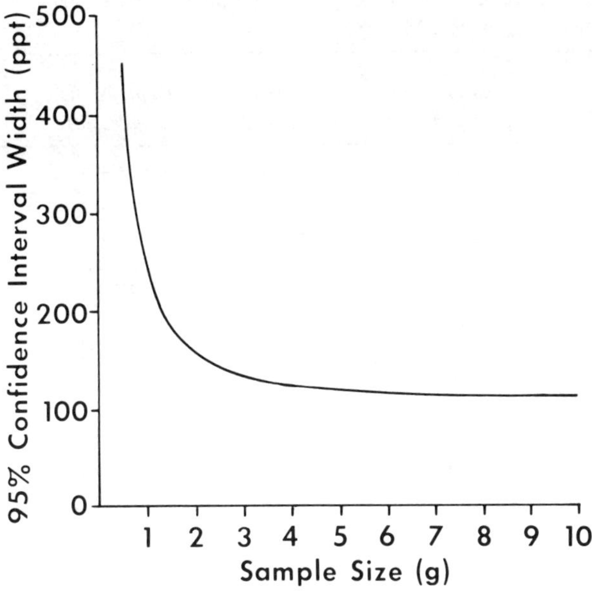

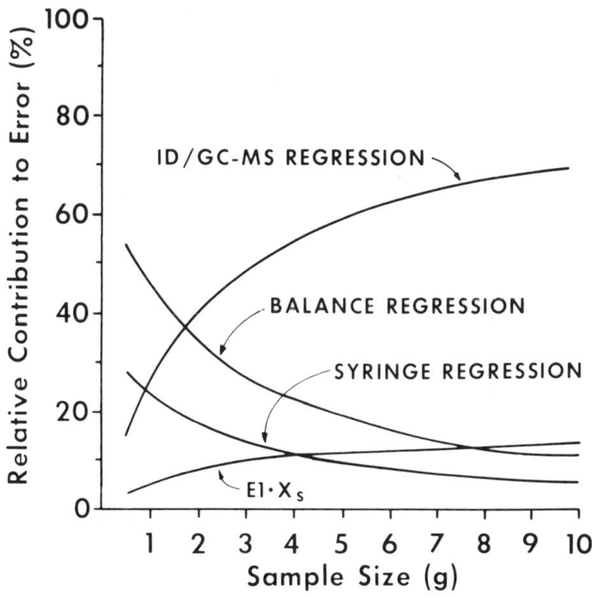

Figure 17.3. Influence of sample size on (a) SSD ID-GC/MS linear regression confidence interval widths and (b) relative contributions to error of four contributors to total error.

Table 17.2. Comparison of 95% Confidence Limits (Parts-per-Trillion) from Four ID-GC/MS Linear Regression Methods

Standards	*Poisson Regression*	*SSD Regression*	*Repl. Meas. Regression*	*Unweighted Regression*
2±2	2±1	1±2	1±5	2±18
5±1	6±2	5±3	5±1	6±18
10±1	10±3	10±3	9±1	10±18
19±1	18±4	17±3	17±3	18±18
24±1	25±5	24±4	25±2	25±18
48±3	◦±11	50±8	51±22	50±18
72±4	68±10	68±6	70±12	67±18
96±5	98±14	98±9	100±26	96±18
144±8	138±21	139±14	142±12	136±18
239±13	226±32	228±23	233±11	223±18
479±26	494±68	500±56	510±53	486±20

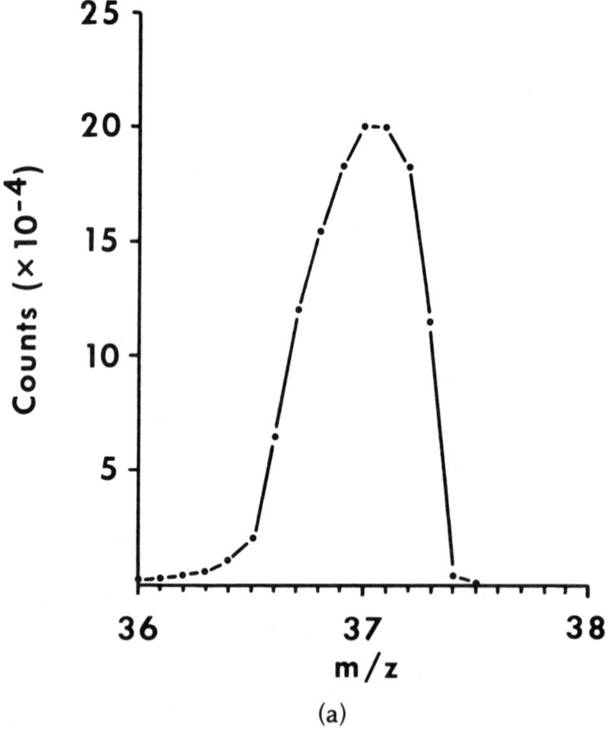

Figure 17.3. (a) Reconstructed positive ion mass spectrum from m/z 36.0 to 37.9. Counts are means of 100 consecutive measurements.

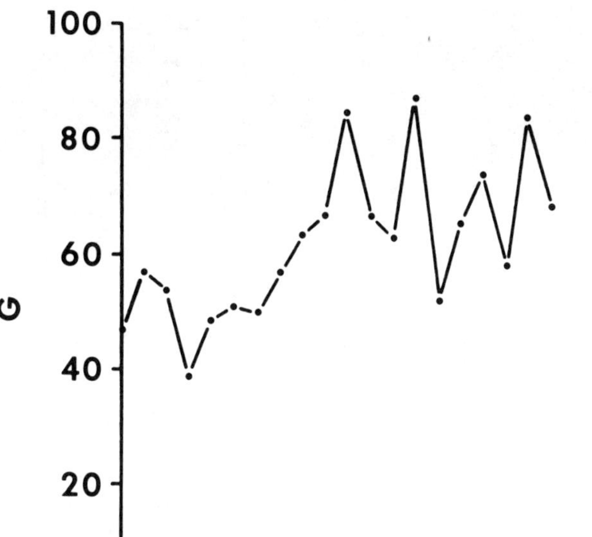

(b)

Figure 17.3. *(continued)* (a) Reconstructed positive ion mass spectrum from m/z 36.0 to 37.9. Counts are means of 100 consecutive measurements. (b) G (variance/mean of counts) vs. m/z.

17.3b shows G (the variance of the intensity divided by the mean of the intensity, Equations 33 and 34) vs. m/z for the data of Figure 17.3a. Figure 17.4a shows a reconstructed positive ion mass spectrum between m/z 27.0 and 32.8. The counts are the means of 100 consecutive measurements. Figure 17.4b shows G vs. m/z for the data of Figure 17.4a.

Cumulative Probability Distribution

Figures 17.5 through 17.8 show CPD of Student's (dashed lines) and of the ID-GC/MS standardized variable U (Equation 42). Table 17.3 reports KS goodness of fit probabilities (p) calculated from Beyer [12].

Figure 17.5a shows the CPD of U from unweighted linear regressions using

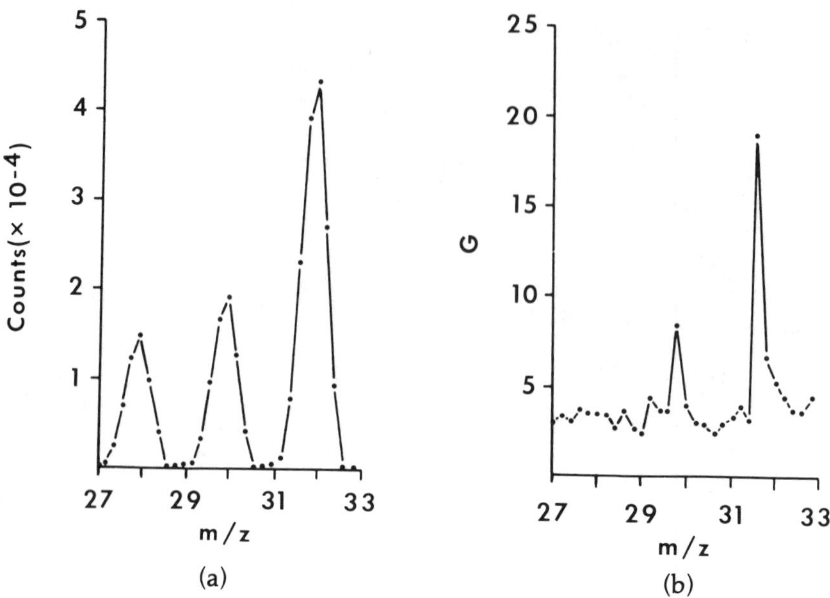

Figure 17.4. (a) Reconstructed positive ion mass spectrum from m/z 27.0 to m/z 32.8. Counts are means of 100 consecutive measurements. (b) G (variance/mean of counts) vs. m/z.

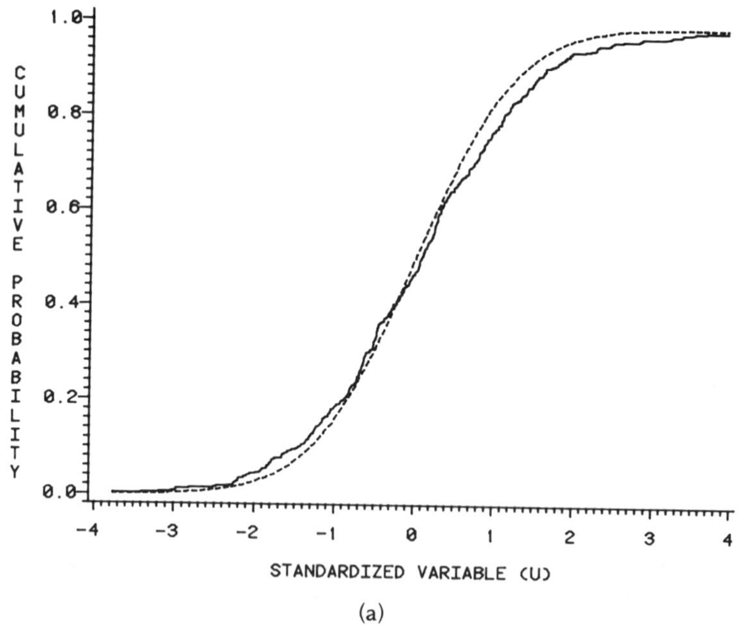

(a)

Figure 17.5. Unweighted linear regression CPD of U (solid line) and Student's t (dashed line) using one point per weight ratio. (a) $0 \leq X_t \leq 1.5$. (b) $0 \leq X_t \leq 0.5$.

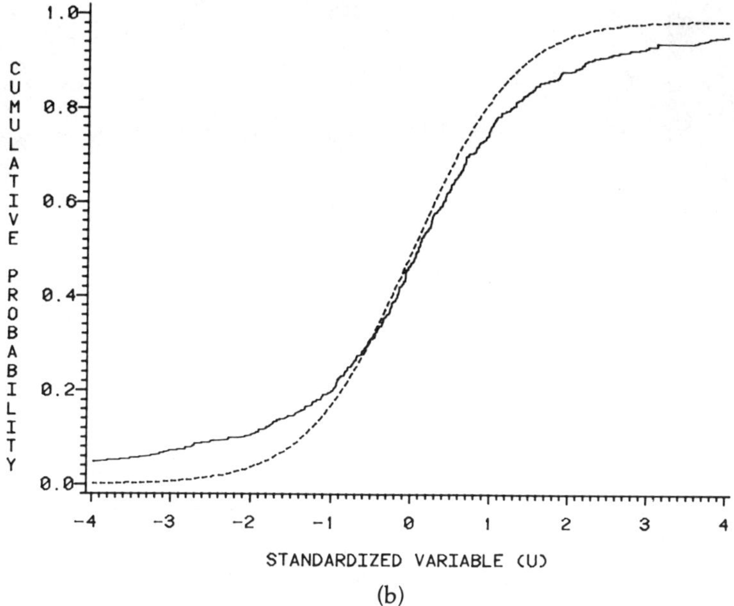

(b)

Figure 17.5. *(continued)* Unweighted linear regression CPD of U (solid line) and Student's t (dashed line) using one point per weight ratio. (b) $0 \leq X_t \leq 0.5$.

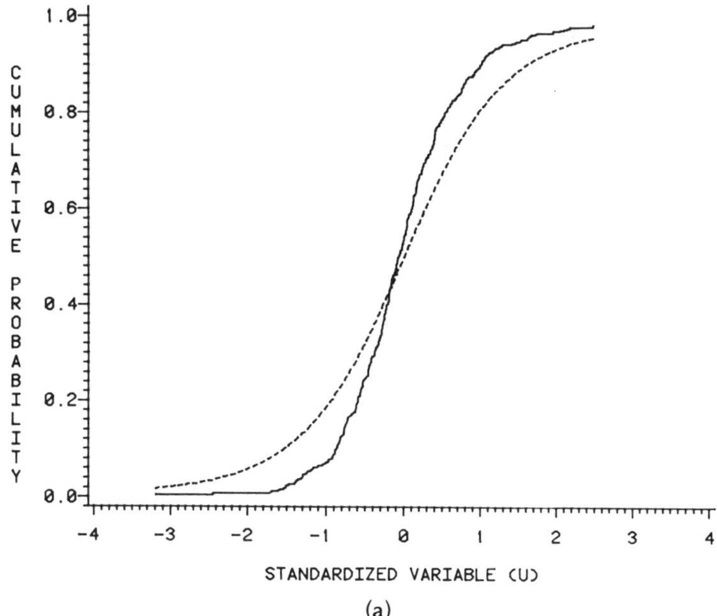

(a)

Figure 17.6. Replicate measurement linear regression CPD of U (solid line) and Student's t (dashed line); $0 \leq X_t \leq 1.5$: (a) two points per weight ratio; (b) five points per weight ratio.

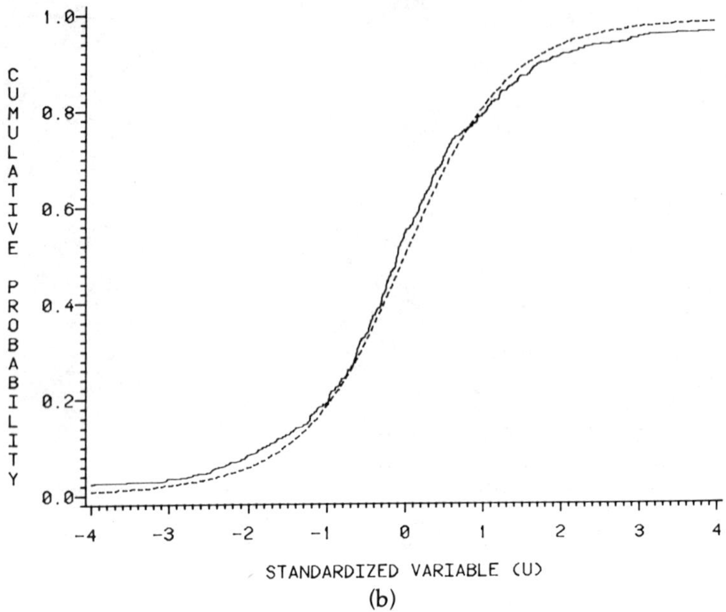

(b)

Figure 17.6. (*continued*) Replicate measurement linear regression CPD of U (solid line) and Student's t (dashed line); $0 \leq X_t \leq 1.5$: (a) two points per weight ratio; (b) five points per weight ratio.

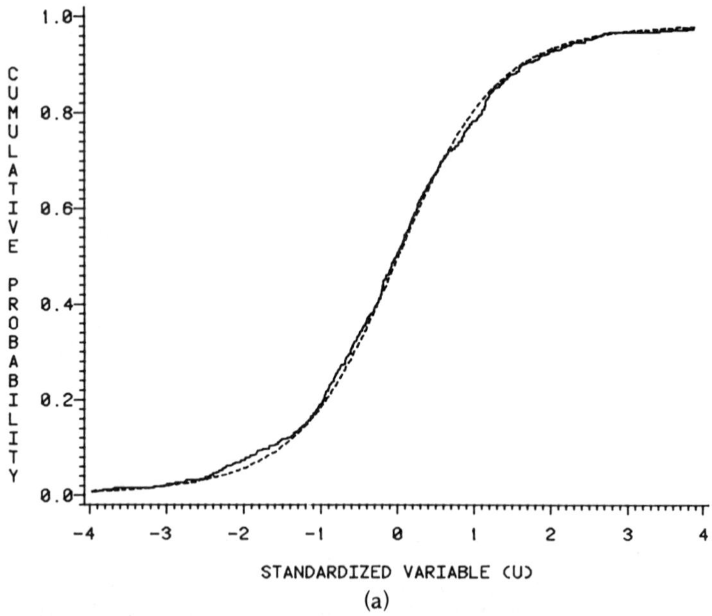

(a)

Figure 17.7. SSD linear regression CPD of U (solid line) and Student's t (dashed line); $0 \leq X_t \leq 1.5$: (a) two points per weight ratio; (b) five points per weight ratio.

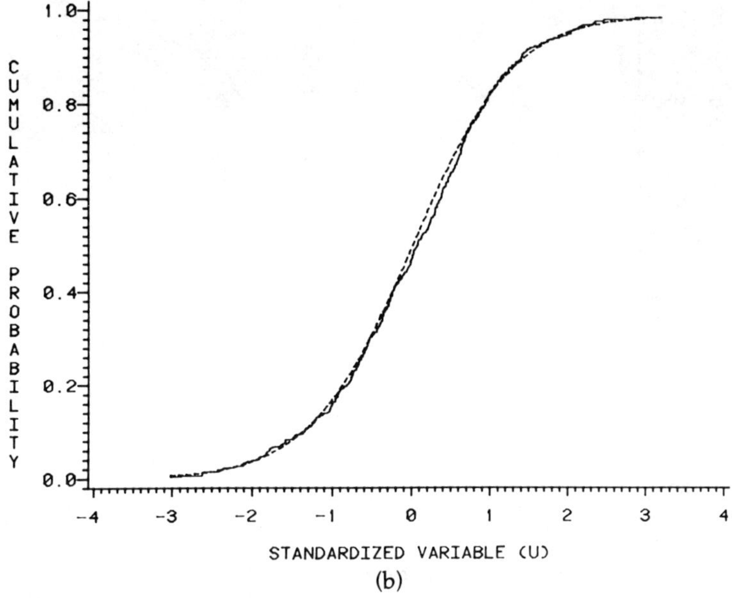

(b)

Figure 17.7 *(continued)* SSD linear regression CPD of U (solid line) and Student's t (dashed line); $0 \leq X_t \leq 1.5$: (a) two points per weight ratio: (b) five points per weight ratio.

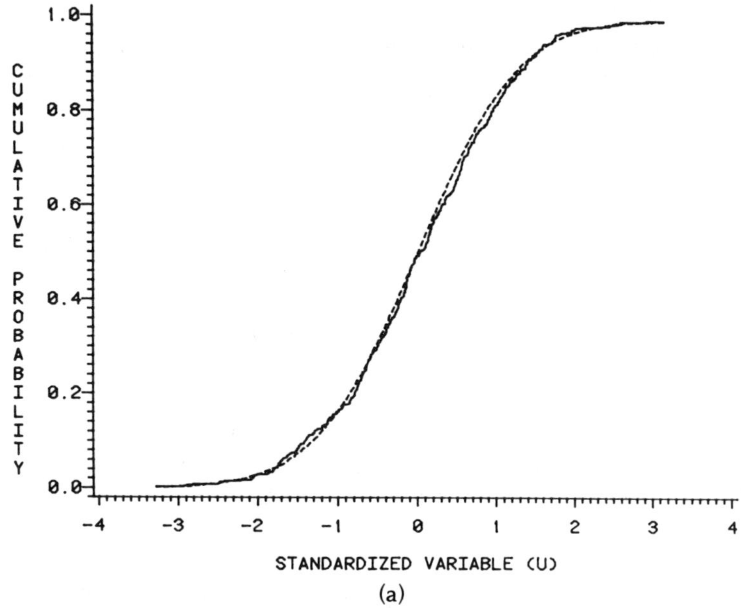

(a)

Figure 17.8. SSD linear regression CPD of U (solid line) and Student's t (dashed line): (a) $0 \leq X_t \leq 1.5$.

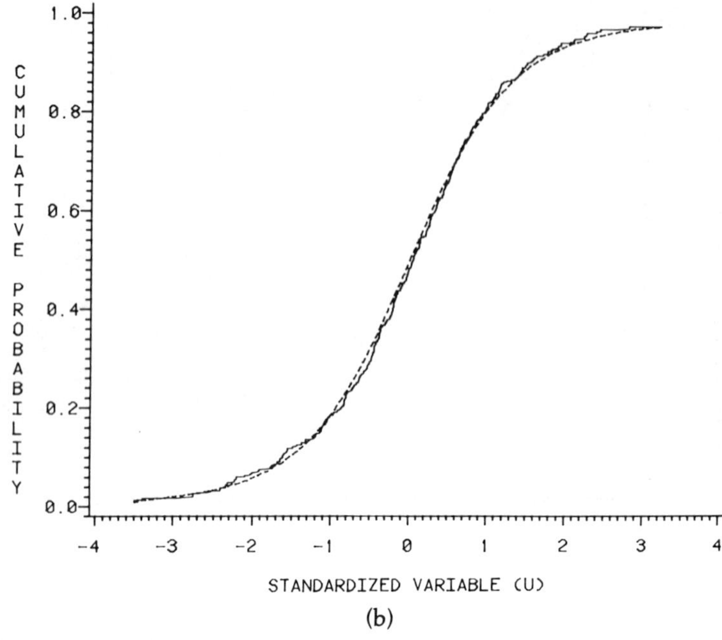

(b)

Figure 17.8. *(continued)* SSD linear regression CPD of U (solid line) and Student's t (dashed line): (b) $0 \leq X_t \leq 0.5$.

Table 17.3. KS Goodness-of-Fit Probabilities (p) for CPD Experiments

X_t	Points/ Wt. Ratio	Unweighted Regression	Repl. Meas. Regression	SSD Regression
0–1.5	1	$0.15 < p < 0.20$		$p >> 0.20$
0–0.5	1	$p << 0.01$		$p >> 0.20$
0–1.5	2		$p << 0.01$	$p > 0.20$
0–1.5	5		$0.01 < p < 0.05$	$p > 0.20$

one point per weight ratio (no replicate measurements). Data of Figure 17.5a were computed from sample weight ratios X_t, uniformly randomly distributed over the full range of weight ratios of the points that produced the standard curves: $0 \leq X_t \leq 1.5$. In Figure 17.5b, X_t was uniformly randomly distributed over only the lower third of the standard curve range: $0 \leq X_t \leq 0.5$.

Figure 17.6 shows the CPD of U from replicate measurement linear regressions. In Figure 17.6a the regressions used two points per weight ratio, and Figure 17.6b the regressions used five points per weight ratio ($0 \leq X_t \leq 1.5$).

Figure 17.7 shows the CPD of U from SSD linear regressions using two points per weight ratio and five points per weight ratio ($0 \leq X_t \leq 1.5$).

Figure 17.8 shows the CPD of U from SSD linear regressions using only one point per weight ratio. In Figure 17.8a, $0 \leq X_t \leq 1.5$. In Figure 17.8b, $0 \leq X_t \leq 0.5$.

DISCUSSION

Relative Contributions to Error

The results presented in Figure 17.2 illustrate effects of changing sample size on confidence limits and on RCE. One would usually expect that the largest RCE would be that of the ID-GC/MS standard curve, with the other RCE being significantly smaller, as is the case with the 10.0-g sample.

As the sample size is reduced, however, the RCE of the syringe regression and the sample balance regression might no longer be of small significance. One should, therefore, avoid using small samples or use more precise instruments than were used in the present study. For example, the Ohaus Model 2610 balance is an inexpensive general-purpose instrument not designed for precisely weighing small samples. Also, a spike level of 100 ppt for a 1.0-g sample required a spike solution volume of only 6.3 μL. Clearly, the coefficient of variation at such a volume for a 100-μL syringe is high; one should, for such a volume level, use a 10-μL syringe instead. It should not be inferred from the foregoing discussion, however, that the Ohaus Model 2610 balance and the 100-μL syringe were unacceptable; indeed, as long as the sample weight was about 10 g or greater their RCE were less than 20%.

The RCE of $E1X_s$ could easily have been less than they were. The standard deviation of weighings on the Electrobalance in the range 1–5 mg was essentially constant. If the ^{12}C-tetra-CDD standard had weighed 5.0 mg instead of 383 μg, for example, then E1 would have been 4/50,000 = 0.0008 instead of 0.0104 and RCE of $E1X_s$ would have been reduced to insignificance.

ID-GC/MS Linear Regression Methods

Table 17.2 shows a comparison of 95% confidence limits produced by each of the four linear regression methods, with each method using the same 36 points. Because of the limited number of points involved and the fact that the "sample" points were in fact taken from the standard curve points, this was by no means a statistically rigorous experiment, but it did expose some obvious trends.

One conclusion from the data in Table 17.2 is that the means of the confidence limits for all methods were very similar. This indicates that the slopes and y-intercepts calculated by the four methods were in close agreement, and this is confirmed by examining Table 17.1

Another conclusion from Table 17.2 is that confidence limits calculated by the three weighted regression methods are remarkably similar, despite totally different formulas the methods used to estimate weights. This may be due to the fact that for weighted regression accuracy, it is necessary only for the weight to be proportional to the inverse of the variance, because any proportionality constant will drop out of Equations 21, 22 and 29. Weighted regressions, therefore, tolerate a certain amount of inaccuracy in their methods of estimating weights. Table 17.1 shows that the sums of weights W1 and, therefore, the individual weights calculated by the three weighted regression methods are indeed widely different.

Information concerning accuracies of the confidence limits can be deduced from Table 17.3. If points about a regression line are independent and normally distributed, and estimated weights are proportional to the theoretical true weights, then the standardized variable U will be distributed as a Student's t [7]. Independence and normalcy were included in the specifications of the CPD simulations performed in this study, so major differences between the CPD of U and Student's t were due to inaccurate weight estimates (minor differences were expected due to the finite number of standard curves simulated).

Conditions in the CPD simulations were severe. For example, the GC/MS peak areas of the isotope diluent were randomly uniformly distributed between 20,000 and 80,000. This was to simulate variance in the amounts of standards injected and in recoveries of spiked samples. Further, the noise levels of the simulated chromatography peaks were very high (40% of the Weibull function height, Figure 17.1). These conditions were deliberately inserted into the simulations to test the statistical robustness of each regression method tested, and also because in some trace-level ID-GC/MS work such conditions are not unusual.

Unweighted Method

It is not difficult to conclude from the data presented in Table 17.2 that the unweigthed regression produced inaccurate confidence limits. First, the widths of the confidence bands for all but one concentration were essentially the same (after rounding off to integer values), contrary to known behavior of ID-GC/MS data. Second, confidence bands around the four lowest concentration levels include zero, which could allow for possible erroneous classification of ^{12}C-tetra-CDD determinations in these samples as being "not detected." Third, the confidence band around the highest computed concentration (486 ppt) was narrower than the confidence band around the standard, which would imply that the linear regression contributed less variance at this point than was contained in the standard itself.

Figure 17.5 and Table 17.3 show that when X_t was chosen from across the full range of standard curve X values (i.e., 0–1.5), the unweighted regression's confidence limits do not appear exceedingly inaccurate. However, when X_t was chosen from only the lower third of standard curve X values (analogous to the lower concentration values in Table 17.2, accuracy diminished dramatically. Clearly, sampling from only the middle standard curve X values tends to mask inaccurate confidence limits at each end of the standard curve.

Poisson Method

Figures 17.3 and 17.4 show that the simple Poisson hypothesis was not supported on the API/MS. Not only was G always found to be greater than the means of the counts, but G was not a constant. Based on these data and other experiments conducted in this study (the data from which are not summarized in the figures), some trends could be seen for G. G tended to rise as the counts rose. G increased as the mass peak width decreased, and decreased slightly as integration times decreased. From this behavior, one can conclude that the variance of the counts is in part a function of the counts, but it is not a simple relationship.

The Poisson model has been used in certain isotope dilution mass spectrometry applications, such as the measurement of $^{13}CO_2/^{12}CO_2$ ratios from gases bled into a mass spectrometer at a constant rate [13]. Integration times in that application were long, and instruments were stable. Schoeller [14] suggested that the Poisson model might apply to ID-MS standard curves in general, although he cautioned that, for the model to be applied validly, instrumental stability was required.

It is clear that in the API/MS there were sources of significant variance other than Poisson processes. Analog-to-digital converter noise [15], alternating current line noise, instability in the quadrupole, etc., might all have contributed. In addition, the chemical ionization process may have added variance due to reactant and product concentration fluxes. Although the simple Poisson model was not supported on the API/MS, the Poisson regression produced confidence limits similar to those calculated by the other weighted regressions (Table 17.2). As discussed earlier, if Poisson estimates of weights are only proportional to the true weights, the regression will be accurate. Thus, the Poisson method may perform adequately even when its model hypothesis is not strictly supported.

Replicate Measurement Method

As Figure 17.6 and Table 17.3 illustrate, the replicate measurement method produces more accurate confidence limits when the number of points (L, Equation 36) per weight ratio is increased. This is as expected, as the estimator $V_{Y_i}^{Rm}$ improves as L increases.

There are theoretical problems and a practical limitation associated with the use of the replicate measurement method for ID-GC/MS standard curves. The theoretical problems stem from the assumption that $V_{Y_i}^{Rm}$ is an accurate estimator of $\sigma_{Y_i}^2$ for all points within a group, i.e., that all points have the same variance which can be accurately estimated with $V_{Y_i}^{Rm}$. But in ID-GC/MS, the variance of individual points are in fact independent, and are functions of the absolute amounts of material injected and of the GC/MS noise level at the time of injection, as much as being a function simply of weight ratios or peak area ratios.

The practical limitation of using the replicate measurement method for ID-GC/MS standard curves is, of course, due to the requirement for a relatively high number of injections at each weight ratio (at least five injections [16]). Obeying this requirement may prove to be excessively laborious if the GC retention times are high, as is the case with some pesticides. Replicate injections must also be made for the sample in order to compute W_{S_i}. This latter requirement poses a problem with the determination of environmental pollutants at very low concentration levels (e.g., parts per trillion), because in this case it is advantageous, for sensitivity considerations, to inject as large a fraction of the sample extract as possible.

Despite these limitations, the replicate measurement method has two strong points. One, it allows for relatively easy computation of an ID-GC/MS weighted linear regression and can be implemented with a calculator. Two, it provides more accurate confidence limits than does the unweighted method.

SSD Method

Based on the data presented in Figures 17.7 and 17.8, and Table 17.3, and on previously determined [4] characteristics, the SSD method for ID-GC/MS has the following advantages:

1. It produces confidence limits consistently more accurate than either the replicate measurement method or the unweighted regression method.
2. The number of standard curve points per weight ratio has no bearing on the accuracy of its calculated confidence limits, so replicate injections of standards and samples are not necessary.
3. Variances of Y_i are estimated independently, so the method imposes no theoretical requirement for identical injection sizes or for day-to-day instrumental noise stability.
4. In the SSD method it is unnecessary [4] to presuppose a particular distribution of the underlying noise of the mass spectrometer—a presupposition, for example, required of the Poisson method. Indeed, a Poisson noise distribution is merely a subset of distributions accommodated by the SSD method.

The disadvantage of the SSD method is that it requires relatively complex calculations (Equation 37); therefore, an on-line computer is a practical necessity for its implementation.

Other Methods

Other linear regression techniques were not examined in this study. For example, one method [17] involves computing several regression lines from one set of points by selectively deleting subsets of points and choosing the line that produces the narrowest confidence band around the unknown sample concentration. In another technique [18] the replicate measurement method is extended by performing a nonlinear regressionof $V_{Y_i}^{Rm}$ on X; this tends to smooth out erratic values of $V_{Y_i}^{Rm}$ that were computed from few points. These and other methods were not explored in the present study because they either assume a variance structure not applicable to ID-GC/MS data, or because they appeared to offer negligible improvement over the performances of the three conventional regression methods under normal experimental conditions.

Recommendations

The SSD method should be implemented if the necessary computational ability is available. Computer software is available from the authors on request. This software consists of two programs and is supplied in FORTRAN IV source code and in compiled code (for use with the Nova III INCOS Data System). One program, called "STANDARD," produces an SSD weighted linear regression ID-GC/MS standard curve. The other program, called "SAMPLE," uses parameters stored on a computer disk file by STANDARD to produce 95% confidence limits for an unknown concentration; SAMPLE also performs qualitative analysis by analyzing molecular fragment m/z ratios.

If adequate computer facilities for implementing the SSD method are not available, then either the Poisson method or the replicate measurement method should be implemented. However, the Poisson method's hypothesis (Equations 33 and 34) may require confirmation on the particular GC/MS in use if very accurate confidence limits are needed. If the replicate measurement method is used, then the amounts of standards and samples injected into the GC/MS should be tightly controlled to reduce intragroup variances.

The unweighted ID-GC/MS regression will usually produce fair estimates of slope and y-intercept, but inaccurate confidence limits. This method should, therefore, be used only when accurate confidence limits are not required.

Implementation Example

Figure 17.9 shows a tetra-CDD ID-GC/MS chromatogram for a fish sample extract injection, and Figure 17.10 is a report produced by SAMPLE for this sample.

Figure 17.10 illustrates part of a comprehensive implementation of qualitative and quantitative analysis procedures discussed in this paper. All variances reported were calculated by the SSD method. "STD FILE NAME" is the name of the standard curve disk file produced by STANDARD. The "CONF BAND OF 178/176" was computed with Equation 51 and was used as part of a ^{12}C-tetra-CDD confirmation process. Note that, in this example (Figure 17.10), this confidence band included the "STD CURVE 178/176" ($\bar{R}$) value, and

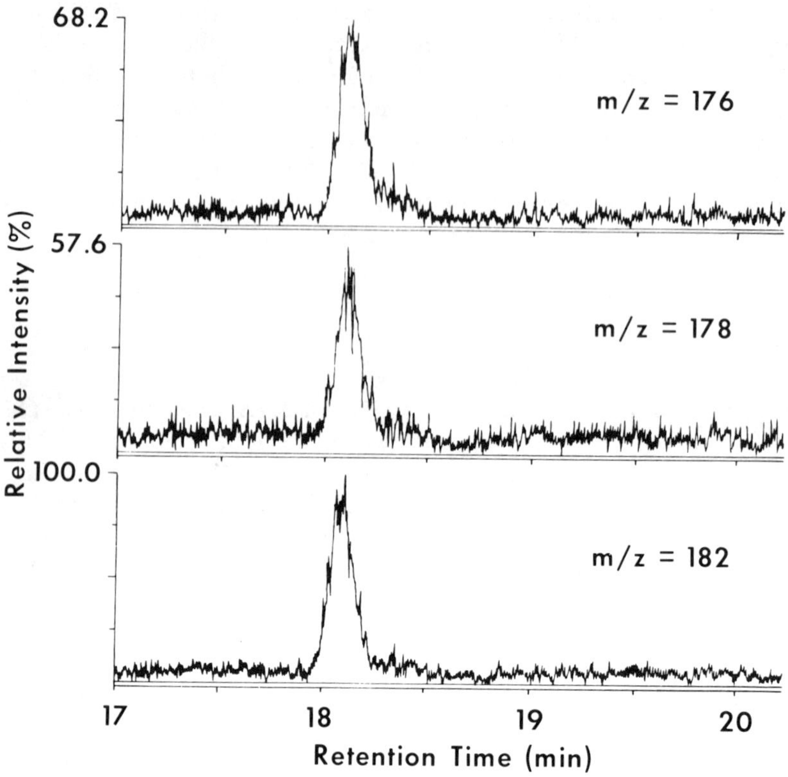

Figure 17.9. Tetra-CDD ID-GC/MS chromatogram of a fish sample extract. The report for this sample is shown in Figure 17.10.

therefore the sample passed the third confirmation criterion listed in the lower part of the figure.

"PK AREA VAR" is V_A^{SSD} (Equation 37) for each of the three tetra-CDD peaks. For example, the area of the peak at m/z 176 is 10,716, and its SSD variance is 76,844. Therefore [4], 95% confidence limits for the area of this peak is estimated as $A_{176} \pm 2(V_A^{SSD})^{1/2} = 10{,}716 \pm 554$.

The final computation reported in Figure 17.10, "MEAN PPT FOR ALL DETERMINATIONS," lists 95% confidence limits for ^{12}C-tetra-CDD confirmed in the sample. These limits incorporate the discussed variances: σ_J^2, σ_M^2, σ_{reg}^2 and $(E1X_s)^2$. SAMPLE handles multiple injections and standard solution injections (for continuous validation of the standard curve).

A detection limit (Equation 52) was computed for every injection. This limit was not computed by the method discussed by Currie [19] and adopted by the ACS Committee on Environmental Improvement [20]. This latter method is

```
OPERATOR:  WK
DATE:  9/21/81
PROJECT:  6009
SAMPLE DESCRIPTION:  CANADIAN FISH SAMPLE "A", RUN 1, 3/5 OF TOTAL
STD FILE NAME:  SWHAT.WT
SPIKE LEVEL (PPT):  99.9
NUMBER OF SAMPLE INJECTIONS:  1

INJECTION 1:

        FILENAME OF BACKGROUND REGION 1:  T917103B1.EP
        FILENAME OF PEAK REGION:          T917103PK.EP
        FILENAME OF BACKGROUND REGION 2:  T917103B2.EP
        176/182 AREA RATIO       0.745342
        VARIANCE OF 176/182      0.000727
        WEIGHT OF 176/182            1373
        178/176 AREA RATIO       0.681399
        VARIANCE OF 178/176      0.000926
        WEIGHT OF 178/176            1078
        CONF BAND OF 178/176     0.681399  ±  0.079459
        STD CURVE 178/176        0.644183

        M/Z                     176            178            182

        BKGD MEANS               23.9           32.7           30.3
        BKGD VAR                189.2          240.0          217.0
        BK MEAN VAR             1.220          1.548          1.400
        THRESHOLD AREA            831            796
        PEAK AREA               10716           7302          14378
        PK AREA VAR             76844          70776         132563

RESULTS:

        INJECTION:  1
        CONFIRMATION CRITERIA:
          1.  176 PEAK AREA > TA (176):            PASS
          2.  178 PEAK AREA > TA (178):            PASS
          3.  178/176 WITHIN 99% CONF LIMITS:      PASS
        INJECTION 1:  TCDD WAS DETERMINED
        DETECTION LIMIT = 3.1 PPT

QUANTITATION:

        MEAN PPT FOR ALL DETERMINATIONS =   54.5  ±  6.6
```

Figure 17.10. Report created by SAMPLE from an ID-GC/MS chromatogram of a fish sample extract. The chromatogram for this sample is shown in Figure 17.9.

based on an implicit assumption that an instrument has an inherent (fixed) detection limit. The noise level of the API/MS was found to be sufficiently variable from day to day, so that no such fixed detection limit could be assigned; therefore, it was computed independently for each injection, automatically taking into account day-to-day noise fluctuations.

Appendix 17A

Derivation of SSD Variance Estimators

Figure 17.11 shows a smooth function F superimposed on a noisy digitized chromatography peak I, the latter consisting of N discrete intensity values. F is the mean of hypothetically infinitely many peaks similar to I that were produced under injection conditions identical to those under which I was produced. F may therefore be called the "expected peak" of I, analogous to the expected value of a data distribution.

If an intensity value I_i is distributed about F_i with variance σ_i^2, and the expected value $< I_i >$ of I_i is equal to F_i,

$$< I_i > = F_i \tag{A1}$$

then

$$< I_i^2 > = F_i^2 + \sigma_i^2 \tag{A2}$$

Let the abscissa interval of I be 1, and let the estimate A of the area under F between i = 1 and i = N be

$$A = \sum_{i=1}^{N} I_i \tag{A3}$$

The variance, σ_A^2, of A is

$$\sigma_A^2 = \sum_{i=1}^{N} \sigma_i^2 \tag{A4}$$

The sum of squared residuals, S_A^2, of A is

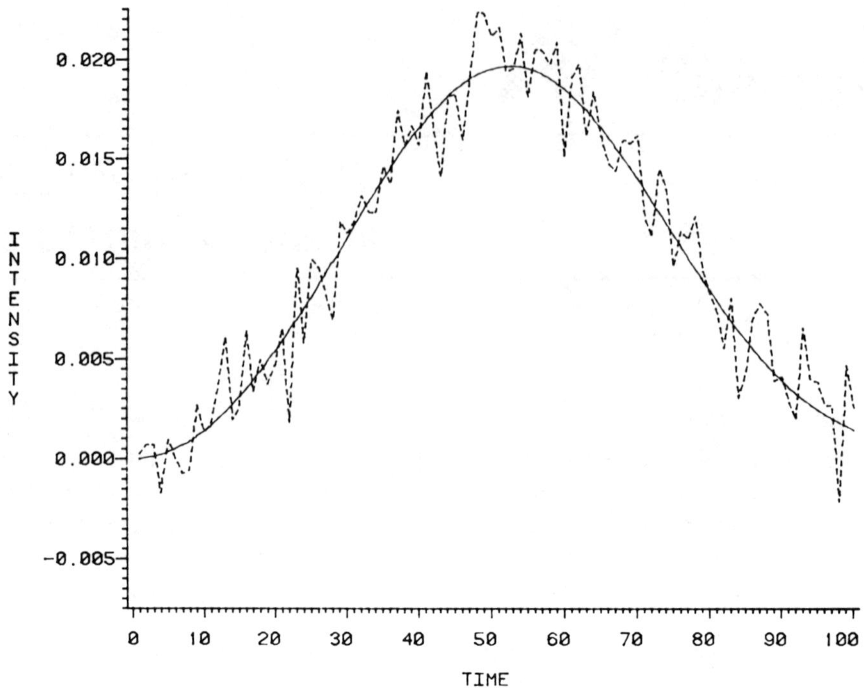

Figure 17.11. Expected peak F (solid line) superimposed on a noisy peak (dashed line). Arbitrary time and intensity units.

$$S_A^2 = \sum_{i=1}^{N} (I_i - F_i)^2 \tag{A5}$$

s_A^2 can be computed only if F is known exactly, as is the case with computer simulations. For real chromatography peaks, F is unobtainable.

Define a function Q_i:

$$Q_i \equiv \frac{(I_i - I_{i+1})^2}{2} \tag{A6}$$

From Equations A1, A2 and A6,

$$< Q_i > = \frac{\sigma_i^2}{2} + \frac{\sigma_{i+1}^2}{2} + \frac{(F_i - F_{i+1})^2}{2} \tag{A7}$$

From Equation A7,

$$< \sum_{i=1}^{N-1} Q_i > = \frac{\sigma_1^2}{2} + \frac{\sigma_N^2}{2} + \sum_{i=2}^{N-1} \sigma_i^2 + \tfrac{1}{2} \sum_{i=1}^{N-1} (F_i - F_{i+1})^2 \tag{A8}$$

Rearranging Equation A8, and using Equations A4 and A6 gives an estimator V_A of σ_A^2:

$$V_A = \tfrac{1}{2} \sum_{i=1}^{N-1} (I_i - I_{i+1})^2 - \tfrac{1}{2} \sum_{i=1}^{N-1} (F_i - F_{i+1})^2 + \tfrac{1}{2}\sigma_1^2 + \tfrac{1}{2}\sigma_N^2 \tag{A9}$$

The contributions of $\sigma_1^2/2$ and $\sigma_N^2/2$ are usually small, so

$$V_A \approx \tfrac{1}{2} \sum_{i=1}^{N-1} (I_i - I_{i+1})^2 - \tfrac{1}{2} \sum_{i=1}^{N-1} (F_i - F_{i+1})^2 \tag{A10}$$

Using a quadratic-cubic smoothing function SF [9] to estimate F in Equation A10 gives the SSD estimator, V_A^{SSD}, for the variance of the area of a peak that does not contain background:

$$V_A^{SSD} = \tfrac{1}{2} \sum_{i=1}^{N-1} (I_i - I_{i+1})^2 - \tfrac{1}{2} \sum_{i=1}^{N-1} (SF_i - SF_{i+1})^2 \tag{A11}$$

Background contained in a chromatography peak is handled as follows. Let $\bar{B}$ be the average of background intensities under the peak. Then the area, A_B, of the peak with background subtracted is

$$A_B = A - N\bar{B} \tag{A12}$$

and the variance σ_{AB}^2 of this area is

$$\sigma_{AB}^2 = \sigma_A^2 + N^2\sigma_{\bar{B}}^2 \tag{A13}$$

Using the estimator, V_B, for the variance of the mean of G background intensities,

$$V_B = \frac{1}{G(G-1)} \sum_{i=1}^{G} (B_i - \bar{B})^2 \tag{A14}$$

the SSD estimator, V_{AB}^{SSD}, for the variance of the area of a chromatography peak which contains background is

$$V_{AB}^{SSD} = \tfrac{1}{2} \sum_{i=1}^{N-1} (I_i - I_{i+1})^2 - \tfrac{1}{2} \sum_{i=1}^{N-1} (SF_i - SF_{i+1})^2 + \frac{N^2}{G(G-1)} \sum_{i=1}^{G} (B_i - \bar{B})^2 \tag{A15}$$

The accuracy and precision of the SSD estimators (Equations A11 and A15) have been examined [4]. It was found that in most circumstances (peak shape, noise levels, etc.) the SSD estimators were accurate to within ±5% of the true variances, as determined by computer simulations.

ACKNOWLEDGMENTS

The authors acknowledge statistical assistance of D.W. Gaylor and R.L. Kodell.

REFERENCES

1. Mitchum, R.K., G.F. Moler and W. A. Korfmacher. *Anal. Chem.* 52:2278 (1980).
2. Mitchum, R.K., W.A. Korfmacher, G.F. Moler and D.L. Stalling. *Anal. Chem.* 54:719 (1982).
3. Moler, G.F., R.R. Delongchamp, W.A. Korfmacher, B.A. Pearce and R.K. Mitchum. *Anal. Chem.* 55:835 (1983).
4. Moler, G.F., R.R. Delongchamp and R.K. Mitchum. *Anal. Chem.* 55:842 (1983).
5. Bevington, P.R. *Data Reduction and Error Analysis in the Physical Sciences* (New York: McGraw-Hill Book Company, 1969).
6. Draper, N.R., and H. Smith. *Applied Regression Analysis* (New York: John Wiley and Sons, Inc., 1966).
7. Brownlee, K.A. *Statistical Theory and Methodology in Science and Engineering,* 2nd. ed. (New York: John Wiley and Sons, Inc., 1965).
8. Snedecor, G.W., and W.G. Cochran. *Statistical Methods,* 7th ed. (Ames, IA: Iowa State University Press, 1980).
9. Savitzky, A., and M.J.E. Golay. *Anal. Chem.* 36–1627 (1964).
10. Hastings, N.A.J. and J.B. Peacock. *Statistical Distributions* (New York: John Wiley and Sons, Inc., 1975).
11. Matthews, D.E., and J.M. Hayes. *Anal. Chem.* 48:1375 (1976).
12. Beyer, W.H. *Handbook of Tables for Probability and Statistics,* 2nd ed. (Cleveland, OH: CRC Press, 1968).
13. Schoeller, D.A., and J.M. Hayes. *Anal. Chem.* 47:408 (1975)
14. Schoeller, D.A. *Biomed. Mass. Spectrom.* 3:265 (1976).
15. Chesler, S.N., and S.P. Cram. *Anal. Chem.* 44:2240 (1972).
16. Kleijnen, J., R. Brent and R. Brouwers. *Kenmerk* 28:355 (1980).
17. Mitchell, D.G., W.N. Mills, J.S. Garden and M. Zdeb. *Anal. Chem.* 49:1655 (1977).
18. Garden, J.S., D.G. Mitchell and W.N. Mills. *Anal. Chem.* 52:2310 (1980).
19. Currie, L.A. *Anal. Chem.* 40:568 (1968).
20. ACS Committee on Environmental Improvement. *Anal. Chem.* 52:2242 (1980).

18

Evaluation of Interferences from Seven Series of Polychlorinated Aromatic Compounds in an Analytical Method for Polychlorinated Dibenzofurans and Dibenzo-*p*-dioxins in Environmental Samples

Lawrence M. Smith and James L. Johnson

A critical aspect of the validation of an analytical method is the assessment of the limitations imposed by interfering substances on the method detection limit [1] and on the degree of assurance of the determination. Although extraordinary enrichment or determinative procedures usually can be applied in instances in which false positive results are suspected, the risk lies in the possibility that an erroneous determination may not be recognized. The problems associated with the presence of compounds that specifically interfere in analyses for chemical residues are especially critical in residue determinations at part-per-trillion levels in which the total analyte may amount to <1 ng. In these situations, analytical options to deal with suspect data are very limited.

The purpose of the study reported here was to determine the degree of interference produced by selected members of seven families of polychlorinated aromatic compounds in a method for the determination of polychlorinated dibenzofurans (PCDF) and dibenzo-*p*-dioxins (PCDD). This method has been employed in this laboratory for the analysis of environmental samples, especially fish, for extremely low levels of PCDF and PCDD.

Analyses for PCDF and PCDD are often performed on environmental samples collected from areas associated with a large number and variety of contaminant sources, such as the Great Lakes, and the Hudson and Ohio Rivers. These analyses are designed to provide congener-specific and, to a limited extent, isomer-specific determinations at a method detection limit of less than 10 ppt and are usually subject to the constraint of extremely small amounts of analytes. For example, at 10 ppt in a 50-g sample, complete recovery of a particular residue will yield only 500 pg. As many as a dozen families of polychlorinated aromatic compounds have been identified as potential specific interferences in the determination of PCDF and PCDD [1–14]. For example, the ubiquitous polychlorinated biphenyls (PCB) and DDE, routinely encountered in environmental samples at concentrations a million-fold greater than those of PCDF and PCDD, exhibit gas chromatography/mass spectrometry (GC/MS) properties similar to those of PCDD. With interfering cocontaminants usually present at levels from parts per billion to parts per million, the reduction of the levels of these compounds in the final analyte to an equivalent of 10 ppt and less requires a preferential enrichment factor for the PCDF and PCDD of 10^3–10^6. Consequently, the detrimental role of interference from cocontaminants is a particularly important consideration in analyses of environmental samples for PCDF and PCDD. In Table 18.1, nine families of polychlorinated aromatic compounds are categorized according to the degree of interference that each poses in determinations of PCDF and PCDD by GC/MS.

These interfering compounds exhibit the following pertinent analytical characteristics:

1. The compounds produce mass spectral fragmentation patterns that overlap to varying degrees with those of PCDF and PCDD.
2. Each chemical family is composed of a large number of isomers and congener groups with GC retention times often similar to those of PCDF and PCDD.
3. Compounds of at least two of the families (methoxy-PCB and benzylphenyl ethers) have the same nominal masses and same number of chlorine substituents as those of the PCDF and PCDD, making the molecular ions indistinguishable by low-resolution mass spectrometry (LRMS).
4. At least five of the chemical families include members that undergo thermal conversions and/or conversion following ionization (during MS fragmentation) to produce PCDF and PCDD.

The most powerful and routinely employed determinative procedures for PCDF and PCDD are high-resolution gas chromatography (HRGC)/LRMS and less frequently with HRGC/HRMS. Under the requirements of a limit of detection of 1–10 ppt, the use of the multiple ion monitoring (MIM) procedure during GC/MS is necessary for attainment of maximum instrumental sensitivity. Unfortunately, the gain in sensitivity is accompanied by a reduction in analytical specificity; only limited portions of the fragmentation pattern of a compound can be observed. The reduced specificity exacerbates the potential for specific interference and false positive determinations.

Table 18.1. Levels of Interferences of Selected Chemical Families in MS Determinations of PCDF and PCDD

	Level of Interference					
	Overlap of Fragmentation Patterns[a]		*Indistinguishable by LRMS*[b]		*Indistinguishable by HRMS*	
	PCDD	*PCDF*	*PCDD*	*PCDF*	*PCDD*	*PCDF*
PCB	++/+++					
Naphthalenes (PCN)		+				
Diphenyl Ethers (DPE)		++/+++		X		X
Methoxy (MeO)-PCB	+++	+++	X	X		X
Hydroxy (HO)-PCB		+++		X		X
MeO-DPE	+++	++	X		X	
HO-DPE	+++		X		X	
Benzylphenyl Ethers (BzPE)	+++		X			
Biphenylenes	++					

[a] + = minor overlap; ++ = major overlap; +++ = complete overlap.
[b] X = interference is observed.

Analyses for PCDF and PCDD in this laboratory include the determination of the members of 10 congener groups (tetra- through octachloro) of the PCDF and PCDD and of two or three isotopically enriched marker compounds. Four to six individual masses of the molecular ion cluster are monitored for each congener group. For example, in the GC window containing tetrachlorodibenzofurans (tetra-CDF) and tetrachlorodibenzo-*p*-dioxins (tetra-CDD) the following masses are monitored: 304, 306, 308 and 310 for tetra-CDF; 320, 322, 324 and 326 for tetra-CDD, 312 for $^{37}C1_4$-tetra-CDF (marker compound), and 332, 334 and 336 for $^{13}C_{12}$-tetra-CDD (marker compound).

Routinely, aliquots of the analyte equivalent to a 5- to 20-g portion of the sample are injected for the HRGC/LRMS-MIM analysis in the electron impact (EI) ionization mode. This corresponds to 50–200 pg of a component present at 10 ppt. Instrumental limits of detection for individual components range 10–30 pg in the MIM mode, compared to near 1 ng for full-scan analysis. The prerequisites adopted at this laboratory for the confirmation by HRGC/LRMS-MIM of a particular PCDF or PCDD isomer are:

1. signal-to-noise ratio of 3;
2. correct and unique relative retention time (within 2 parts in 1000) compared with an authentic sample (in some cases determined on two different liquid phases);
3. correct nominal molecular mass; and

4. correct chlorine isotope abundance ratios for four to six members of the molecular ion cluster.

EXPERIMENTAL

The degrees of interference associated with the first seven families of compounds listed in Table 18.1 were evaluated for the analytical procedure of Stalling et al. [15]. The enrichment procedure employs a series of acid- and base-treated and neutral silica gels, activated carbon, and alumina. Determinations of PCDF and PCDD were carried out in the EI mode with a Finnigan 4023 GC/MS equipped with an INCOS data system. GC was carried out on a 30-m x 0.25-mm DB-5 fused-silica capillary column (J&W Scientific, Inc., Rancho Cordova, California) using helium carrier gas and the following temperature program: 150°C for 2 min, 20°C/min to 210°C, 5°C/min to 290°C, hold 7 min.

Seven families of polychlorinated aromatic compounds were selected for evaluation: PCB, PCN, DPE, HO-PCB, MeO-PCB, HO-DPE and MeO-PCB. Representative isomers of the benzylphenyl ethers were not available. Table 18.2 lists the individual components and mixtures used in the study. Aroclor® 1248 + 1254 (PCB) and Halowax® 1014 (PCN) were applied at levels of 5–50 ppm (0.25–2.5 mg) relative to the usual 50-g sample size. All other compounds listed in Table 18.2 were applied at a level corresponding to 100 ppb (5.0 μg each). The mixture of compounds listed in Table 18.2 (except the Aroclors, which were applied separately) was applied to the enrichment system once in the presence of 2.5 ng each of three marker compounds and once in the absence of the marker compounds. The marker compounds are uniformly labeled (UL) ^{13}C-2,3,7,8-tetra-CDD, UL ^{37}Cl-2,3,7,8-tetra-CDF (plus other tetra-CDF isomers) and UL ^{37}Cl-octa-CDD.

The extent of elimination of the interfering compounds was determined both by full scan and MIM procedures. For GC/MS analysis, the sample was reduced in volume to 100 μL in toluene, and 4-μL injections were made.

RESULTS AND DISCUSSION

The approach used in this study to evaluate the level of interference for each component was to carry out the routine HRGC/LRMS-MM analysis on the enriched sample. Comparisons of signal intensities of sample components with those of an external standard and those of the internal marker compounds permitted determination of the level of sensitivity of the measurements; both yielded a limit of detection of approximately 10 ppt using MIM analysis of the enriched (processed) samples. Figure 18.1 presents the HRGC/LRMS analyses of the mixture of interfering compounds listed in Table 18.2 (excluding the Aroclors) before application of the enrichment procedure. Identification of each component of the fragmentograms was confirmed by full scan analyses (Figure 18.1B) and by com-

Table 18.2. Recoveries of Selected Compounds and Mixtures in the PCDF and PCDD Enrichment Procedure

Compound	*GC/MS No.*	*Recovery (%)*
PCB		
3,4,3′,4′-tetra-	3	48
3,4,5,3′,4′-penta-	9	45
3,4,5,3′,4′,5′-hexa-	13	55
Aroclor 1248 + 1254 (1:1)		0
PCN (Halowax 1014)	3,14	0–50
DPE		
3,4,3′,4′-tetrachloro-	1	0
2,4,5,2′,4′-pentachloro-	4	0
2,3,5,3′,4′,5′-hexachloro-	11	0
2,3,4,5,2′,3′,4′-heptachloro-	15	0
2,3,4,5,6,2′,3′,4′-octachloro-	16	0
Decachloro-	19	0
HO-PCB		
3,2′,3′,4′,5′-pentachloro-2-OH-	10	0
2,3,2′,4′,5′-pentachloro-2-OH-	7	0
MeO-DEP		
3,4,5,2′,4′-pentachloro-2-MeO-	11	0
Nonachloro-2-OMe-	17	0
Nonachloro-3-OMe-	18	0
HO-DPE		
3,4,5,2′,4′-pentachloro-2-OH-	12	0
Nonachloro-3-OH-	20	0
Nonachloro-3-OH-	22	0
MeO-PCB		
3,4,3′,4′-tetrachloro-2-MeO-	6	0
2,4,3′,4′-tetrachloro-3-MeO-	5	0
3,5,3′,4′-tetrachloro-4-MeO-	8	0

parison with GC/MS analyses of individual preparations of the compounds; the assignments are listed by number in Table 18.2. Component 21 was determined to be octa-CDD, believed to arise from thermolysis of nonachloro-2-hydroxy diphenyl ether, No. 22. Figure 18.1A presents a GC/MS analysis of the interference mixture using a MIM procedure (extracted from a full scan analysis) in which each mass is monitored during the entire analysis; no GC elution windows were set. Compound 2, the tetrachlorobiphenyl, is not observed in Figure 18.1A because the molecular weight of this compound is well below that of the members of the lowest congener group, tetra-CDF. In other MIM analyses, compound 2 will be observed because an additional set of descriptors is usually included spe-

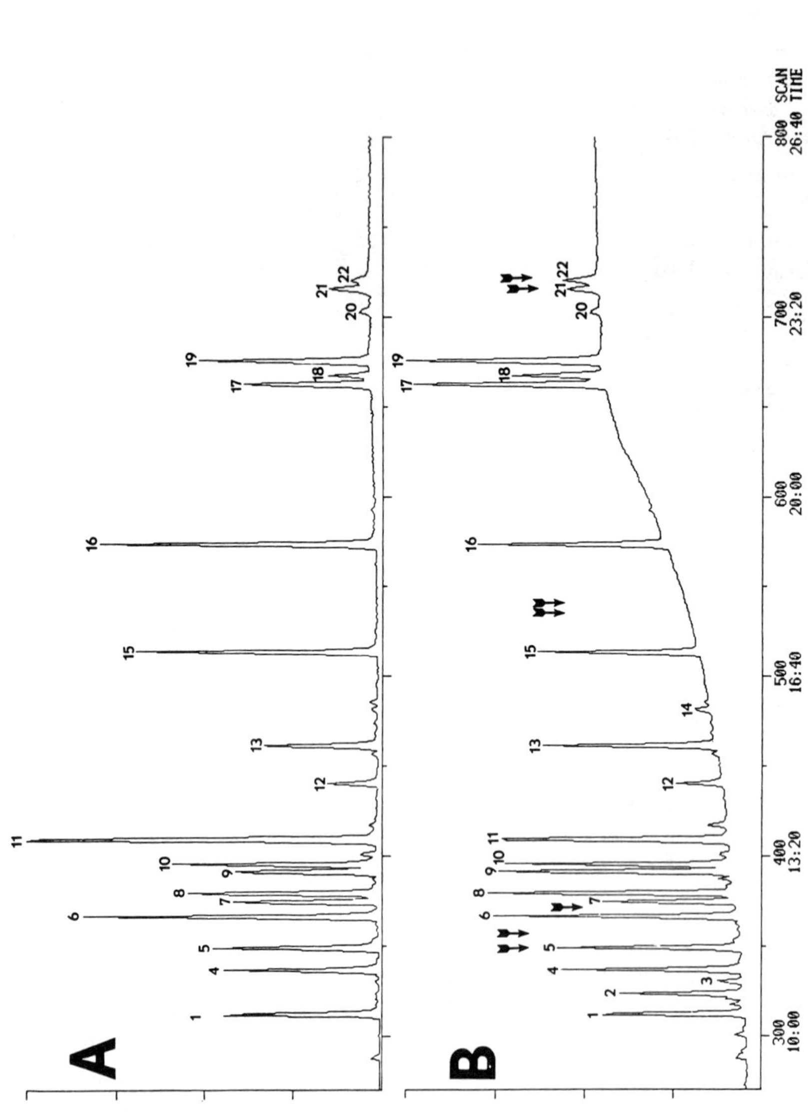

Figure 18.1. (A) HRGC/LRMS Full Scan of Interference Mixture in which only Responses at the Masses of Tetra- through Octa-CDD and -CDF Are Displayed. Arrows Identify Points at which Positive Determinations for PCDF and PCDD Were Later Observed. (B) HRGC/LRMS Full Scan Analysis of Interference Mixture, Including All Responses in the Mass Range 230–530.

cifically to permit the determination of this compound. When the constraints of GC elution "windows" are coupled with the MIM analysis, the complexity of the fragmentogram of the interference mixture (Figure 18.2) is reduced. Under these conditions, only nine major components are observed by HRGC/LRMS-MIM. An important consideration here is that components apparently eliminated from detection by the MIM procedure may produce low-level fragmentations that could elicit positive responses during the determinations at part-per-trillion levels.

The mass fragmentograms of the processed samples (Figure 18.3) were dominated by three or four isomers of PCB and by six to eight isomers of PCN. The three predominant PCB isomers (b, h and k in Figure 18.3, corresponding to 2, 9 and 13 in Table 18.2) were recovered at approximately the 50% level. The majority of PCN components were eliminated by the enrichment. A small proportion of PCN isomers persisting in the final analyte were recovered, mostly below the 50% level. The PCB and PCN responses can obscure PCDF and PCDD components, but do not generate false positive determinations. Those PCB isomers containing no chlorine substitution in the four *ortho* positions and certain PCN isomers are retained by the procedure and are frequently observed by this procedure at sub-part-per-billion concentrations in fish samples. The analytical behavior of these compounds in this procedure is well defined and their presence has not adversely affected the accuracy of the analyses.

HRGC/LRMS-MIM analyses of both samples of the processed spiking mixture (i.e., the analytes produced by application of the enrichment procedure) contained nine significant responses in the tetrachloro congener region, two in the pentachloro region, two in the hexachloro region, zero in the heptachloro region, and two in the octachloro region (Figure 18.3). Of these, the following numbers of positive determinations for PCDF and PCDD were observed: three tetra-CDF, two hexa-CDF, octa-CDF and octa-CDD. The other responses in the fragmentogram (Figure 18.3) did not meet the requirements of correct nominal mass or molecular ion cluster pattern. The retention times of these seven components, which produce positive responses, are designated by the arrows in the fragmentograms. Only two of the seven positive responses exhibit the same retention time as any one of the compounds in the interference mixture listed in Table 18.2. Compound 21, observed in the interferences mixture (but not listed in Table 18.2), was identified as octa-CDD and most probably arises largely from thermolysis of compound 22, the predioxin nonachloro-2-hydroxydiphenyl ether. The predioxin compound 22 was eliminated during the enrichment procedure. The residual octa-CDD detected in the final analyte amounts to less than 30 ppt and is suspected to originate as a trace impurity in the preparation of the predioxin 22 [13].

The positive response for a tetra-CDF (component c in Figure 18.3) observed at the same retention time as compound 5 (the tetrachloromethoxybiphenyl) did not exhibit the characteristic mass spectrum of compound 5 under full scan analysis. The same observation was made with regard to the positive response of component o in Figure 18.3 at the same retention time as compound 22. This

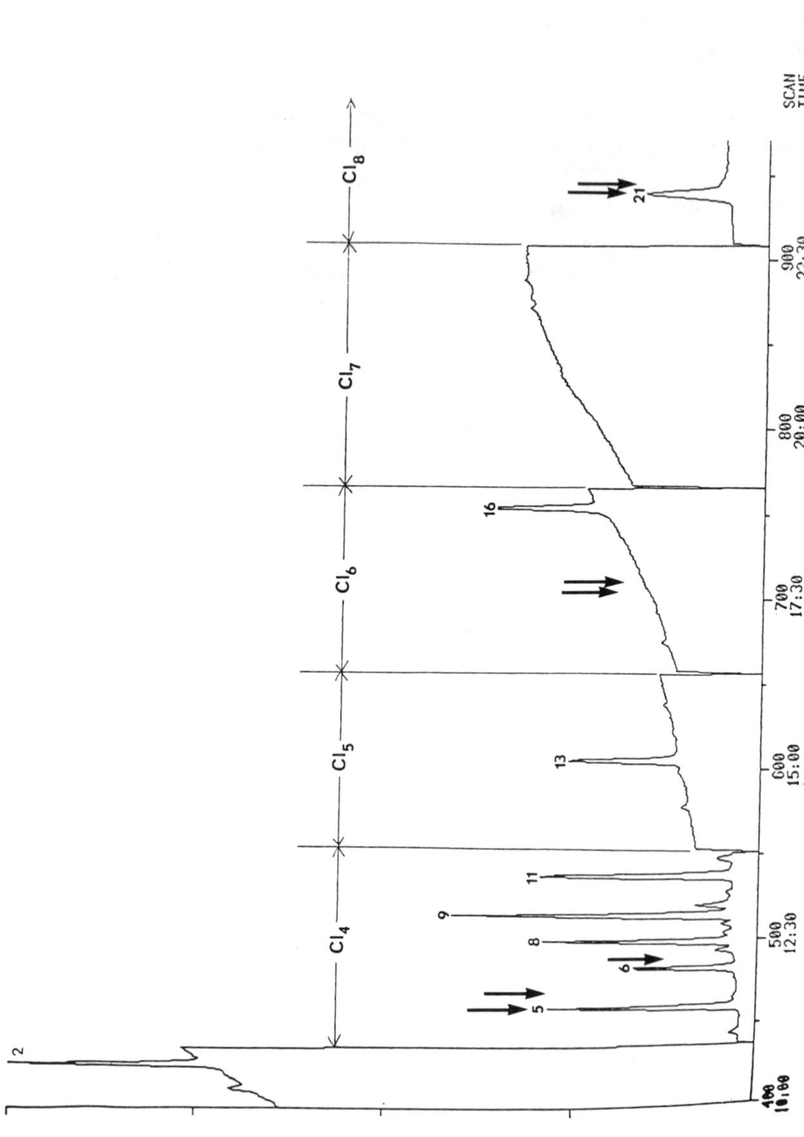

Figure 18.2. HRGC/LRMS-MIM Analysis of Interference under Analytical Conditions Normally Used for Determination of Tetra- through Octa-CDD and -CDF.

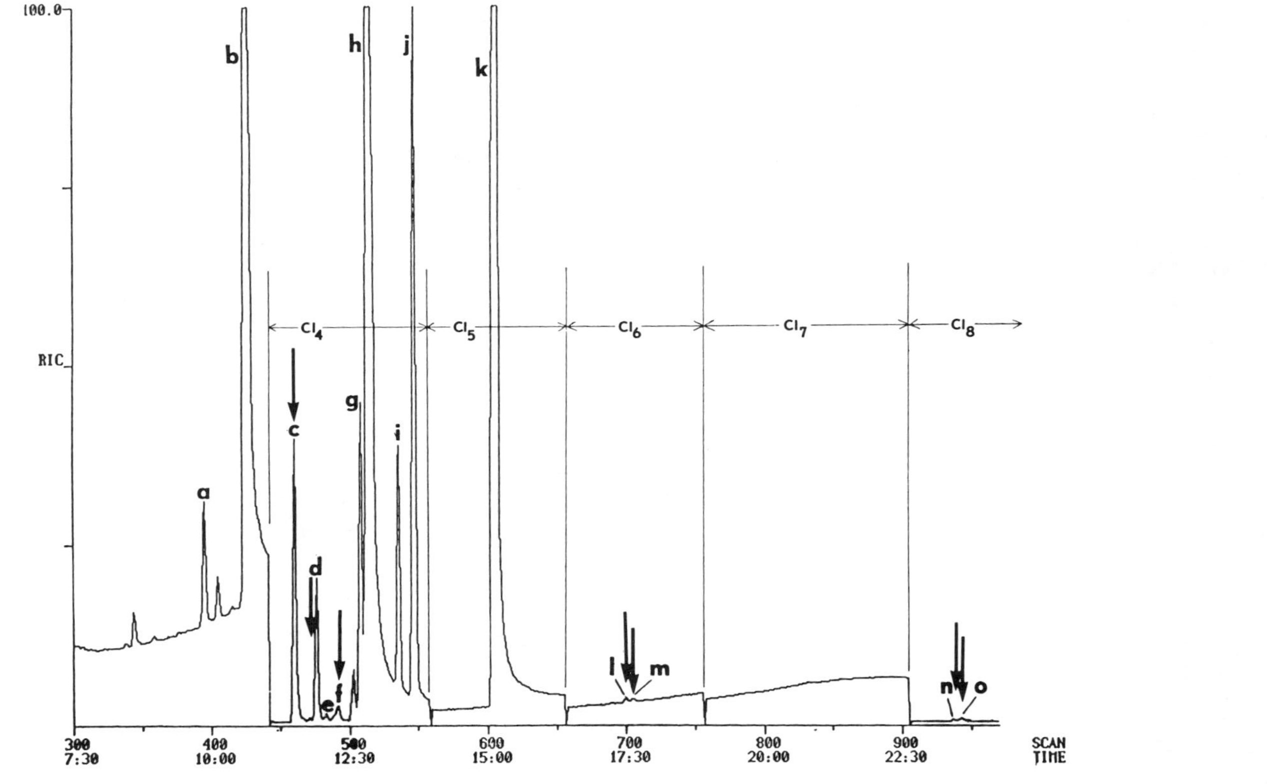

Figure 18.3. HRGC/LRMS Analysis of Interference Mixture Following Enrichment (a) Penta-PCN; (b) No. 2, Table 18.2; (c) Tetra-CDF; (d) Penta-CBP; (e) Tetra-CDF; (f) Tetra-CDF and ^{37}C1-Tetra-CDF; (g) Hexa-PCN; (h) No. 9, Table 18.2; (i) Hexa-PCN; (j) Hexa-PCN; (k) No. 13, Table 18.2; (l) Hexa-CDF; (m) Hexa-CDF; (n) Octa- and ^{37}C1-Octa-CDD; (o) Octa-CDF.

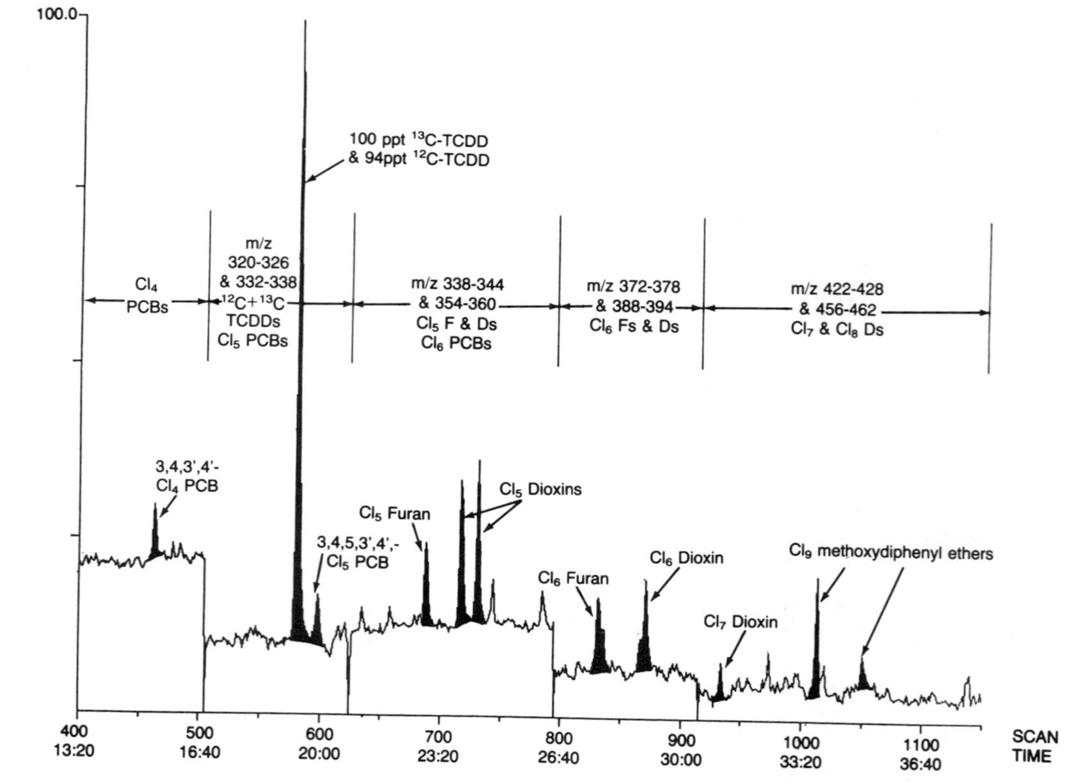

Figure 18.4. Analysis of a Fish Sample from Saginaw Bay for PCDF, PCDD and Related Compounds.

component of the final analyte is apparently octa-CDF, but compound 22 does not produce the octa-CDF molecular cluster. Consequently, all seven positive responses observed in the processed sample are not produced by any of the original components of the interference mixture. The seven positive-response compounds are all believed to be PCDF and PCDD that most likely originate as trace impurities in the chemicals used to prepare the interference mixture (Table 18.2).

The elimination of all PCB except the nonortho isomers was demonstrated in similar spiking studies, being effectively eliminated by use of activated carbon [16]. Furthermore, the elimination of high levels of PCB and DDE present in fish samples from contaminated environments has been repeatedly demonstrated. Figure 18.4 shows the HRGC/LRMS-MIM analysis of a fish sample taken from Saginaw Bay in Lake Huron, a sample that is probably typical of those containing a large variety of chemical contaminants.

This study demonstrates that the series of compounds listed in Table 18.2 do not produce significant levels of false positive interferences in the analytical procedure employed. Significant levels of obscuring interference can be produced by high concentrations of PCN (greater than 1 ppm) and nonortho PCB (greater than several parts per billion), but experience in this laboratory with more than 100 environmental fish samples indicates that the presence of these cocontaminants is generally not detrimental to the determinations of PCDF and PCDD at low parts-per-trillion levels. The analysis of a sample of soot produced in accidental pyrolysis of a PCB preparation from a transformer showed that polychlorinated biphenylenes are largely recovered in the final analyte of this procedure. Polychlorinated biphenylenes have not been observed in fish or sediment samples analyzed in this laboratory. The collection of compounds listed in Table 18.2 and used in this study cannot be considered universally representative of all the members of the particular chemical families, especially because of the large number of possible isomers. Significant variations in the retention times and fragmentation patterns can be expected. Indeed, in a small number of fish samples analyzed for PCDF and PCDD at this laboratory, nonachloromethoxydiphenyl ethers have been tentatively identified. On the other hand, many of the compounds selected for this study are worst-case representatives, those lacking ortho-substitution (known to be most strongly adsorbed by the activated carbon employed) [17] and those most prone to ring closure to form PCDF and PCDD.

REFERENCES

1. Dolphin, R.J., and F. Willmott. *J. Chromatog.* 149:161 (1978).
2. Hummel, R.A. *J. Agric. Food Chem.* 25:1049 (1977).
3. Shadoff, L.A., W.W. Blaser, C.W. Kocher and H.G. Fravel. *Anal. Chem.* 50:1586 (1978).
4. Harless, R.L., E.O. Oswald, M.K. Wilkinson, A.E. Dupesy, Jr., D.D. McDaniel and H. Tai. *Anal. Chem.* 52:1239 (1980).
5. Lamparski, L.L., T.J. Nestrick and R.H. Stehl. *Anal. Chem.* 51:1453 (1979).

6. Hummel, R.A., and L.A. Shadoff. *Anal. Chem.* 52:191 (1980).
7. Lamparski, L.L., and T.J. Nestrick. *Anal. Chem.* 52:2045 (1980).
8. Langhorst, M.L., and L.A. Shadoff. *Anal. Chem.* 52:2037 (1980).
9. Phillipson, D.W., and B.J. Puma. *Anal. Chem.* 52:2328 (1980).
10. Albro, P.W., and C.E. Parker. *J. Chromatog.* 197:155 (1980).
11. Mitchum, R.K., W.A. Korfmacher, G.F. Molar and D.L. Stalling. *Anal. Chem.* 54:719 (1982).
12. Harless, R.L., and R.G. Lewis. Proceedings of the 28th Annual Conference on Mass Spectrometry and Allied Topics (1980), pp. 592–593.
13. Rappe, C., and C.A. Nilsson. *J. Chromatog.* 67:247 (1972).
14. Baker, P.G., R.A. Hoodless and J.F.C. Tyler. *Pest. Sci.* 12:297 (1981).
15. Stalling, D.L., J.D. Petty, L.M. Smith, C. Rappe and H.R. Buser. In: *Chlorinated Dioxins and Related Compounds,* O. Hutzinger, Ed. (New York: Pergamon Press, 1982), pp. 77–85.
16. Smith, L.M. *Anal. Chem.* 53:2152 (1981).
17. Huckins, J.N., D.L. Stalling and J.D. Petty. *J. Assoc. Off. Anal. Chem.* 63:750 (1980).

V

Occupational Exposure and Effects

19

Occupational Exposure to Polychlorinated Dibenzo-*p*-dioxins and Dibenzofurans: A Perspective

Gangadhar Choudhary

Recently, two classes of closely related tricyclic aromatic compounds, polychlorinated dibenzo-*p*-dioxins (PCDD) and dibenzofurans (PCDF), have been a subject of major concern due to their acute toxicological effects and associated adverse health implications. These two classes of compounds, like most halogenated aryl hydrocarbons, have a number of common chemical, physical, biological and toxicological properties, and have relatively stable aromatic nuclei. They also have relative inertness to acids, bases, oxidation, reduction and heat. The chemical inertness tends to increase with chlorination. Both classes of compounds have planar chemical structures. The chemical structures of these compounds are shown in Figure 19.1.

The growing concern with exposure to these two classes of compounds arises primarily from their potential toxicity and their presence as contaminants in various commercial products. The purpose of this chapter is to collect information from literature sources as to what is currently known about the occurrence of

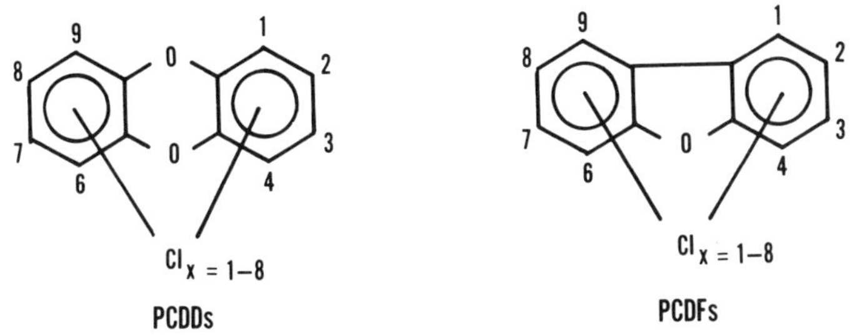

PCDDs ($Cl_{x = 1-8}$) PCDFs ($Cl_{x = 1-8}$)

Figure 19.1. Structures of PCDD and PCDF

these compounds in the workplace atmosphere and how the exposure can correctly be assessed.

SOURCES OF OCCUPATIONAL EXPOSURES

Occupational exposure to PCDD and PCDF can be attributed to a variety of chemical products manufactured or used in the workplace. Some chemicals considered as the source of PCDD and PCDF are mono-, tri-, tetra- and pentachlorophenols (CP) and their salts, tetra- and hexa-chlorinated benzenes, chlorinated phenoxyphenols (predioxins), chlorinated diphenyl ethers, and polychlorinated biphenyls (PCB). PCDD and PCDF are present in these chemicals as trace contaminants or are produced as unwanted by-products during their industrial manufacture and use. One such reaction is the production of PCDD and PCDF from industrially important chlorobenzenes [1] as shown in Figure 19.2.

Also, PCB, which are commonly used as dielectrics in electrical transformers and capacitors (and less commonly used as lubricators in cutting oils, pesticides, heat exchanger fluids, plasticizers, adhesives and sealants), contain PCDD and PCDF as contaminants [2]. PCB also produce PCDF on heating [3,4] as shown in Figure 19.3.

Similarly, workers in a CP plant or in facilities using CP may be exposed to PCDD or PCDF, depending on the prevailing ambient conditions. Dioxins found in 2,4,5-trichlorophenoxyacetic acid (2,4,5-T), penta-CP and hexachlorophene are formed during the manufacturing process as exemplified in Figure 19.4.

Classical Ullmann dimerization [5] of chlorophenate may produce highly toxic 2,3,7,8-tetra-CDD from trichlorochlorophenate as per Figure 19.5.

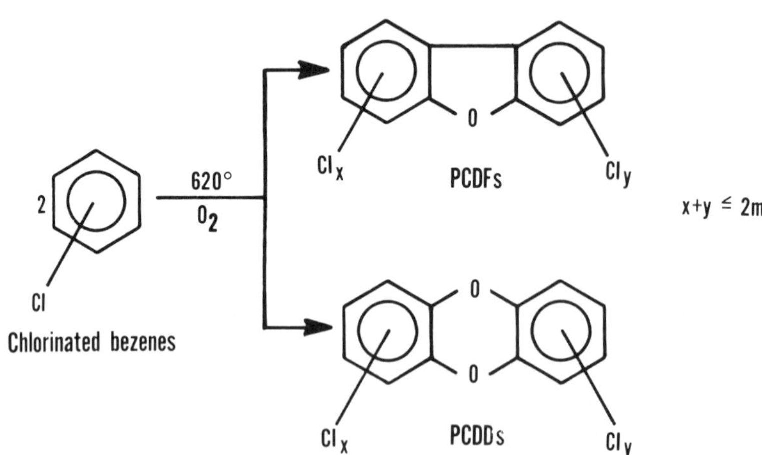

Figure 19.2. Production of PCDD and PCDF from Chlorinated Benzenes.

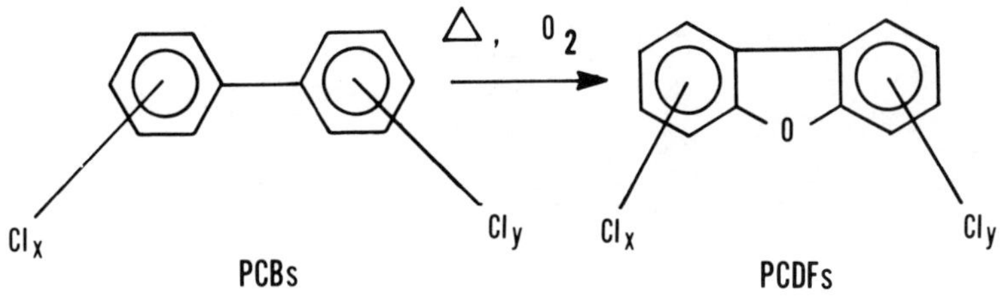

Cl$_x$ PCBs Cl$_y$ Δ, O_2 Cl$_x$ O PCDFs Cl$_y$

Figure 19.3. Production of PCDF from PCB.

Trichlorochlorophenate may also produce PCDF by reaction with common contaminants present in the CP, namely, polychlorinated diphenyl ether (PDPE) and chlorobenzenes [6] as shown in Figure 19.6. $FeC1_3$-catalyzed dimerization reactions usually take place. PDPE itself converts to PCDD and PCDF under heat conditions at 600°C according to Figure 19.7 [7].

Chlorinated dioxins and dibenzofurans are also formed in combustion sources and often detected in incinerators. Two important reviews have appeared in the literature [8,9].

Some potential workplaces where occupational exposure to PCDD and PCDF may be suspected due to the use of their chemical precursors are:

1. CP-related facilities

- 2,4,5-T phenoxy herbicide production facilities,
- sawmills,
- wood-treatment plants,
- manufacture, shipping and formulation facilities of industrial CP,
- leather and tanning, including gelatin production,
- carpentry and other timber and wood-working,
- termite control spraying,
- agricultural pesticide application, including greenhouses,
- industrial cooling towers and evaporative condensers,
- treatment and handling of wool,
- treatment and handling of burlap, canvas and rope,
- pulp and paper plants,
- petroleum and other drilling,
- paint and adhesive manufacture and use,
- telephone and electrical line work,
- dyeing and cleaning of garments (textile industry),
- sewing facilities (related to garment impregnation),
- health-related facilities (using hexachlorophene bactericide),
- nursing (using hexachlorophene as bactericide),
- cutting fluids and hydraulic fluids, and
- handling latex rubber;

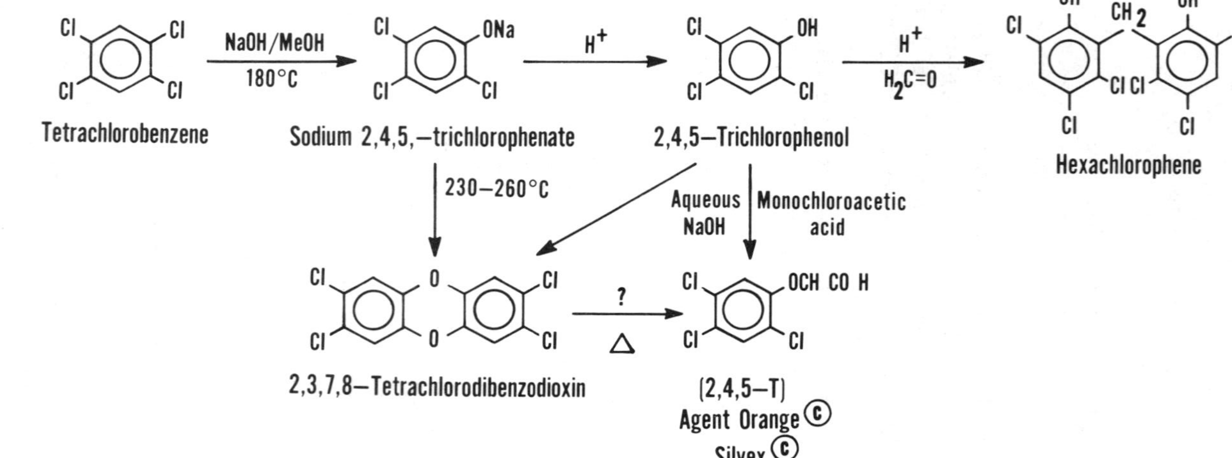

Figure 19.4. Industrial Production of Trichlorophenol, Hexachlorophene, 2,4,5-T and By-Product, 2,3,7,8-Tetra-CDD.

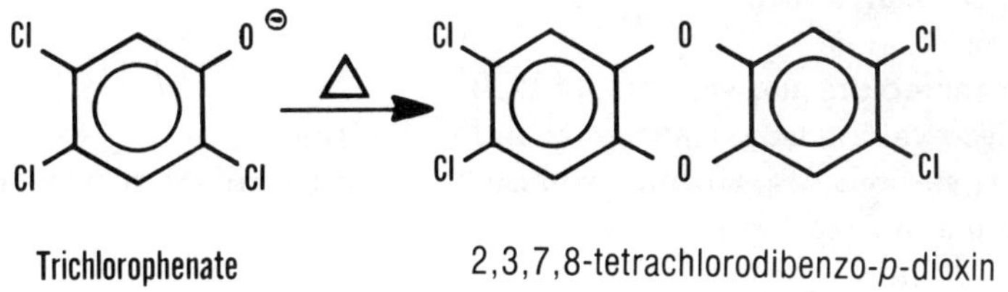

Figure 19.5. Production of 2,3,7,8-Tetra-CDD from Trichlorophenate.

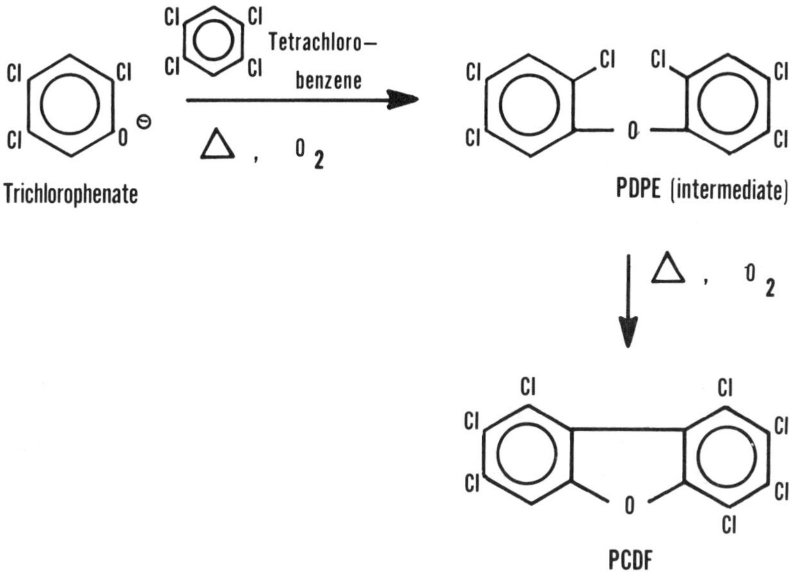

Figure 19.6. Production of PCDF via PDPE from Trichlorophenate.

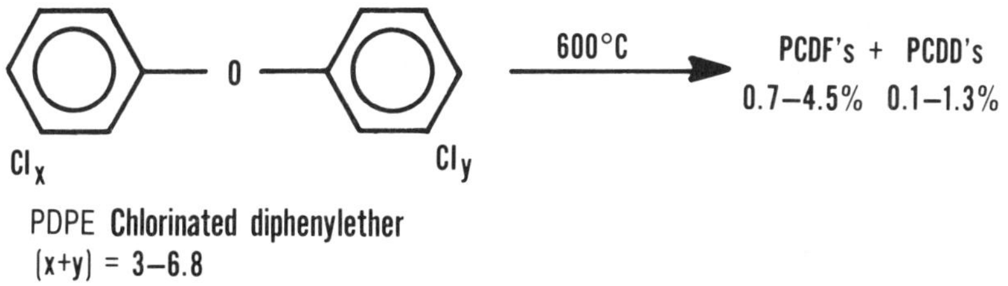

Figure 19.7. Production of PCDD and PCDF from Chlorinated Diphenylether.

2. PCB-related facilities

- manufacture and shipping of PCB,
- electrical industry (capacitors and transformers),
- fire stations (transformer explosions and use of flame retardants),
- contaminated food supplies,
- paint industry,
- foundries and industries using cutting fluids,
- adhesives and sealants,
- pesticides (spraying facilities),
- copying paper handling,
- plastics, and
- utilizing casting waxes; and

3. combination of PCDD and PCDF precursor = related facilities

- waste disposal sites (dump sites), and
- municipal incinerators.

COMPLEXITIES OF EXPOSURE ASSESSMENT

Chemically, PCDD and PCDF occur in various positional isomeric forms and various levels of chlorine substitution. These isomers have been found to possess different degrees of toxicity. Consequently, an investigation of the toxicological properties of these compounds without regard to isomeric specificity may lead to erroneous conclusions. Determination of the exact isomer and its toxicity may be of vital importance during any unequivocal assessment of the exposure hazard to these compounds. Therefore, it is crucial to any health hazard evaluation that both questions of exact analysis and exact toxicity be addressed and the relation between toxicity and analytical isomer be established.

Analytical Considerations

During any evaluation of the problem, it should be asked how certain one is that the analyte is correctly identified and accurately measured. The problem becomes more complicated while determining the amount of PCDD or PCDF in an environmental sample from a workplace.

The number of chlorine atoms in these compounds varies between one and eight and, because of various levels of chlorine substitution, there are 75 possible positional isomers for PCDD and 135 possible positional isomers for PCDF. Table 19.1 lists the number of isomers at each level of chlorine substitution for these compounds.

Table 19.1. Possible Number of Positional PCDD and PCDF Isomers

Level of Chlorine Substitution	*Number of Positional Isomers*	
	PCDD	*PCDF*
Mono-	2	4
Di-	10	16
Tri-	14	28
Tetra-	22	38
Penta-	14	28
Hexa-	10	16
Hepta-	2	4
Octa-	1	1
Total	75	135

Of the 75 possible PCDD isomers and 135 PCDF isomers, only a few have been investigated at any length with regard to their isomer-specific analysis. Only recently has some attention been paid to isomeric separation of PCDD and PCDF homologs. To elucidate the point, one such compound that has received considerable attention in the recent past is the highly toxic congener 2,3,7,8-tetra-CDD. Unfortunately, in considering dioxin exposure, 2,3,7,8-tetra-CDD has often been equated to the gross combined tetra- isomers of tetra-CDD. The determined exposure may not be to the highly toxic 2,3,7,8-tetra-CDD. Rather, it may be to some other tetra- isomers whose toxicities are less severe or uncertain. All the congeners of hepta-CDF, tetra-CDD [10,11], hexa-CDD, hepta-CDD, and tetra-CDF [12] have been separated and identified chromatographically. Figures 19.8 through 19.11 show some of the efforts that have been made for isomeric separation of congeners. Table 19.2 provides the peak identification key to Figures 19.8 and 19.9. More efforts are being made in the scientific community to improve the analytical methodology, synthesize the standards, achieve the isomer separation and thereby provide a meaningful measurement in terms of sampling and analysis. To date, the problem of validated sampling and analysis is far from being solved. Unavailability of appropriate standards at various levels of chlorination of these compounds adds to the complexities of any measurement efforts.

One of the major analytical problems associated with PCDD and PCDF determination is measurement near the limit of detection, as required by their occurrence as trace contaminants (parts per billion or lower) with other chemical species. Hence, any measurement has some uncertainty, and validating the analyses at these low levels becomes difficult. Extreme caution should be exercised if the correct exposure assessment is to be made.

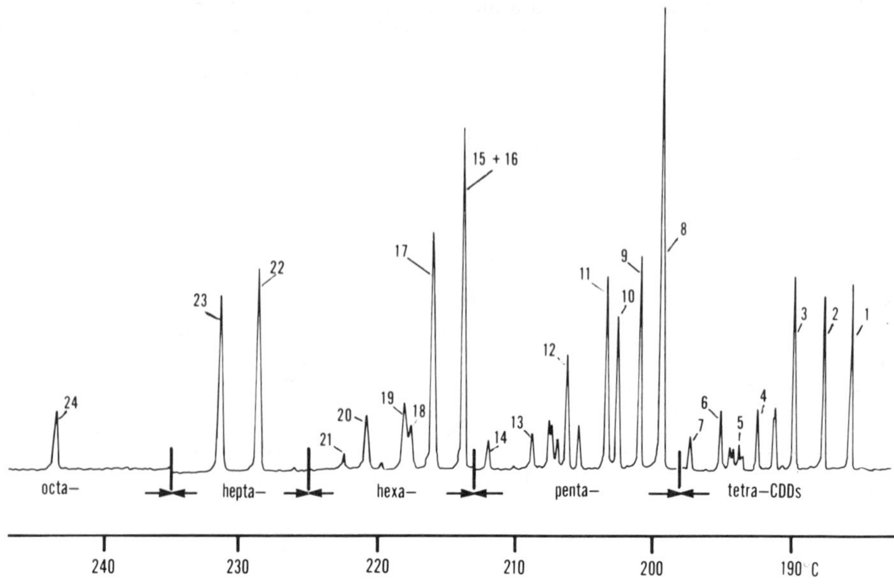

Figure 19.8. Mass fragmentograms (50-m Silar-10c glass capillary column, m/e 320, 354, 388, 424 and 460) showing elution of tetra-, penta-, hexa-, hepta- and octa-CDD in a combined chlorobenzene pyrolyzate. Peak identifications are given in Table 19.2. Column conditions: 100°C for 2 min, 10°C/min to 140°C, and then 50°C/min to 240°C. Helium carrier gas [1].

Toxicity Considerations

Similar to analytical considerations, isomer-specific toxicity investigations of PCDD and PCDF have not been addressed adequately in the literature. In fact, these investigations are much more limited than are the analytical investigations. Animal toxicity using highly toxic 2,3,7,8-tetra-CDD has been investigated abundantly, and a clear picture is beginning to emerge with respect to toxicological effects of 2,3,7,8-tetra-CDD in animals. However, no definitive epidemiological data involving a human population are available. Data pertaining to congeners or homologs of PCDF are almost nonexistent in the literature. A study [13] of animal exposure to a mixture of 31% 1,2,3,6,7,8-hexa-CDD and 67% 1,2,3,7,8,9-hexa-CDD has been made. A summary of available toxicological data for some congeners is given in Table 19.3.

The potential for toxicity investigation exists for 75 congeners of PCDD and 135 congeners of PCDF, but it is impractical, and probably impossible, to assess the potential hazards of even a small number of these congeners through classical testing procedures. Understanding the mechanisms by which these compounds

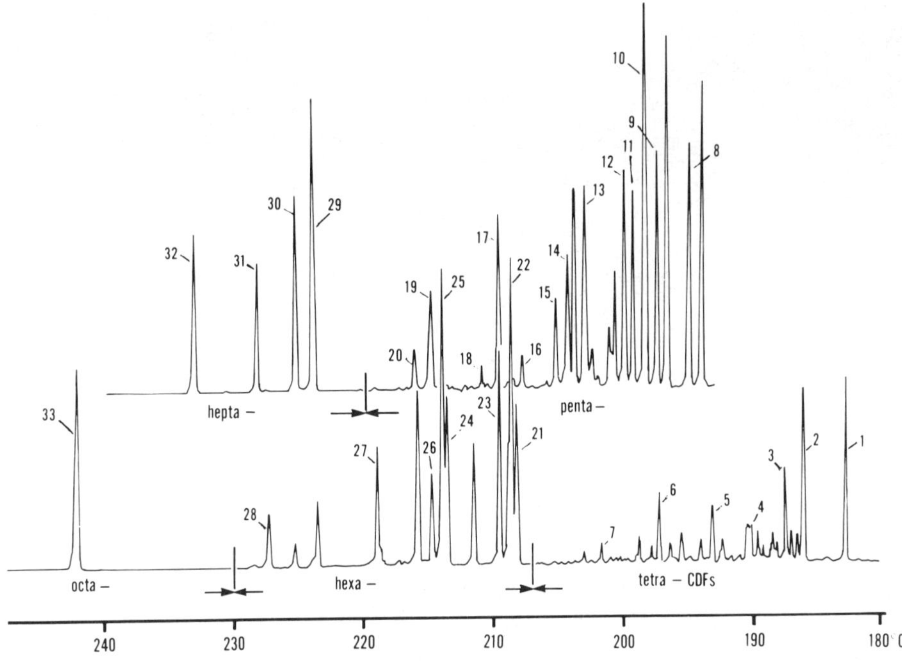

Figure 19.9. Mass fragmentograms (50-m Silar-10c glass capillary column, m/e 306, 340, 374, 410 and 444) showing elution of tetra-, penta-, hexa-, hepta- and octa-CDF in a combined chlorobenzene pyrolyzate. Peak identifications are given in Table 19.2. Column conditions: 100°C for 2 min, 10°C/min to 140°C, and then, 50°C/min to 240°C. Helium carrier gas [1].

produce their toxic responses will, hopefully, uncover some commonality in their mode(s) of action. This may lead to development of more rapid and concise tests, such as a structure-activity relation study, to identify those congeners of most concern, and subsequent testing of such congeners.

Based on the induction of δ-aminolevulinic acid synthetase (ALAS) and aryl hydrocarbon hydroxylase (AHH) enzymes by a series of halogenated PCDD, Greenlie and Poland [15] concluded that isomers with halogen atoms on three of the four lateral ring positions and with at least one unoccupied ring carbon are found to be inducers. Such isomers are supposed to be toxic. Toxicity will be higher if all the four lateral ring positions are substituted with chlorine. This hypothesis is consistent with the fact that octa-CDD, where there is no unoccupied ring carbon, is found nontoxic, whereas 2,3,7,8-tetra-CDD is highly toxic. No actual toxicological data are available for PCDF. Moore [16] predicted that the toxicological properties of PCDF should be similar to those of PCDD. Figure 19.12 lists some highly toxic congeners of PCDD and PCDF.

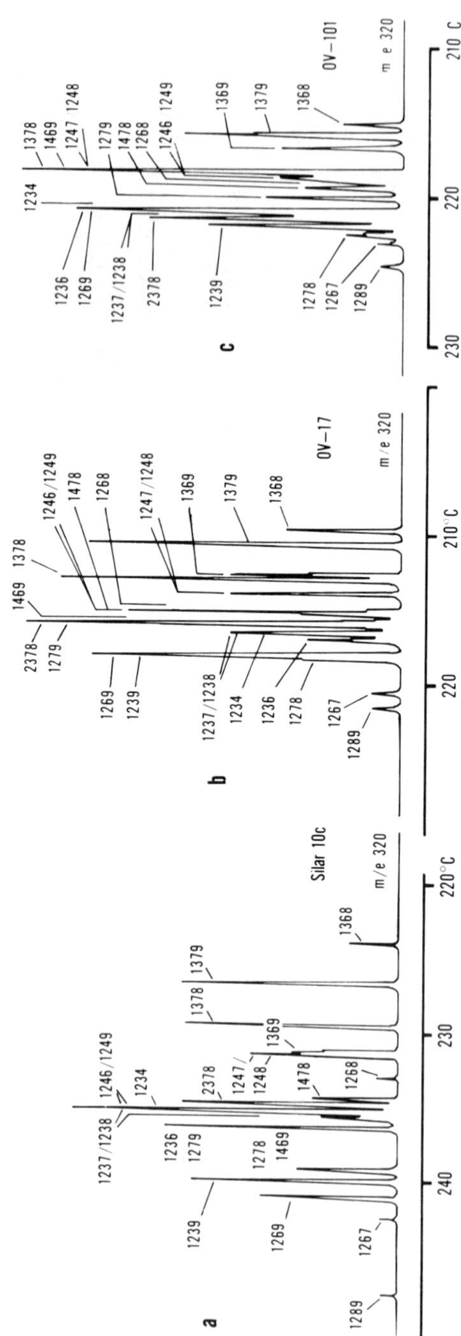

Figure 19.10. Mass fragmentograms (m/e 320) of a composite sample showing elution of all 22 tetra-CDD isomers on (a) 55-m Silar-10c; (b) 50-m OV-17; and (c) 50-m OV-101 HRGC columns. Column conditions: 100°C for 3 min, 20°C/min to 180° and then 20°C/min to 240°C. Helium carrier gas [10].

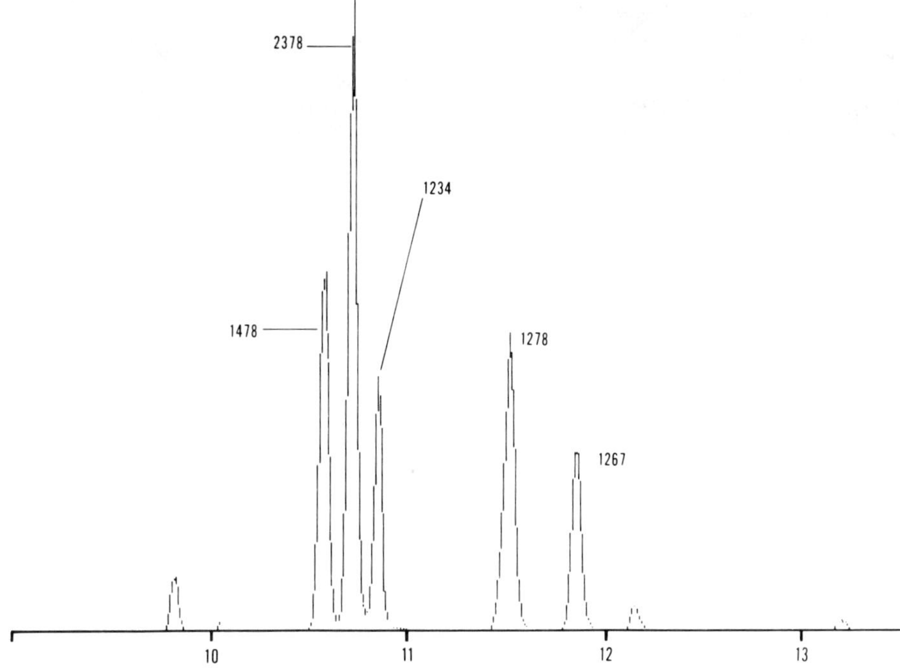

Figure 19.11. Performance evaluation standard mixture of tetra-CDD analyzed using 54-m fused silica SP-2330 column. Column conditions: 200°C for 1 min, 12°C/min to 265°C. Hydrogen carrier gas [12].

Table 19.2. Identification of PCDF and PCDD Isomers in the Combined Chlorobenzene Pyrolyzate (Figures 19.8 and 19.9) [1]

Peak No. (Figure 19.3)	*PCDF Isomer*	*Peak No. (Figure 19.2)*	*PCDD Isomer*
	Tetra-CDF		Tetra-CDD
1	1,3,6,8-	1	1,3,6,8-
2	1,3,7,9-	2	1,3,7,9-
3	1,3,6,7-	3	1,3,7,8-
4	1,2,4,6-	4	1,3,6,7-
5	2,4,6,8-	5	2,3,7,8-
6	2,3,6,8-	6	1,3,8,9-
7	2,3,7,8-	7	1,2,7,8-
	Penta-CDF		Penta-CDD
8	1,2,4,6,8-	8	1,2,4,6,8-
9	1,3,4,7,9-	9	1,2,3,6,8-
10	1,3,4,7,8-	10	1,2,4,7,8-
11	1,2,4,7,8-	11	1,2,3,7,9-
12	1,2,4,7,9-	12	1,2,3,7,8-

Table 19.2. *(continued)* Identification of PCDF and PCDD Isomers in the Combined Chlorobenzene Pyrolyzate (Figures 19.8 and 19.9)

Peak No. (Figure 19.3)	*PCDF Isomer*	*Peak No. (Figure 19.2)*	*PCDD Isomer*
13	1,2,3,7,8- + 1,2,3,4,8-	13	1,2,3,6,7-
14	1,2,3,6,7-	14	1,2,3,8,9-
15	1,2,6,7,8-		Hexa-CDD
16	1,3,4,8,9-	15	1,2,4,6,7,9- or 1,2,4,6,8,9-
17	2,3,4,6,8-	16	1,2,3,4,6,8-
18	1,2,4,8,9-	17	1,2,3,6,8,9- or 1,2,3,6,7,9-
19	2,3,4,7,8-	18	1,2,3,4,7,8-
20	2,3,4,6,7-	19	1,2,3,6,7,8-
	Hexa-CDF	20	1,2,3,7,8,9-
21	1,2,3,4,6,8-	21	1,2,3,4,6,7-
22	1,3,4,6,7,8-		Hepta-CDD
23	1,2,4,6,7,8-	22	1,2,3,4,6,7,9-
24	1,2,3,4,7,8-	23	1,2,3,4,6,7,8-
25	1,2,3,6,7,8-		
26	1,2,4,6,8,9- + 1,2,3,4,6,7-	24	Octa-CDD
27	1,2,3,6,8,9-		
28	2,3,4,6,7,8-		
	Hepta-CDF		
29	1,2,3,4,6,7,8-		
30	1,2,3,4,6,7,9-		
31	1,2,3,4,6,8,9-		
32	1,2,3,4,7,8,9-		
33	Octa-CDF		

Table 19.3. Summary of Toxicological Data of PCDD in Terms of LD50-30 [13,14]

	Guinea Pigs		*Mice*	
Chlorination	*μg*	*μmol/kg*	*μg/kg*	*μmol/kg*
2,7-	> 2,000,000			33,000
2,8-	> 300,000	>1180	>150,000	
2,3,7-	29,444	102.41	>3,000	10
2,3,7,8-	2	0.006	283.7	0.88
1,2,3,4-	>1,000,000			
1,2,3,7,8-	3.1	0.009	337.5	0.94
1,2,4,7,8-	1,125	3.15	>5,000	>14
1,2,3,6,7,8-	70–100	0.178–0.255	1,250	3.19
1,2,3,7,8,9-	60–100	0.153–0.255	1,440	3.67
1,2,3,4,6,7,8-	180	0.423		

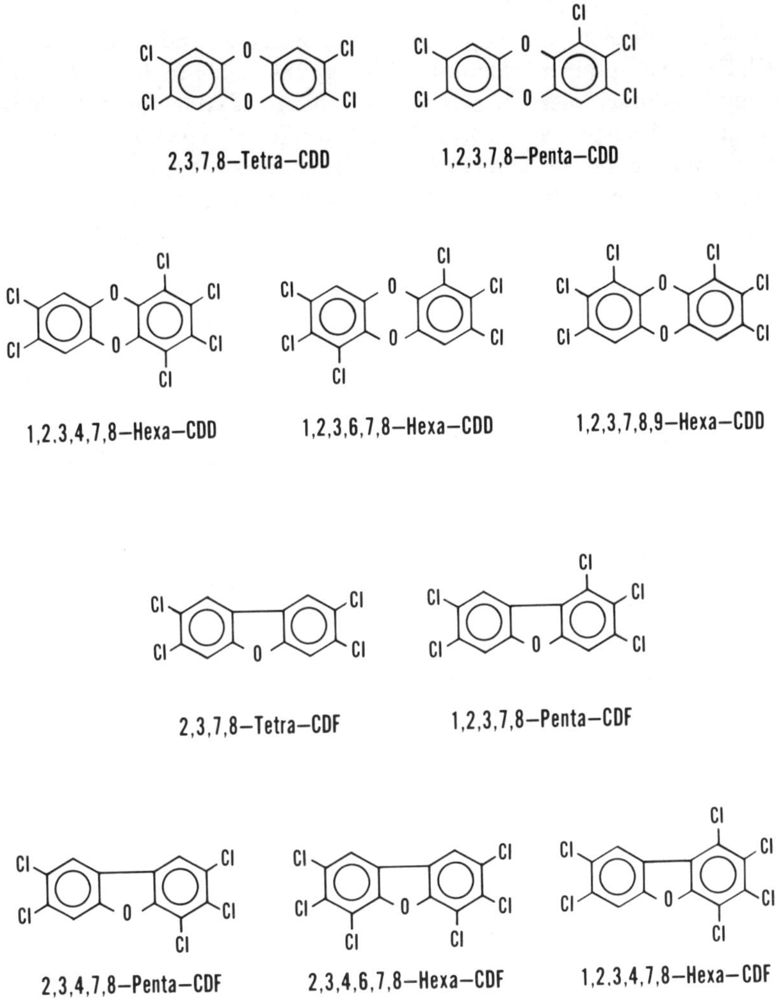

2,3,7,8–Tetra–CDD 1,2,3,7,8–Penta–CDD

1,2,3,4,7,8–Hexa–CDD 1,2,3,6,7,8–Hexa–CDD 1,2,3,7,8,9–Hexa–CDD

2,3,7,8–Tetra–CDF 1,2,3,7,8–Penta–CDF

2,3,4,7,8–Penta–CDF 2,3,4,6,7,8–Hexa–CDF 1,2,3,4,7,8–Hexa–CDF

Figure 19.12. Some Highly Toxic PCDD and PCDF Isomers.

FIELD ANALYTICAL RESULTS

In 1975 a plant in Seveso, Italy, producing hexachlorophene from tri-CP, had an explosion [17]. Soil from the surrounding areas has indicated the presence of 2,3,7,8-tetra-CDD. Some toxic PCDF congeners have been found in the clinical samples from yusho patients in Japan [18,19]. Their diet was contaminated with PCB. Sludge and other wastes from the wastewater disposal stream in a wood preserving facility in the United States have shown the presence of 200–6000 ng/g of 2,3,7,8-tetra-CDD [20]. Rappe [21] analyzed the clinical samples from the workers of a textile plant, a sawmill and a tanning facility, and congeners of PCDD and PCDF were found in all samples. Although the highly toxic 2,3,7,8-congener is not found in the samples, the presence of other toxic components in these samples indicates the suspected hazard. Rappe's results are given in Table 19.4 and 19.5. Buser and Rappe [10] and Lamparski and Nestrick [11] found low amounts (parts per billion) of 2,3,7,8-tetra-CDD in fly ash from a municipal incinerator. Table 19.6 gives PCDD found in fly ash by Lamparski and Nestrick [11].

Table 19.4. Levels of PCDD and PCDF in Blood Samples from Workers in the Textile and Tannery Industries after Exposure to Penta-CP or Its Derivatives [21]

Occupation	Chlorophenols in Urine (μg/mL)	PCDD (pg/g blood)				PCDF (pg/g blood)			
		O_8	H_7[a]	H_6	P_5	O_8	H_7[a]	H_6	P_5
Textile	3.12	304	59	<1	<1	10	33	<1	<1
	<0.01	3	<1	<1	<1	<2	<1	<1	<1
	<0.01	10	1	<1	<1	<2	<1	<1	10
	0.42	105	15	<1	<1	<2	<1	<1	<1
	0.16	30	6	<1	<1	<2	<1	<1	1
Tannery[c]	0.55[d]	20	7	<3		<3	7	<3	
	0.04[d]	80	30	3		7	18	3	
	0.03[d]	12	4	<3		3	3	<3	
		7	2	<2		3	3	<3	

[a]Major isomer 1,2,3,4,6,7,8-hepta-CDD.
[b]Major isomer 1,2,3,4,6,7,8-hepta-CDF.
[c]Blood sampling eight months after last exposure.
[d]Urine sampling six months after exposure.

Table 19.5. Levels of PCDD and PCDF in Blood Samples from Workers in Saw Mill Industry after Exposure to 2,3,4,6-Tetrachlorophenate [21]

Person	Profession	Chlorophenols in Urine (μg/mL)	PCDD (pg/g blood)				PCDF (pg/g blood)			
			Octa-	Hepta-[a]	Hexa-	Penta-	Octa-	Hepta-[b]	Hexa-	Penta-
Blank A			7	< 2	< 3		< 3	2	< 3	
Blank B			<2	< 1	< 1	< 1	< 2	< 1	< 1	<1
1	Loader[c]	0.04	5	< 2	< 3		< 3	40	< 3	
	Loader[d]	5.2	7	< 1	< 1	< 1	< 2	22	< 1	<1
2	Cleaner[c]	<0.02	5	2	< 3		< 3	30	< 3	
	Cleaner[d]	0.23	22	8	< 2	< 2	< 3	17	< 1	5

Table 19.5. *(continued)* Levels of PCDD and PCDF in Blood Samples from Workers in Saw Mill Industry after Exposure to 2,3,4,6-Tetrachlorophenate [21]

Person	*Profession*	*Chlorophenols in Urine (μg/mL)*	*PCDD (pg/g blood)*				*PCDF (pg/g blood)*			
			Octa-	*Hepta-*[a]	*Hexa-*	*Penta-*	*Octa-*	*Hepta-*[b]	*Hexa-*	*Penta-*
3	Loader[c]	0.03	18	10	3		<3	18	<3	
	Loader[d]	0.83	3	<1	<1	<1	<2	17	<1	1
5	Package[c]	<0.05	<3	<2	<3		<3	7	<3	
	Package[d]	0.11	4	<1	<1	2	<2	12	<1	<1
7	Control	<0.01	3	<1	<1	<1	<2	3	<1	<1

[a]Major isomer 1,2,3,4,6,7,8-hepta-CDD.
[b]Major isomer 1,2,3,4,6,7,8-hepta-CDF.
[c]Sampling six months after latest exposure.
[d]Sampling after one month of exposure.

Table 19.6. Tetra-CDD Isomer Concentrations (ng/kg) Detected in U.S. and European Fly Ash [11]

Isomer	*U.S.*	*European*
1,2,6,9-	190	1,000
1,4,6,9-	ND (50)[b]	250
1,2,6,7-/1,2,8,9-[a]	220	800
1,2,6,8-/1,2,7,9-[a]	500	2,500
1,3,6,9-/1,4,7,8-[c]		—
1,2,7,8-	ND (80)[b]	3,100
1,2,3,6-/1,2,3,9-[a]	430	2,300
1,2,3,7-/1,2,3,8-[a]	720[d]	8,500[d]
1,2,4,6-/1,2,4,9-[a]	730[d]	3,500[d]
1,2,4,7-/1,2,4,8-[a]	310	6,900
1,3,7,8-	1,370	13,200
1,3,7,9-	1,160	7,000
1,3,6,8-	1,320	16,200
1,2,3,4-	370	2,100
2,3,7,8-	430	2,300
Total	7,750	69,650

[a]Summation of two retention times.
[b]Not detected (at detection limit).
[c]Not recovered.
[d]Possible isomer interference.

Rappe et al. [22] and Rappe [23] found considerable (parts-per-million range) amounts of toxic PCDF in the soots from accidental PCB fires in Skovde, Sweden, and Binghampton, New York, where PCB-filled capacitors burned.

In all above cases the results were based on isomer-specific determinations using HRGC/MS. The positive findings of toxic components of PCDD and PCDF in workplace samples suggest that the exposure to PCDD and PCDF can occur, and assessment of exposure in the workplace should be made carefully.

DECONTAMINATION

One of several means to free the workplace from PCDD and PCDF is to use decontamination procedures. Although no control technologies have been perfected, there are some new efforts being made in this direction. They are briefly mentioned below. In most cases, incineration of the species above 800°C or, sometimes, repeated washing of the contaminated area with phosphate detergent is used.

Thermal Degradation

Temperatures higher than 800°C are used to incinerate PCDD and PCDF. This is the most common method of destruction of PCDD and PCDF.

Photodegradation

Photodegradation of PCDD and a PCDF by ultraviolet (UV) light is one of the most frequent decontamination measures mentioned in the literature. Thus far, efforts have been only partially successful. The process appears to be stepwise degradation of the polychlorinated entities by UV light [24–26] and takes places in solution in the presence of a hydrogen donor solvent.

Chemical Degradation

Oxidative degradation of PCDD (particularly 2,3,7,8-tetra-CDD) has been tried on several occasions. A recent reference is made to using ruthenium oxide as a potential chemical oxidizing agent to degrade tetra-CDD [27].

Biodegradation

A great deal of effort is underway in the scientific community to detoxicate PCDD and PCDF by stripping these compounds of halogens by the action of microorganisms. To date, no successful dechlorination of 2,3,7,8-tetra-CDD has been reported. Prospects are in favor of some breakthroughs in the future. Klecka and Gibon [28] presented a state-of-the-art analysis of biodegradation of PCDD.

Electrochemical Degradation

Efforts have been made to destroy toxic PCDD and PCDF by redox procedures on electrodes in electrochemical cells [29,30]. Although decontamination efforts are being made, not much success has been reported in this regard particularly in the workplace [13,31]. It is hoped that in the near future some breakthroughs will occur in freeing the workplace from the contamination of PCDD and PCDF.

CONCLUSION

A clear understanding of the chemistry, anticipated occurrence, transport properties and specific toxicological properties is necessary to correctly assess the health implications of exposure to these classes of compounds. A search of the literature shows that insufficient analytical and toxicological data are available to evaluate correctly the exposure and hazards of these compounds. To establish unequivocally the hazards from exposure to a specific isomer, an isomer-specific analysis of the sample must be made. Similarly, isomer-specific toxicological studies must be carried out for which only scant data are available. Isomer-specific toxicological studies beyond 2,3,7,8-tetra-CDD are almost nonexistent. Currently, most exposure evaluation studies are based on animal studies and gross (isomer-nonspecific) tetra-CDD. Besides animal studies, human toxicological studies as well as isomer-specific toxicological investigations at various levels of chlorine substitution of both classes of compound will be helpful in determining the relationship between gross exposure and health hazard potential.

Available field sample analytical (isomer-specific) data suggest that exposure to toxic PCDD and PCDF in the workplace is real in many cases. Hence, extreme caution should be used while handling the anticipated precursors of these compounds or chemicals contaminated by PCDD and/or PCDF in the workplace environment. As no federal or state exposure standards are available, even the presence of ultratrace amount of these compounds in the workplace should be construed as dangerous.

Usually, because of the extreme toxicological properties of some of these compounds, the suspected presence of PCDD and PCDF in the workplace environment becomes a matter of concern. However, the myth of the suspected exposure must be differentiated from the reality of exposure through authentic specific analytical and toxicological data for these compounds. Consideration of the following simplified questions may provide a guideline to assess exposure to PCDD and PCDF.

1. Has isomer-specific measurement of the samples been made?
2. Are toxicological data available for the determined isomer?
3. If not available, can the available toxicological data be estimated based on toxicity-structure relationship?
4. Does the measured amount exceed the minimum level of government guidelines (if available)?

If the answer to all the questions is "yes," there is definite exposure to PCDD or PCDF. If the answers are "no" or a combination of "yes" and "no," the correct assessment of exposure may be in doubt and reconsideration of the health hazard evaluation procedure of the workplace may be necessary. In want of governmental guidelines, the presence of even ultratrace amounts of toxic isomers of PCDD and PCDF must not be considered safe.

DISCLAIMER

Mention of a specific product or company does not constitute endorsement by the National Institute for Occupational Safety and Health.

REFERENCES

1. Buser, H.R., *Chemosphere* 6:415 (1979).
2. Schneider, M.-J. In: *Persistent Poisons* (New York: New York Academy of Sciences, 1979), p. 13.
3. Choudhry, G.G. In: *Chlorinated Dioxins and Related Compounds,* O. Hutzinger, R.W. Frei, E. Meriam and F. Pocchiari, Eds. (New York: Pergamon Press, 1982), p. 286.
4. Buser, H.R., and C. Rappe. *Chemosphere* 3:157 (1979).
5. Ullmann, R. *Ber. Deutch. Chem. Gesells.* 36:2382 (1903).
6. Nilsson, C.A., A. Norstrom, K. Anderson and C. Rappe. In: *Pentachlorophenol Chemistry, Pharmacology and Environmental Toxicology,* K. Ranga Rao, Ed. (New York: Plenum Press, 1978), p. 313.
7. Lindahl, R., C. Rappe and H.R. Buser. *Chemosphere* 9:351 (1980).
8. Harris, J.C., R.C. Anderson, B.E. Goodwin and C.E. Rechsteiner. In: *A.D. Little Review Report to Research Committee on Industrial Municipal Wastes* (New York: American Society of Mechanical Engineers, 1981).
9. Shih, C., D. Alkerman, L. Scinto and B. Johnson. "Emissions of PCDDs and PCDFs from the Combustion of Fossil Fuels," U.S. EPA Contract No. 68-02-3138 (1980).
10. Buser, H.R., and C. Rappe. *Anal. Chem.* 52:2257 (1980).
11. Lamparski, L.L., and T.J. Nestrick. *Anal. Chem.* 52:2045 (1980).
12. Hileman, F., Monsanto Research Corporation. Personal communication (1982).
13. "Polychlorinated Dibenzo-*p*-dioxins," NRCC No. 18576, National Research Council of Canada (1981).
14. McConnell, E.E., and J.A. Moore, *Ann. N.Y. Acad. Sci.* 320:138 (1979).

15. Greenlie, W.F., and A. Poland. In: *Dioxin Toxicological and Chemical Aspects,* F. Cattabenis, Al Cavallaro and Giovanni Galli, Eds. (New York: S.P. Medical and Scientific Books, 1976) p. 122.
16. Moore, J.A. *Environ. Health Perspec.* 5:313 (1973).
17. Buser, H.R. *Anal. Chem.* 49:918 (1977).
18. Rappe, C., University of Umeå, Sweden. Personal communication (1981).
19. Buser, H.R. *Chemosphere* 5:231 (1977).
20. Wilson, D., U.S. EPA Contract Report 68–03–2567 and 68–03–3028, Cincinnati, OH (1981).
21. Rappe, C., and H.R. Buser. In: *Chemical Hazards in the Workplace,* G. Choudhary, Ed. ACS Symposium series 149 (Washington, DC: American Chemical Society, 1981), p. 328.
22. Rappe, C., S. Marklund, P.-A. Bergqvist and M. Hansson. *Chemica Scripta* 20:56 (1982).
23. Rappe, C., S. Marklund, P.-A. Bergqvist and M. Hansson. "Polychlorinated Dioxins, Dibenzofurans and Other Polychlorinated Polynuclear Aromatics Formed During Incineration and Polychlorinated Biphenyl Fires," Chapter 7, this volume.
24. Liberti, A., D. Brocco, I. Allegrini and G. Bertoni. In: *Dioxin,* F. Cattabeni et al., Eds., (New York: S.P. Medical and Scientific Books, 1976) p. 195.
25. Zepp, R.G., and D.M. Cline. *Environ. Sci. Technol.* 11:359 (1977).
26. Dobbs, A.J., and C. Grant. *Nature* 278:163 (1978).
27. Ayres, D.C. *Nature* 290:323 (1981).
28. Klecka, G.M., and D.T. Gibson. *Appl. Environ. Microbiol.* 39:288 (1980).
29. Harrison, J.M., T.D. Inch and R.G. Wilkinson. *Nature* (in preparation).
30. Fleet, B., Environmental and Monitoring, Inc., Rexdale, Ontario. Personal communication (1982).
31. "Dioxins," U.S. EPA-600/2-80-197 (1980).

20

Human Exposure to Polychlorinated Dibenzo-*p*-Dioxins and Dibenzofurans

C. Rappe, M. Nygren and G. Gustafsson

Some of the polychlorinated dibenzo-*p*-dioxins (PCDD) and dibenzofurans (PCDF) have extraordinarily toxic properties, and have been the subject of much concern. They are known as highly stable contaminants in chlorinated phenols (CP), phenoxy acids, hexachlorophene and polychlorinated biphenyls (PCB). They have also been identified in fly ash, flue gas condensates and other products from municipal incinerators and accidental fires.

There is a pronounced difference in biological effects between different PCDD and PCDF isomers. A 10^3–10^4 difference in toxicity can be found for such closely related isomers as 2,3,7,8- and 1,2,3,8-tetrachlorodibenzo-*p*-dioxin (tetra-CDD). The isomers with the highest acute toxicities have 4–6 chlorine atoms, and they have their four lateral positions substituted for chlorine. They all have LD_{50} values in the range 1–100 μg/kg for the most sensitive animal species.

Human exposure to PCDD or PCDF may be through specific exposure (mainly of occupational origin), accidental exposure or general exposure of the public. A general exposure of the public to 2,3,7,8-tetra-CDD has been discussed in relation to the herbicide spraying program in Vietnam in the 1960s and the explosion in a chemical plant near Seveso, Italy, in July 1976. The Yusho oil accidents in Japan in 1968 and in Taiwan in 1979 also resulted in an exposure of the public to various PCDF.

Because of the extreme toxicity of some of the PCDD and PCDF isomers, very sensitive and highly specific analytical techniques are required. The different isomers of PCDD and PCDF vary greatly in their biological and toxicological properties; consequently, the separation, identification and quantification of the most toxic isomers become important.

In recent years a number of analytical methods have been developed for trace analyses of PCDD and PCDF in a variety of matrices. The most specific, sensitive and selective of these methods use high-resolution gas chromatography/

mass spectrometry (HRGC/MS). A cleanup procedure resulting in one single fraction to be analyzed is a major advantage.

This chapter discusses the levels of PCDD and PCDF in human blood plasma and tissue samples. Special attention is paid to variation in individual levels and the reasons for these variations.

EXPERIMENTAL

Cleanup Procedures

Blood Plasma

Cleanup of blood plasma samples was based on partitioning between *n*-hexane and acetonitrile. *n*-Hexane (10 mL) was added to 10 mL of blood plasma. Each sample was spiked with 0.1 ng of ^{13}C-2,3,7,8-tetra-CDD or ^{37}Cl-2,3,7,8-tetra-CDF, extracted with three 20-mL portions of acetonitrile saturated with *n*-hexane, and centrifuged. The combined acetonitrile extracts were treated with 10 mL of 1% aqueous sodium sulfate and extracted three times with 20 mL of *n*-hexane. The combined *n*-hexane layers were dried with sodium sulfate and concentrated to 2 mL and were added to a 1-g alumina column. The column was eluted first with 10 mL of 2% methylene chloride in *n*-hexane and thereafter with 10 mL of 50% methylene chloride in *n*-hexane. The first eluate was discarded, and the second was concentrated and analyzed by GC/MS.

Tissue Samples

Cleanup of the tissue samples was performed using methods described elsewhere [1,2].

GC/MS Analyses

The purified extracts of each sample yielding one fraction were used directly for the final analyses by a Finnigan 4021 GC/MS system equipped with a PROMIM multiple ion monitoring unit and negative chemical ionization options. Aliquots (1–2 μg) corresponding to up to 1 g of blood plasma or tissue sample were injected splitlessly onto a glass capillary column (OV-17, Silar-10C and Supelco SP-2330) or a fused-silica column (SP-2100, SE-54) leading into the ion source [3]. Mass-specific detection was used by selective monitoring of M, M+2 and/or M+4 ions, and some typical fragmentations [4]. The recovery efficiency of the cleanup procedure was determined using the internal standards.

Quantification was based on peak area measurements using external standards and a calibration curve of the actual PCDD and PCDF isomer. Isomer identification was based on retention time studies using mainly polar (but also nonpolar) capillary columns and the available qualitative standards (about 40 PCDD and 70 PCDF). 2,3,7,8-Tetra-CDD was separated from other tetra-CDD isomers using a 55-m narrow-bore Silar-10C glass capillary column [3].

RESULTS AND DISCUSSION

Influence on Individual Levels Due to Nature of Exposure

Earlier studies investigated the levels of PCDD and PCDF in samples of blood plasma taken from workers occupationally exposed to chlorinated phenols in the sawmill, leather and textile industries [5,6]. The levels were normally below 100 ppt (100 pg/g) in blood plasma. Table 20.1 shows values from this investigation that illustrate the variation of individual levels due to different types of exposure. The results indicate that the highest values for both CP in urine and PCDD and PCDF in blood were found for the workers directly exposed to liquid CP formulations (textile industry and tannery) or aqueous solutions of chlorophenates (sawmills). In the textile industry there was a 100-fold difference between workers in different job categories. For epidemiological studies it must be of great importance to differentiate between highly exposed workers and those who were only slightly exposed.

Influence in Individual Levels Due to Difference in Duration of Exposure

We have also performed a study on samples from a group of workers in a CP formulation facility. The analytical values are summarized in table 20.2. This table also includes the number of years the workers were employed in this facility. A good relationship was found for these two parameters; the highest levels were found for those workers having the longest exposure.

Variation in Individual Levels Due to Different Chemical Exposure

In examining Tables 20.1 and 20.2 a difference can be found between the sawmill and chemical workers and the textile and tannery workers. The sawmill and chemical workers were exposed to a 2,3,4,6-tetrachlorophenate formulation where PCDF were the major impurities [7]. These workers generally had higher levels of PCDF than of PCDD; the major component was 1,2,3,4,7,8-hepta-CDF. This isomer is also the major hepta-CDF isomer in the formulation used.

In the textile and leather industry, PCP were used, and the hepta- and octa-CDD were the major impurities in these formulations [7]. The blood plasma

Table 20.1. Levels of CP in Urine and PCDD and PCDF in Blood Plasma from Workers Exposed to CP [5,6]

	CP in Urine (μg/mL)	PCDD in Blood Plasma (pg/g)			PCDF in Blood Plasma (pg/g)		
		Octa-	*1,2,3,4,6,7,8-*	*1,2,3,4,6,7,9-*	*Octa-*	*1,2,3,4,6,7,8-*	*1,2,3,4,6,8,9-*
Sawmill							
Loader	0.04	5	< 2	<2	< 3	40	< 1
Loader[a,b]	0.03	18	7	3	<3	18	< 1
Packer[a]	< 0.05	<3	< 2	< 2	<3	7	< 1
Cleaner [a]	<0.02	5	2	< 2	<3	30	< 1
Textile, Mixer[b]	3.12	304	59	< 2	10	25	8
Textile	0.16	30	6	< 2	<1	3	<1
Textile	< 0.01	3	< 1	< 2	<1	<1	<1
Textile[b]	0.42	105	15	< 2	<1	<1	<1
Tannery[a,b]	0.04	80	32	< 2	7	18	<1
Tannery	0.03	12	4	< 2	< 3	3	<1

[a]Sampling of blood five months after last exposure to CP.
[b]Exposure to solutions of CP.

Table 20.2. Levels of PCDD and PCDF in Blood Plasma of Workers Exposed During Manufacture of CP

	PCDD in Blood Plasma (pg/g)		*PCDF in Blood Plasma (pg/g)*	
Years	*Octa-*	*1,2,3,4,6,7,8-*	*Octa-*	*1,2,3,4,6,7,8-*
3	ND[a]	ND	ND	47
3	ND	ND	ND	46
13	ND	ND	ND	132
6 + 3[b]	ND	ND	ND	184
< 0.5	ND	ND	ND	3
18	ND	ND	ND	197
5	ND	ND	ND	20
1	ND	ND	ND	191
2	ND	ND	ND	19
4	ND	ND	ND	42
0.5	ND	ND	ND	5

[a]ND = not detected. Detection level ≃ 1 pg/g.
[b]Two separate work periods of six and three years.

samples of these workers were mainly contaminated with 1,2,3,4,6,7,8-hepta-and octa-CDD.

Variation in Individual Levels Due to Metabolism or Elimination

Textile Industry

We have studied workers in the textile industry for a longer time. Samples of blood plasma were taken on two occasions—5 and 16 months after exposure to the contaminated penta-CP formulations was stopped (Table 20.3). It was found that the reduction of PCDD and PCDF in blood levels during the 11-month period between the two sampling occasions ranged 60–90%; the values for the most exposed workers (numbers 3 and 6 in the table) were 88% and 76%, respectively.

Yusho Episodes Japan, 1968, and Taiwan, 1979

In 1968 more than 1500 persons in southwest Japan were intoxicated by consuming a commercial rice oil accidentally contaminated by polychlorinated biphenyls (PCB), PCDF and polychlorinated quarterphenyls (PCQ). In 1979 a similar episode was reported from Taiwan, where the number of persons involved approached 2000.

Table 20.3. Levels of PCDD and PCDF (pg/g) in Blood Plasma from Textile Workers

		PCDD			PCDF		
Worker		*Octa-*	*1,2,3,4,6,7,9-*	*1,2,3,4,6,7,8-*	*Octa-*	*1,2,3,4,6,7,8-*	*1,2,3,4,6,8,9-*
1	A[a]	14	1	3	1	1	ND[b]
	B[c]	6	ND	1	ND	1	ND
2	A	6	1	2	1	1	ND
	B	2	ND	ND	ND	ND	ND
3	A	304	ND	59	10	25	8
	B	35	ND	7	ND	5	ND
4	A	3	ND	ND	ND	ND	ND
	B	3	ND	1	ND	ND	ND
5	A	10	ND	1	ND	ND	ND
	B	1	ND	ND	ND	ND	ND
6	A	105	ND	15	ND	ND	ND
	B	25	ND	3	ND	1	ND
7	A	30	ND	6	ND	3	1
	B	5	ND	1	ND	1	ND

[a]A = five months after exposure to contaminated product.
[b]ND = not detected. Detection level 1 pg/g.
[c]B = 16 months after exposure to contaminated product.

Earlier analyses have proven that the Japanese rice oil contained more than 40 PCDF isomers, ranging from tri- to hexa-CDF [8]. Analysis of liver samples taken from the Japanese patients about 18 months after the exposure showed a dramatic decrease in the number of PCDF isomers (Figure 20.1) [9]. Apparently, most of the PCDF isomers were metabolized or excreted during the period between exposure and sampling.

A comparison between the PCDF isomers found in the Yusho oil and in the

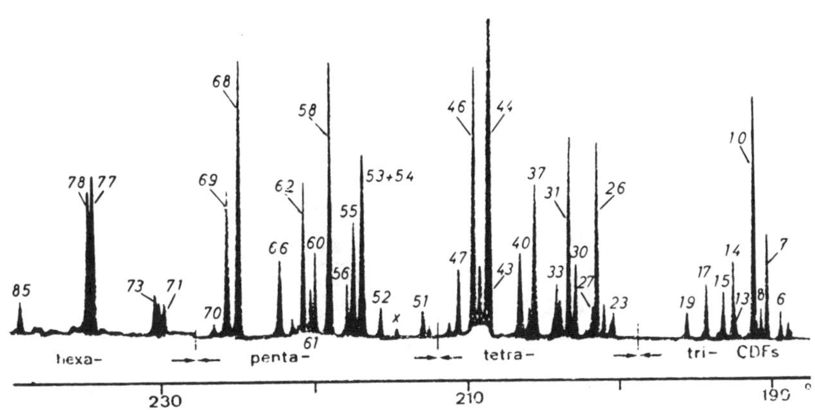

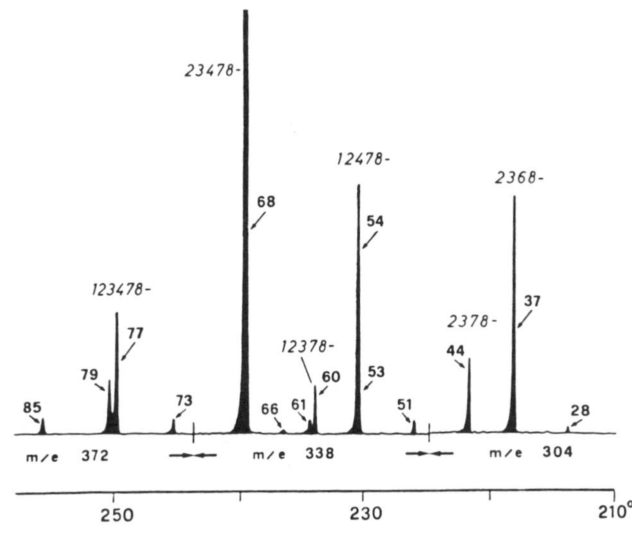

Figure 20.1. Mass Fragmentogram of Yusho Oil (Top) and a Liver Sample (Bottom).

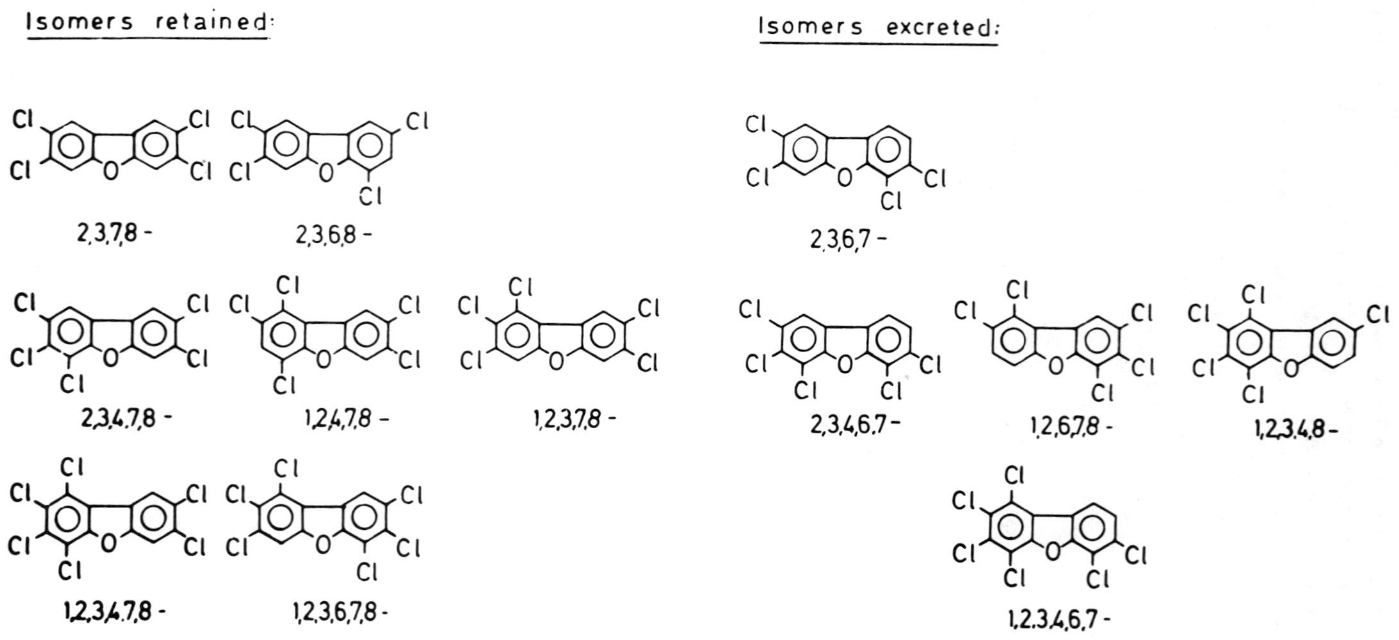

Isomers retained:

2,3,7,8-

2,3,6,8-

2,3,4,7,8-

1,2,4,7,8-

1,2,3,7,8-

1,2,3,4,7,8-

1,2,3,6,7,8-

Isomers excreted:

2,3,6,7-

2,3,4,6,7-

1,2,6,7,8-

1,2,3,4,8-

1,2,3,4,6,7-

Figure 20.2. PCDF Isomers Retained (Left) and Excreted (Right) in Yusho Oil.

liver samples revealed an interesting relationship between the isomers retained in the liver and the isomers excreted (Figure 20.2). All the PCDF isomers *excreted* had two vicinal unchlorinated C atoms in at least one of the two aromatic rings of the PCDF system. However, none of the isomers *retained* had two vicinal unchlorinated C atoms in any of the two aromatic rings. Most of the latter isomers had all lateral (2-, 3-, 7-, and 8-) positions substituted for chlorine [9].

We also analyzed a set of four blood samples taken from Yusho patients; two collected in Japan in 1979 (11 years after exposure), and the two collected in Taiwan in 1980 (one year after exposure). The results are shown in Table 20.4. The highly toxic 2,3,4,7,8-penta-CDF could be identified in all samples, although the levels in the Japanese samples were just above the detection limit. Two additional penta-CDF and 1,2,3,4,7,8-hexa-CDF were also found in the Taiwanese samples.

We also have preliminary data from blood samples from Taiwanese Yusho patients [10]. These samples were collected in 1980 and 1981, and the reduction of PCDF during the year between the two sampling occasions was unexpectedly low. For 2,3,4,7,8-penta-CDF we found an average 20% reduction; whereas for 1,2,3,4,7,8-hexa-CDF only a 15% reduction was found.

The data reported here indicate a dramatic difference in half-life between the hepta- and octa- congeners and the highly toxic penta- and hexa- isomers. The latter have a half-life of more than one year. This observation is in accord with data reported by McNulty et al. [11]. They found the half-life in fat of 2,3,7,8-tetra-CDD for a rhesus monkey to be 1 year.

Variation in Isomers in Different Organs

We have analyzed samples of the liver and kidney from a man working in the production of phenoxy acid herbicides such as 2,4-D, methyl-2-chlorophenoxy-

Table 20.4. Levels of PCDF (pg/g) in Blood Plasma from Yusho Patients

	Japan		*Taiwan*	
Isomer	*A*	*B*	*C*	*D*
Tetra-CDF	<3	<3	<30[a]	<30[a]
1,2,4,7,8-Penta-CDF	ND[b]	ND	60	40
1,2,3,7,8-Penta-CDF	ND	ND	30	20
2,3,4,7,8-Penta-CDF	3[c]	3[c]	120	80
1,2,3,4,7,8-Hexa-CDF	<6	<6	150	60

[a]High detection limits due to large amounts of overlapping PCB and other chlorinated compounds.

[b]ND = not detected. Detection level ≃1 pg/g.

[c]Detection limit = 3 pg/g.

acetic acid (MCPA) and 2,4,6-trichlorophenol (2,4,6-tri-CP). Exposure to the herbicides ceased 3–4 years before death from a pancreatic tumor. The results of our analyses are given in table 20.5. In general, the levels were much higher in the liver than in the kidney; however, for the highly toxic 2,3,7,8-tetra-CDD the ratio of the contaminant in liver vs kidney was reversed.

We have also analyzed various organs of a "Yusho baby" from Taiwan. Table 20.6 presents the results of these PCDF analyses together with the PCBH levels. The PCB levels seem to parallel the fat levels in the tissue. However, PCDF could only be detected in the adipose tissue and liver. The highest PCDF values were found in the liver, but amazingly here we found the lowest PCB values. In the liver sample, the dominating isomer is 1,2,3,7,8-penta-CDF, while in adipose tissue the highest value was found for 2,3,4,7,8-penta-CDF.

Table 20.5. Levels of PCDD and PCDF (pg/g) in Tissue Samples (values in parenthesis are the detection limits)

Isomer	*Liver*	*Kidney*
2,3,7,8-Tetra-CDD	2 (1)	5 (1)
1,3,6,8-Tetra-CDD	1 (1)	1 (1)
Penta-CDD	<2	<2
1,2,3,6,7,8-Hexa-CDD	4 (4)	2
1,2,3,4,6,7,8-Hepta-CDD	72 (5)	8 (4)
Octa-CDD	350 (8)	15 (4)
Tetra-CDF	<1	<1
2,3,4,7,8-Penta-CDF	10 (2)	7 (2)
Hexa-CDF	55 (2)	8 (2)
1,2,3,4,6,7,8-Hepta-CDF	100 (4)	6 (4)
Octa-CDF	<3	<3

Table 20.6. Levels of PCDF (pg/g) and PCB (ng/g) in Tissue Samples of a Yusho Baby from Taiwan

Isomer	*Adipose*	*Liver*	*Muscle*	*Omentum*	*Diaphragm*
2,3,7,8-Tetra-CDF	17	60	NA[a]	ND	ND
1,2,4,7,8-Penta-CDF	14	42	ND	ND	ND
1,2,3,7,8-Penta-CDF	44	194	ND	ND	ND
2,3,4,7,8-Penta-CDF	68	91	ND	ND	ND
1,2,3,4,7,8-Hexa-CDF	88	193	ND	ND	ND
PCB	316	27	38	64	46

[a]ND = not detected. Detection level 1 pg/g.

SUMMARY

Analytical methods are available for identification and quantification of PCDD and PCDF in human blood and tissue samples. These methods have been used to investigate occupational exposure in the sawmill, leather, textile and chemical industries. The yusho oil accidents in Japan in 1968 and Taiwan in 1979 were also investigated. The rate of elimination indicates that the most toxic isomers are also the slowest to be eliminated.

REFERENCES

1. Stalling, D.L., J.D. Petty, L.M. Smith, C. Rappe and H.R. Buser. "Isolation and Analysis of Polychlorinated Dibenzofurans in Aquatic Samples," in *Chlorinated Dioxins and Related Compounds*, O. Hutzinger, R.W. Frei, E. Merian and F. Pocchiari, Eds. (Oxford: Pergamon Press, Ltd., 1982), pp. 77–85.
2. Rappe, C., H.R. Buser, D. Stalling, L.M. Smith and R.C. Dougherty. "Identification of Polychlorinated Dibenzofurans in Environmental Samples," *Nature* 292:524–526 (1981).
3. Buser, H.R., and C. Rappe. "High-Resolution Gas-Chromatography of the 22 Tetrachlorodibenzo-*p*-dioxin Isomers," *Anal. Chem.* 52:2257–2262 (1980).
4. Rappe, C., S. Marklund, M. Nygren and A. Garå. "Parameters for Identification and Confirmation in Trace Analyses of Polychlorinated Dioxins and Dibenzofurans," Chapter 15, this volume.
5. Rappe, C., and H.R. Buser. "Occupational Exposure to Polychlorinated Dioxins and Dibenzofurans," in *Chemical Hazards in the Workplace, Measurements and Control*, G. Choudhary, Ed. ACS Symp. Ser. No. 149 (Washington, DC: American Chemical Society, 1981), pp. 319–342.
6. Rappe, C., M. Nygren, H.R. Buser and T. Kauppinen. "Occupational Exposure to Polychlorinated Dioxins and Dibenzofurans," in *Chlorinated Dioxins and Related Compounds*, O. Hutzinger, R.W. Frei, E. Merian and F. Pocchiari, Eds. (Oxford: Pergamon Press, Ltd. 1982), pp. 495–513.
7. Rappe, C., A. Gara and H.R. Buser. "Identification of Polychlorinated Dibenzofurans (PCDFs) in Commercial Chlorophenol Formulations," *Chemosphere* 7:981–991 (1978).
8. Buser, H.R., C. Rappe and A. Garå. "Polychlorinated Dibenzofurans (PCDFs) in Yusho Oil and Used Japanese PCB," *Chemosphere* 7:439–449 (1978).
9. Rappe, C., H.R. Buser, H. Kuroki and Y. Masuda. "Identification of Polychlorinated Dibenzofurans (PCDFs) Retained in Patients with Yusho," *Chemosphere* 8:259–266 (1979).
10. Rappe, C., M. Nygren, P. Chen. (in preparation).
11. McNulty, W.P., K.A. Nielsen-Smith, J.O. Lay, D.L. Liffstreu, N.L. Kangas, P.A. Lyon and M.L. Gross. Persistence of 2,3,7,8-Tetrachlorodibenzo-*p*-dioxin in Fat of a Rhesus Monkey (*Macaca mulatta*)," *Food Chem.* (in press).

21

Epidemiological Studies on Soft-Tissue Sarcoma, Malignant Lymphoma, Nasal and Nasopharyngeal Cancer, and Their Relation to Phenoxy Acid or Chlorophenol Exposure

Lennart Hardell

Chemical pesticides were introduced in agriculture and forestry in Sweden at the end of the 1940s. The use of herbicides has increased considerably, with phenoxy acids as the dominant weed killers. In 1947 less than 1 ton of phenoxy acids was used; the consumption is now roughly 3000 ton/yr. In forestry, about 300 ton/yr of phenoxy acids have been used for defoliation and control of shrubbery since the early 1950s. The use of 2,4,5-trichlorophenoxyacetic acid (2,4,5-T) was prohibited in Sweden in 1977 because of contamination by polychlorinated dibenzo-*p*-dioxins (PCDD) and dibenzofurans (PCDF). Table 21.1 shows levels of 2,3,7,8-tetrachlorodibenzo-*p*-dioxin (tetra-CDD) in different samples of 2,4,5-T.

The phenoxy acids now available on the Swedish market are 4-chloro-2-methyl phenoxyacetic acid (MCPA), 2,4-dichlorophenoxyacetic acid (2,4-D), and the analogous phenoxypropionic acids, mecoprop and dichloroprop. In comparison to 2,4,5-T, these phenoxy acids are less prone to contain PCDD and PCDF as impurities. In 1978, 1,3,6,8- and 1,3,7,9-tetra-CDD were identified as impurities in 2,4-D by Rappe et al. [1,3]; these findings have been confirmed by Cochrane et al. [4].

The phenoxy acids used as herbicides are formed from chlorinated phenols (CP) and chloroacetic acid. The major identified sources of PCDD and PCDF in the environment are associated with the use of CP in the wood industry (Table 21.2) [3,5].

About 90% of the use of CP has been in sawmills as impregnates or for protection against blue stain. High levels of PCDD and PCDF have been found in the

Table 21.1. Levels of 2,3,7,8-Tetra-CDD in 2,4,5-T Acid and Ester Formulations [1,2]

Sample	*Origin*	*2,3,7,8-Tetra-CDD ($\mu g/g$)*
2,4,5-T Acid	Sweden, 1952	1.10
2,4,5-T Ester	Sweden, 1960	0.40
2,4,5-T Ester	Finland, 1962	0.95
2,4,5-T Ester	Finland, 1967	0.22
2,4,5-T Acid	U.S., 1964	4.8
2,4,5-T Acid	U.S., 1969	6.0
Agent Orange	U.S., unknown	0.12
Agent Orange	U.S., unknown	5.1

Table 21.2. Levels of PCDD and PCDF in Commercial Chlorinated Phenols ($\mu g/g$) [3,5]

	PCDD	*PCDF*
2,4,6-Tri-CP, Sweden	3	60
2,4,6-Tri-CP, U.S.	0.3	4.6
2,3,4,6-Tetra-CP, Finland	12	160
Penta-CP, U.S.	2625	190
Penta-CP, U.S.	1900	790
Penta-CP, Germany	6.8	2.1

sawdust from sawmills [6]. Pentachlorophenol (penta-CP) has been used for slime control in the manufacture of paper pulp and a variety of other purposes, such as in cutting oils and fluids, as wood preservatives in paints, and to waterproof leather and textiles. 2,4-Di- and 2,4,5-tri-CP are used to produce 2,4-D and 2,4,5-T herbicides, and hexachlorophene. The use of CP was banned in Sweden in 1978, except for impregnating textiles.

During the past few years, exposure to phenoxy acids has attracted an increasing interest as a possible cause of malignant diseases. In an investigation of Swedish railroad workers, an increased incidence of malignant tumors was found, related to exposure to amitrol and phenoxy acid preparations [7,8]. In East Germany, a high incidence of lung cancer was observed among pesticide workers, but, again, the subjects were exposed to several different pesticides [9].

Based on clinical observations, soft-tissue sarcomas and malignant lymphoma were reported as possibly related to exposure to phenoxy acids or CP [10,11]. These observations initiated two subsequent case-control studies of soft-tissue sarcoma that demonstrated a roughly sixfold increase in the risk after exposure to phenoxy acids or CP [12,13]. A similar matched case-control study of malignant lymphoma (both Hodgkin's disease and non-Hodgkin lymphoma) indicated an association between that disease and exposure to phenoxy acids or CP and to organic solvents [14]. Exposure to various other agents was equal in cases and controls. In case-control studies the results might be biased if information on exposure is obtained in a noncomparable manner from cases and controls. Thus, despite precautions in the study methodology, observational bias might still exist. The validity of the assessment of exposure to phenoxy acids and CP in the previous studies was further analyzed by using a fairly common cancer type, colon cancer, for comparison. No association was found between exposure to phenoxy acids or CP and colon cancer [15]. Consequently, the earlier findings could not be explained by a systematic observational bias.

A strong association between occupational exposure to wood dust, mainly from hardwoods, and adenocarcinoma of the nose and paranasal sinuses was reported by Acheson et al. [16]. The particle size distribution of wood dust has a maximum at 6–10 μm according to Andersen et al. [17]. Particles of 5 μm or more in diameter are deposited in the nose, whereas smaller particles tend to pass further down into the respiratory tract [18]. During knapsack spraying of phenoxy herbicides, the median diameter of the droplets has been calculated as 300 μm [19,20]. Only 2% of the droplets are 60 μm or less [21]. Consequently, the droplets are trapped in the upper respiratory tract and then are swallowed and resorbed. Hence, exposure through inhalation might be expected to strike particularly the nose and nasopharynx, although skin absorption rather than inhalation and ingestion seems to be the major exposure route [22,23]. It was thus of interest to see if CP or phenoxy acids contaminating sawdust exert local carcinogenic effects in the nose, paranasal sinuses and nasopharynx. This study followed the methodological design of the series of studies referred to above. Exposure to phenoxy acids gave formally a doubled (but still insignificant) risk for nasal and nasopharyngeal cancer. Exposure to CP, as present particularly in wood work, was related to an approximately sevenfold and significant increase in the risk for both cancer types [24]. In wood workers without exposure to CP, there was an approximately normal risk, but cabinetmakers, even without exposure to CP, had a nearly doubled (but still insignificant) risk for nasal cancer. The results of these investigations are summarized in Table 21.3.

A Swedish study of causes of death among 375 lumberjacks showed a slight numerical excess of deaths due to ischemic heart diseases, cerebrovascular diseases and all other circulatory diseases. There were fewer deaths than expected from cancer and other causes. A case-referent analysis showed an increased standard mortality ratio for kidney cancer and cancer of the lymphatic and hematopoietic system [26]. Another cohort study of 142 lumberjacks exposed to phenoxy acids and 244 unexposed showed an increased cancer mortality among the foremen, who were most exposed to phenoxy acids [27].

A Finnish cohort study of 1926 workers who had sprayed brush vegetation with phenoxy herbicides during 1955–1971 showed no increased mortality from

Table 21.3. Risk Ratios (RR) in Cases Exposed to Phenoxy Acids or CP in Swedish Case-Control Studies Referred to in the Text

	Exposure		
Cancer Type	*Phenoxy Acids*	*High-Grade Chlorophenols*	*Reference*
Soft-Tissue Sarcoma			
Northern Sweden	5.3	6.6	12
Southern Sweden	6.8	3.3[a]	13
Malignant Lymphoma[b]	4.8	7.6	14
Hodgkin's Disease[c]	5.0	6.5	25
Colon Cancer	1.3[d]	1.8[d]	15
Nasal, Nasopharyngeal Cancer	2.1[d]	6.7	24

[a] $p < 0.01$.
[b] An association was found with exposure to organic solvents, RR = 3.1.
[c] Organic solvents, RR = 3.0.
[d] Not significant. For all other RR, $p < 0.001$.

cancer [28]. Because of the brief followup period, the small size of the cohort and the low past exposure, the results must be interpreted with caution.

There are now several case reports and cohort studies on workers exposed to CP, phenoxy acids and their impurities in accidents during the processing of these chemicals. In West Germany, 74 persons were exposed to tri-CP and PCDD in an accident in 1953. Of the 21 deceased, 7 had cancer, compared with 4.1 expected. In addition, two other cases of cancer (one bronchial and one prostate carcinoma) are still alive. Three deaths due to stomach cancer were found, compared with 0.6 expected [29].

A review of four U.S. studies of industrial workers exposed to CP and phenoxy acids showed that 3 of the 105 deaths had been due to soft-tissue sarcomas [30]. Only approximately 0.07% total soft-tissue sarcoma deaths were expected. A fourth case of soft-tissue sarcoma has since then been reported in one of these groups of men [31]. In addition, three cases of soft-tissue sarcoma in subjects exposed to phenoxy acids or CP during the manufacture of these compounds have been reported [32,33]. Among Vietnam war veterans exposed to the mixture of phenoxy acids known as Agent Orange, three cases of thoracic soft-tissue sarcomas have lately been reported [34]. In Vietnam an increase was reported in the number of persons with primary liver cancer in proportion to the number of all cancer in North Vietnam during the period when herbicides were used in South Vietnam [35].

One U.S. cohort study of mortality among workers exposed to tri-CP yielded a suspicious increase of lymphatic and hematopoietic malignancies, i.e., 3 deaths as compared to 0.88 expected [36]. In a British plant manufacturing penta-

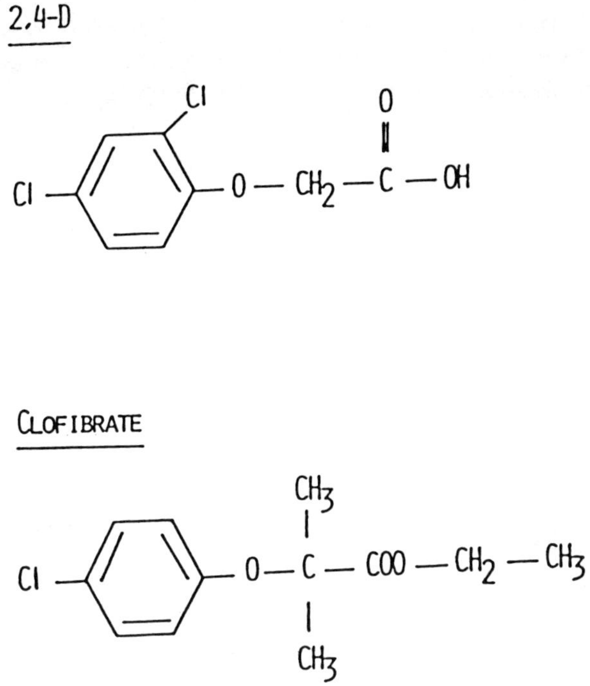

2,4-D

Cl — O — CH_2 — C(=O) — OH

CLOFIBRATE

Cl — O — C(CH_3)(CH_3) — COO — CH_2 — CH_3

Figure 21.1. Structures of 2,4-D and Clofibrate.

CP, 2 men with non-Hodgkin lymphoma of the scalp were reported vs. 0.28 cases expected [37]. Also, a study by Cantor [38] is in agreement with the Swedish study showing an elevated risk among younger farmers for histiocytic lymphoma in counties with high consumption of herbicides. Several studies have revealed an increased incidence of Hodgkin's disease in lumberjacks and wood workers [39,40].

Animal data demonstrate that PCDD are carcinogenic in rats and mice [41–44]. Tetra-CDD has been shown to act as a promoter [45]. Whether PCDD act as promoters, initiators or both is unclear.

It is of some interest that blood fat–lowering drugs of the clofibrate type contain a phenoxy acid derivative [2-(4-chlorophenoxy)-2-methylpropionic acid ethyl-ester] (Figure 21.1). There has been some concern regarding their possible carcinogenic effects. In a double-blind trial of 15,745 healthy men assigned to three groups on the basis of their blood cholesterol concentrations, there was, among other diseases, an overall excess of deaths from gastrointestinal cancer among patients treated with clofibrate [46]. Clofibrate produces hydrogen peroxide and proliferation of hepatic peroxisomes, and has been suggested to be a carcinogen not possessing mutagenic activity in bacterial mutagenesis assays [47]. It has been demonstrated that the phenoxy herbicides 2,4-D and MCPA are able to increase either the size or the amount of peroxisomes in the liver cells of Chinese hamsters [48]. Thus, the possibility exists that neoplastic transformation by phenoxy acids could be the result of continued production of DNA-damaging oxygen radicals as a result of the persistent proliferation of peroxisomes.

REFERENCES

1. Rappe, C., H.R. Buser and H.P. Bosshardt. "Identification and Quantitation of Polychlorinated Dibenzo-*p*-dioxins (PCDDs) and Dibenzofurans (PCDFs) in 2,4,5-T-Ester Formulations and Herbicide Orange," *Chemosphere* 7:431–438 (1978).
2. Norström, A., C. Rappe, R. Lindahl and H.R. Buser. "Analysis of Some Older Scandinavian Formulations of 2,4-D (2.4-Dichlorophenoxyacetic Acid) and 2,4,5-T (2,4,5-Trichlorophenoxyacetic Acid) for Contents of Chlorinated Dibenzo-dioxins and Dibenzofurans," *Scand. J. Work, Environ. Health* 5:375–378 (1980).
3. Rappe, C., A Garå and H.R. Buser. "Identification of Polychlorinated Dibenzofurans (PCDFs) in Commercial Chlorophenol Formulations," *Chemosphere* 7:981–991 (1978).
4. Cochrane, W.P., J. Singh, B. Miles, B. Wakeford and J. Scott. "Analysis of Technical and Formulated Products of 2,4-Dichlorophenoxyacetic Acids for the Presence of Chlorinated Dibenzo-*p*-dioxins," Laboratory Services Division, Food Production and Inspection Branch, Canada Agriculture, Ottawa, Ontario (1981).
5. Rappe, C., M. Nygren, H.R. Buser and T. Kauppinen. "Occupational Exposure to Polychlorinated Dioxins and Dibenzofurans," in *Impact of Chlorinated Dioxins and Related Compounds on the Environment,* O. Hutzinger, R.W. Frie, E. Merian and F. Pocchiari, Eds. (Oxford: Pergamon Press, 1981), pp. 495–513.
6. Levin, J.O., C. Rappe and C.A. Nilsson. "Use of Chlorophenols as Fungicides in Sawmills," *Scand. J. Work Environ. Health* 2:71–81 (1976).
7. Axelson, O., L. Sundell, K. Andersson, C. Edling, C. Hogstedt and H. Kling. "Herbicide Exposure and Tumor Mortality: An Updated Epidemiological Investigation on Swedish Railroad Workers," *Scand. J. Work Environ. Health* 6:73–79 (1980).
8. Axelson, O., and L. Sundell. "Herbicide Exposure, Mortality and Tumor Incidence: An Epidemiological Investigation on Swedish Railroad Workers," *Work Environ. Health* 11:21–28 (1974).
9. Barthel, E. "Increased Risk of Lung Cancer in Pesticide-Exposed Male Agricultural Workers," *J. Toxicol. Environ. Health* 8:1027–1040 (1981).
10. Hardell, L. "Soft-Tissue Sarcomas and Exposure to Phenoxy Acids: A Clinical Observation," *Lakartidningen* 74:2753–2754 (1977).
11. Hardell, L. "Malignant Lymphoma of Histiocytic Type and Exposure to Phenoxyacetic Acids of Chlorophenols," *Lancet* i:55–56 (1979).
12. Hardell, L., and A. Sandstrom. "Case-Control Study: Soft-Tissue Sarcomas and Exposure to Phenoxyacetic Acids or Chlorophenols," *Brit. J. Cancer* 39:711–717 (1978).
13. Erikson, M., L. Hardell, N.O. Berg, T. Moller and O. Axelson. "Soft-Tissue Sarcomas and Exposure to Chemical Substances: A Case-Referent Study," *Brit. J. Ind. Med.* 38:27–33 (1981).
14. Hardell, L., M. Eriksson, P. Lenner and E. Lundgren. "Malignent Lymphoma and

Exposure to Chemicals, Especially Organic Solvents, Chlorophenols and Phenoxy Acids: A Case-Control Study," *Brit. J. Cancer* 43:169–176 (1981).

15. Hardell, L. "Relation of Soft-Tissue Sarcoma, Malignant Lymphoma and Colon Cancer to Phenoxy Acids, Chlorophenols and Other Agents," *Scand. J. Work Environ. Health* 7:119–130 (1981).
16. Acheson, E.D., E.H. Hadfield and R.G. Macbeth. "Carcinoma of the Nasal Cavity and Accessory Sinuses in Woodworkers," *Lancet* i:311–312 (1967).
17. Andersen, H.C., I. Andersen and J. Solgaard. "Nasal Cancers, Symptoms and Upper Airway Function in Woodworkers," *Brit. J. Ind. Med.* 34:201–207 (1977).
18. Hilding, A.C. "Nasal Filtration," in *Scientific Foundations of Otolaryngology*, E. Perkins and D.W. Hill, Eds. (Chicago, IL: Year Book Publishers, 1977), pp. 502–512.
19. Permin, O. "Problems of Wind Drift Related to the Technique of Spraying and Powdering," *Ungeskr. Agron.* 114:308–314, 339–347 (1969).
20. Svensson, K. "The Technical and Biological Background of Spraying," Landbrukshogskolan, Uppsala (1969).
21. Akeson, N.B., and W.E. Yates. "Pesticides in the Air Environment," unpublished (1978).
22. Kolmodin-Hedman, B., and K. Erne. "Estimation of Occupational Exposure to Phenoxy Acids (2,3–D and 2,4,5–T)," *Arch. Toxicol.* Suppl. 4 (1980), pp. 318–321.
23. Kolmodin-Hedman, B., S. Hoglund and M. Akerblom. "Exposure to Phenoxy Acids (MCPA, Dichlorprop, and Mecoprop) in Agricultural Work," *J. Occup. Med.* (in press).
24. Hardell, L., B. Johansson and O. Axelson. "Epidemiological Study on Nasal and Nasopharyngeal Cancer and Their Relation to Phenoxy Acid or Chlorophenol Exposure," *Am. J. Ind. Med.* 3:247–257 (1982).
25. Hardell, L., and N.O. Bengtsson. "Chemical Exposure, Socioeconomic Factors and Clinical Findings in Hodgkin's Disease" (submitted to *Brit. J. Cancer*)
26. Edling, C., and S. Granstam. "Causes of Death Among Lumberjacks—A Pilot Study," *J. Occup. Med.* 22:403–406 (1980).
27. Hogstedt, C., and B. Westerlund. "Kohortstudie av Dodsorsaker for Skogsarbetare med och utan Exposition for Fenoxisyrepreparat," *Lakartidningen* 77:1828–1831 (1980).
28. Riihimaki, V., S. Asp and S. Hernberg. "Mortality of 2,4-Dichlorophenoxyacetic Acid and 2,4,5-Trichlorophenoxyacetic Acid Herbicide Applicators in Finland: First Report of an Ongoing Prospective Cohort Study," *Scand. J. Work Environ. Health* 8:37–42 (1982).
29. Thiess, A.M., R. Frenze-Beyme and R. Link. "Mortality Study of Persons Exposed to Dioxin in a Trichlorophenol Process Accident That Occurred in the BASF AG on November 17, 1953," *Am. J. Ind. Med.* 3:179–189 (1982).
30. Honchar, P.A., and W.E. Halperin. "2,4,5–T, Trichlorophenol and Soft-Tissue Sarcomas," *Lancet* i.268–269 (1981).
31. Cook, R.R. "Dioxin, Chloracne, and Soft-Tissue Sarcoma," *Lancet* i:618–619 (1981).
32. Moses, M., and I.J. Selikoff. "Soft-Tissue Sarcomas, Phenoxy Herbicides and Chlorinated Phenols," *Lancet* i:1370 (1981).
33. Johnson, F.E., M.A. Kugler and S.M. Brown. "Soft-Tissue Sarcomas and Chlorinated Phenols," *Lancet* i:1370 (1981).
34. Sarma, P.R., and J. Jacobs. "Thoracic Soft-Tissue Sarcoma in Vietnam Veterans Exposed to Agent Orange," *New Engl. J. Med.* 306:1109 (1982).

35. Tung, T.T. "Le Cancer Primaire du Foie du Viet-Nam," *Chirurgie* 99:427–436 (1973).
36. Zack, J.A., and R.R. Suskind. "The Mortality Experience of Workers Exposed to Tetrachlorodibenzo-*p*-dioxin in a Trichlorophenol Process Accident," *J. Occup. Med.* 22:11–14 (1980).
37. Bishop, C.M., and A.H. Jones. "Non-Hodgkin's Lymphoma of the Scalp in Workers Exposed to Dioxins," *Lancet* ii:369 (1981).
38. Cantor, K.P. "Farming and Mortality from Non-Hodgkin's Lymphoma: A Case-Control Study," *Int. J. Cancer* 29:239–247 (1982).
39. Milham, S., and J.E. Hesser. "Hodgkin's Disease in Wood-Workers," *Lancet* ii:136–137 (1967).
40. Greene, M.H., L.A. Brinton, J.F. Fraumeni and R. D'Amico. "Familial and Sporadic Hodgkin's Disease Associated with Occupational Wood Exposure," *Lancet* ii:626–627 (1978).
41. Van Miller, J.P., J.J. Lalich and J.R. Allen. "Increased Incidence of Neoplasms in Rats Exposed to Low Levels of 2, 3, 7, 8-Tetrachlorodibenzo-*p*-dioxin," *Chemosphere* 10:625–632 (1977).
42. Kociba, R.J., D.G. Keyes, J.E. Beyer, R.M. Carreon, C.E. Wade, D.A. Dittenber, R.P. Kalnins, L.E. Frauson, D.N. Park, S.D. Barnard, R.A. Hummel and C.G. Humiston. "Results of a Two-Year Chronic Toxicity and Oncogenicity Study of 2,3,7,8-Tetrachlorodibenzo-*p*-dioxin in Rats," *Toxicol. Appl. Pharmacol.* 46:279–303 (1978).
43. "Bioassay of 2,3,7,8-Tetrachlorodibenzo-*p*-dioxin (Gavage Study)," DHHS Publ. No. (NIH)80-1765, Carcinogenesis Testing Program, National Cancer Institute.
44. "Bioassay of 2,3,7,8-Tetrachlorodibenzo-*p*-dioxin for Possible Carcinogenicity," Department of Health and Human Services Publ. No. (NIH)80-1757, Tech. Rep. 201, Carcinogenesis Testing Program, National Cancer Institute.
45. Pitot, H.C., T. Goldsworthy and H. Poland. "Promotion by 2,3,7,8-Tetrachlorodibenzo-*p*-dioxin of Hepatocarcinogenesis from Diethylnitrosamine," *Cancer Res.* 40:3616–3620 (1980).
46. Oliver, M.F., J.A. Heady, J.N. Morris and J. Cooper. "W.H.O. Cooperative Trial on Primary Prevention of Ischemic Heart Disease Using Clofibrate to Lower Serum Cholesterol: Mortality Follow-up," *Lancet* ii:379–385 (1980).
47. Reddy, J.K., and D.L. Zaranoff. "Hypolipidaemic Hepatic Peroxisome Proliferators from a Novel Class of Chemical Carcinogens," *Nature* 283:397–398 (1980).
48. Vainio, H., J. Nickels and K. Kinnainmaa. "Phenoxy Acid Herbicides Cause Peroxisome Proliferation in Chinese Hamsters," *Scand. J. Work Environ. Health* 8:70–73 (1982).

22

Polychlorinated Dibenzofurans in the Tissues of Patients with Yusho and Their Enzyme-Inducing Activities on Aryl Hydrocarbon Hydroxylase

Yoshito Masuda, Hiroaki Kuroki and Junya Nagayama

A mass food poisoning called "yusho" occurred in western Japan in 1968 by ingestion of rice oil contaminated with Kanechlor, a brand of polychlorinated biphenyl (PCB) [1]. The patients have had various chronic symptoms of acneform eruption, hypersecretion of the Meibomian glands, pigmentation of the face, eyelids, gingiva and nails, and others [2]. The rice oil was determined to contain not only PCB but also polychlorinated dibenzofurans (PCDF) and polychlorinated quaterphenyls (PCQ) [3–5], which were considered to have been produced from PCB during the use of Kanechlor as a heat-transfer medium for the production of the rice oil. The toxicity of PCDF is very high as compared with those of PCB and PCQ [6–8]. PCDF consist of various isomers, and their toxicities and biological behaviors are inevitably different from one another. We therefore analyzed the liver and adipose tissue from the deceased patients with yusho for PCDF isomers. Enzyme-inducing effects of these PCDF isomers on benzo[a]pyrene (B[a]P) 3-hydroxylase were also examined for understanding the etiology of yusho.

EXPERIMENTAL

Materials Analyzed

Specimens of the liver and mesenteric adipose tissue of yusho patients were supplied by the pathology departments of Kyushu University, Yamaguchi University and Fukuoka University. The tissues were preserved in a 4% formaldehyde solution after resection.

Chemicals

PCDF isomers were synthesized from corresponding chlorophenols (CP) and chloronitrobenzenes. The chloronitrodiphenylether produced from the starting materials was reduced to form chloroaminodiphenylether, which was cyclized to chlorodibenzofuran after diazotization with amyl nitrite. The chlorodibenzofuran produced was purified by column chromatography on silica gel, recrystallization and, if necessary, high-pressure liquid chromatography (HPLC) on μ-Bondapaks C-18 [9–12]. The synthetic precursors and physical properties of PCDF isomers are listed in Table 22.1. All the other chemicals and solvents used were of the purest grade available.

Analytical Procedures

The liver and adipose tissue were analyzed by a method described previously [3,10,11]. The samples were saponified with 1 *N* NaOH ethanol solution, extracted with *n*-hexane, fractionated on a silica gel column eluting with *n*-hexane, and then chromatographed on a column of alumina eluting successively with *n*-hexane: methylene chloride at 49:1 and at 4:1. The final eluate was analyzed for PCDF by gas chromatography (GC) with electron capture detection (ECD), mass spectrometry (MS) and mass fragmentography (MF). The GC/ECD of the PCDF fractions was performed on three different 3-mm × 5-m glass columns of 1.5% Apiezon L, 1.5% OV-17 and 4% SF-96 and on a 50-m fused-silica capillary of OV-101.

Animal Treatment and Enzyme Assay

The PCDF isomers were dissolved in olive oil and injected into male Wistar rats intraperitoneally (i.p.) at a single dose of 5 μg/kg. Three days after the injection, the animals were killed by cervical dislocation. The dissected liver and lung were separately homogenized in a solution of 0.15 *M* KCl and 0.02 *M* N-2-hydroxyethylpiperazine-N′-2-ethanesulphonic acid (Hepes). The homogenate was centrifuged at 9000 g and the resultant supernatant was used for B[a]P 3-hydroxylase assay. The supernatant was mixed with solutions of Hepes, NADPH, $MgCl_2$ and B[a]P and then incubated at 37°C for 5 min (liver) or 20 min (lung). After the addition of acetone, the mixture was extracted with *n*-hexane. The *n*-hexane layer was extracted with 1 *N* NaOH. The NaOH solution was determined by a fluorophotometer at 395 nm for excitation and 522 nm for emission wavelength. Quinine sulfate and 3-hydroxy B[a]P (donated by N. Kinoshita, School of

Table 22.1. Synthetic Precursors and Physical Properties of PCDF Isomers

	Synthetic Precursor		*mp*		*PMR ($CDCl_3$)*	
PCDF	*Chlorophenol*	*Chloronitrobenzene*	*(°C)*	*MS M*$^+$	*(ppm)*	*J (Hz)*
2,4,8-Tri-	2,4-Di-	2,5-Di-	155–156	270	7.87 dd	1.9, 0.6
					7.77 d	2.0
					7.56 dd	8.8, 0.9
					7.49 d	1.8
					7.48 dd	8.8, 1.8
Tetra-						
1,2,7,8-	3,4-Di-	2,4,5-Tri-	210–211	304	8.44 s	
					7.71 s	
					7.57 d	8.8
					7.43 d	8.8
1,3,6,7-	3,5-Di-	2,3,4-Tri-	177–178.5	304	8.08 d	8.6
					7.54 d	1.7
					7.46 d	8.4
					7.36 d	1.5
2,3,6,7-	2,3-Di-	2,4,5-Tri-	195–196	304	7.99 s	
					7.77 s	
					7.71 d	8.6
					7.47 d	8.4
2,3,6,8-	a	a	197–198	304	7.98 s	
					7.77 d	1.5
					7.76 s	
					7.51 d	1.8
2,3,7,8-	3,4-Di-	2,4,5-Tri-	219–221	304	7.97 s	
					7.70 s	

Table 22.1. *(continued)* Synthetic Precursors and Physical Properties of PCDF Isomers

	Synthetic Precursor		*mp*		*PMR ($CDCl_3$)*	
PCDF	*Chlorophenol*	*Chloronitrobenzene*	*(°C)*	*MS* M^+	*(ppm)*	*J (Hz)*
Penta-						
1,2,3,6,7-	3,4,5-Tri-	2,3,4-Tri-	205–207	338	8.15 d	8.6
					7.72 s	
					7.52 d	8.6
1,2,3,7,8-	3,4,5-Tri-	2,4,5-Tri-	225–227	338	8.40 s	
					7.71 s	
					7.65 s	
1,2,4,7,8-	2,4,5-Tri-	2,4,5-Tri-	236–238	338	8.45 s	
					7.79 s	
					7.63 s	
1,2,6,7,8-	3,4-Di-	2,3,4,5-Tetra-	220–221	338	8.38 s	
					7.61 d	8.8
					7.52 d	8.8
2,3,4,6,7-	2,3-Di-	2,3,4,5-Tetra-	201.5–202	338	7.93 s	
					7.71 d	8.4
					7.50 d	7.9
2,3,4,7,8-	2,3,4-Tri-	2,4,5-Tri-	196–196.5	338	7.97 s	
					7.91 s	
					7.77 s	
Hexa-						
1,2,3,4,6,7-	2,3,4,5-Tetra-	2,3,4-Tri-	227–228	372	8.16 d	8.2
					7.55 d	8.1
1,2,3,4,7,8-	2,3,4,5-Tetra-	2,4,5-Tri-	225.5–226.5	372	8.41 s	
					7.79 s	
1,2,3,6,7,8-	3,4,5-Tri-	2,3,4,5-Tetra-	232–234	372	8.35 s	
					7.74 s	

[a]Donated by Pomerantz.

Health Science, Kyushu University, Fukuoka, Japan) were used as standards for the determination.

RESULTS

Figure 22.1 shows GC/ECD of the PCDF fraction from the liver of a yusho patient. The retention time of each peak was identical with that of 2,3,6,8-tetra-, 2,3,7,8-tetra-, 1,2,4,7,8-penta-, 2,3,4,7,8-penta-, 1,2,3,4,7,8-hexa- or 1,2,3,6,7,8-hexa-CDF. The presence of these PCDF in the tissues of yusho patients was confirmed by GC/MS (Figure 22.2) and MF (Figure 22.3). The concentrations of individual PCDF isomers in the liver and adipose tissue are shown in Table 22.2. Whereas the PCB concentrations in the adipose tissue were much higher than those in the liver, the PCDF concentrations in both the tissues were at similar levels. This pattern of distribution indicates that PCDF are very accumulative in the liver, in contrast to PCB. While 2,3,7,8-tetra-CDF was detected in the liver

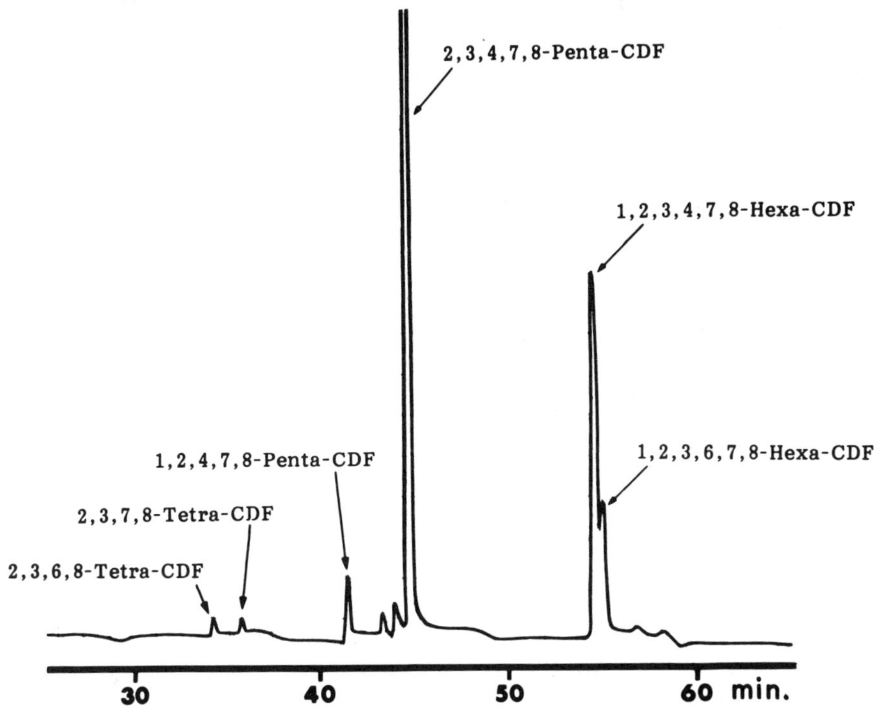

Figure 22.1. GC/ECD of the Fraction from the Liver of a Yusho Patient. Column: 50-m fused-silica glass capillary, OV-101; oven temperature: 180–250°C at 2°C/min.

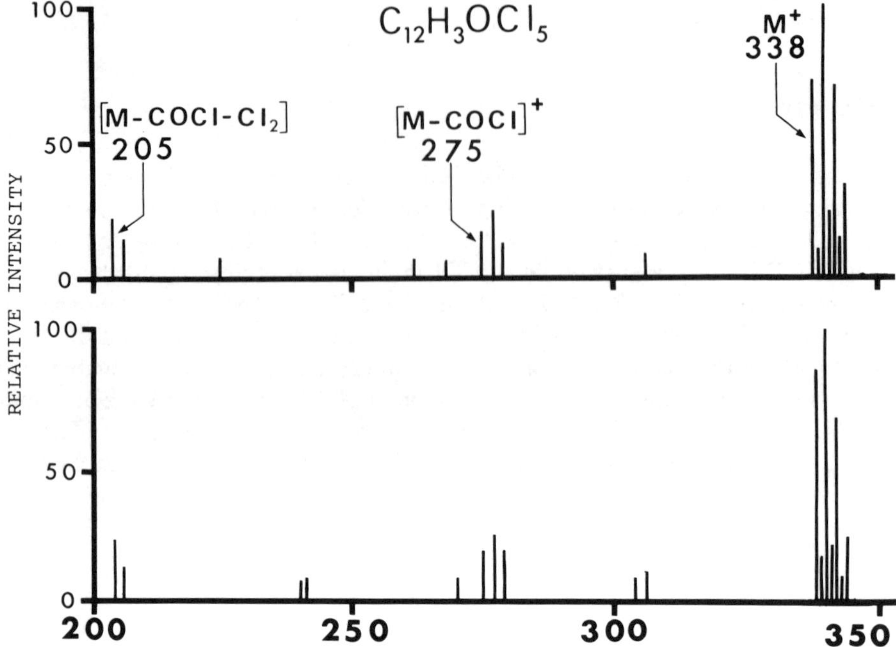

Figure 22.2. GC/MS (penta-CDF) of the PCDF Fraction from the Liver of a Yusho Patient (top) and Synthesized Penta-CDF (bottom).

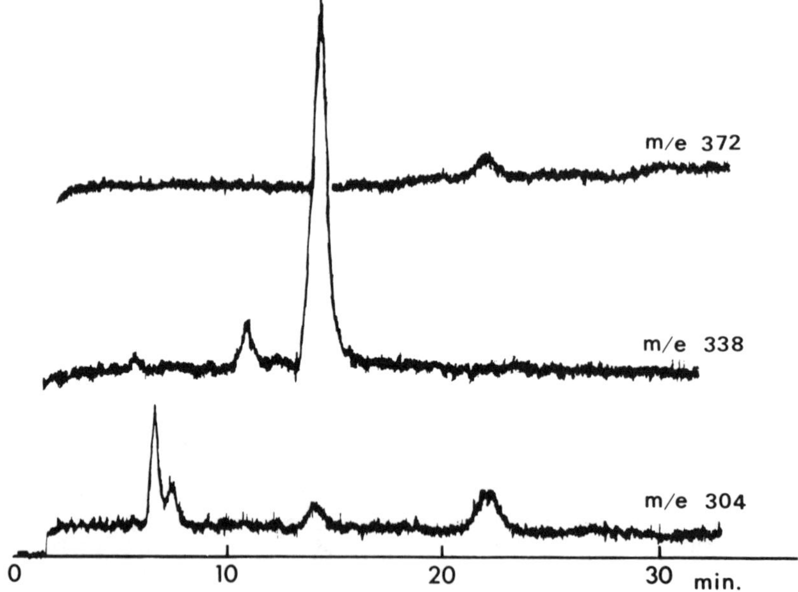

Figure 22.3. Mass Fragmentograms of the PCDF Fraction from the Liver of a Yusho Patient. Column: 2-m glass column, 3% Apiezon L; Ionizing energy: 25 eV; Ionizing current: 300 μA.

Table 22.2. Concentrations of PCB and PCDF Isomers in the Liver and Adipose Tissue of Yusho Patients

				PCDF Conc. (ppb)					
Case	*Time of Death*	*Tissue*	*PCB Conc. (ppm)*	*2,3,6,8-Tetra-*	*2,3,7,8-Tetra-*	*1,2,4,7,8-Penta-*	*2,3,4,7,8-Penta-*	*1,2,3,4,7,8- and 1,2,3,6,7,8-Hexa-*	*Total*
1	Jul 1969	Liver	0.14	0.7	0.3	7.1	6.9	2.6	17.6
2	Jul 1969	Liver	0.2	0.08	0.02	0.4	1.2	0.3	2.0
		Adipose	2.8	0.6	0.3	1.0	5.7	1.7	9.3
3	May 1972	Liver	0.08	0.03	0.005–0.01	0.09	0.3	0.03	0.45
		Adipose	4.3	0.08	<0.005	0.2	0.8	0.2	1.28
4	Apr 1975	Adipose	0.2	0.4	<0.005	0.8	0.1	0.5	1.8
5	Mar 1977	Liver	0.006	<0.005	<0.005	0.02	0.1	0.04	0.16
		Adipose	1.2	<0.005	<0.005	0.2	0.5	<0.005	0.7
6	Sep 1977	Liver	0.02	<0.005	<0.005	<0.005	<0.005	<0.005	
		Adipose	0.8	<0.005	<0.005	<0.005	<0.005	<0.005	

and adipose tissue at relatively low concentrations, 2,3,4,7,8-penta-CDF was determined as a major component of PCDF even in the patient who died nine years after onset [11].

The aryl hydrocarbon hydroxylase (AHH) activities in the liver and lung of the PCDF treated rats are shown in Table 22.3. B[a]P 3-hydroxylase in the liver was significantly induced by pretreatment with 2,3,7,8-tetra- and 2,3,4,7,8-penta-CDF, while in the lung 2,3,6,7-tetra-, 2,3,7,8-tetra-, 1,2,3,7,8-penta-, 2,3,4,6,7-penta-, 2,3,4,7,8-penta-, 1,2,3,4,5,6-hexa-, 1,2,3,4,7,8-hexa- and 1,2,3,6,7,8-hexa-CDF enhanced B[a]P 3-hydroxylase activity. When 2,3,7,8-tetra-CDF is considered as a prototype, 1,2,3,7,8-penta-CDF reduced the enzyme-inducing activity in the lung by introducing a chlorine atom to position 1 of the prototype, whereas 2,3,4,7,8-penta-CDF slightly enhanced the activity by introducing a chlorine atom to position 4. 1,2,3,4,7,8-Hexa-CDF substituted two chlorine atoms at positions 1 and 4 of the prototype, and intermediately induced the activity between the two penta-CDF. Similarly, the enzyme-inducing activity was changed by introducing chlorine atoms to positions 1, 4 or both of the other prototype of 2,3,6,7-tetra-CDF.

DISCUSSION

As shown in Table 22.3, all the PCDF isomers tentatively identified in the patients with yusho have three or four chlorine atoms at the lateral positions (2-, 3-, 7- and 8-) and do not have vicinal hydrogens in the dibenzofuran ring. PCDF having vicinal hydrogens were probably easily metabolized and excreted from the patients [12]. On the other hand, all the PCDF isomers that showed strong enzyme-inducing activities (Table 22.3) also have three or four chlorine atoms at the lateral positions, but some of them have vicinal hydrogens in the dibenzofuran ring such as 2,3,6,7-tetra-, 2,3,4,6,7-penta- and 1,2,3,4,6,7-hexa-CDF. For enhancement of enzyme activity, it seems necessary to have three or four chlorine atoms at the lateral positions, but it is not necessary to have no vicinal hydrogens in the dibenzofuran ring. The PCDF isomers with four chlorines in the lateral positions enhanced the enzyme activity in the lung. Above all, 2,3,7,8-tetra- and 2,3,4,7,8-penta-CDF significantly induced the enzyme in the liver (Table 22.3). Strong toxic potencies of these two isomers were observed on thymic atrophy and liver hypertrophy of the treated rats even at low doses of 1–10 μg/kg [7]. On the other hand, the levels of 2,3,4,7,8-penta-CDF in the liver and adipose tissue of the patients were much higher than those of 2,3,7,8-tetra-CDF (Table 22.2). The high accumulation property of penta-CDF in the liver was also observed in monkeys and rats [10]. Considering all of these aspects, among various toxic PCDF isomers, 2,3,4,7,8-penta-CDF would be the most important compound for understanding the etiology of yusho.

Table 22.3. Structural Characteristics of PCDF Isomers and Their Effects on B[a]P 3-Hydroxylase Activity in the Lung and Liver of Treated Rats

Isomer	*Occurrence in the Patients*	*Vicinal Hydrogen in the Ring*	*Number of Chlorines in the Lateral Positions*	*B[a]P Hydroxylase Activity*[a] *(pmol/min/mg protein)*	
				Lung	*Liver*
Control				1.1±0.2	188±16
2,4,8-Tri-		Yes		1.1±0.2	181±24
Tetra-					
1,2,7,8-		Yes	3	1.1±0.2	170±9
1,3,6,7-		Yes		1.3±0.1	209±16
2,3,6,7-		Yes	3	12.2±2.2[b]	184±14
2,3,6,8-	+	No	3	2.5±0.5	250±10
2,3,7,8-	+	No	4	23.2±4.7[b]	783±181[c]
Penta-					
1,2,3,6,7-		Yes	3	4.3±1.5	197±3
1,2,3,7,8-		No	4	5.7±1.4[d]	204±31
1,2,4,7,8-	++	No	3	1.2±0.5	261±92
1,2,6,7,8-		Yes	3	2.3±0.2	178±29
2,3,4,6,7-		Yes	3	10.1±1.2[b]	193±41
2,3,4,7,8-	+++	No	4	33.1±2.7[b]	391±47[c]
Hexa-					
1,2,3,4,6,7-		Yes	3	8.4±2.5[d]	208±21
1,2,3,4,7,8-	++	No	4	13.5±4.2[c]	252±29
1,2,3,6,7,8-	+	No	4	11.3±1.5[b]	167±19

[a]Each value represents the mean ± standard error of four rats.
[b]Significantly different from the control group, $p < 0.001$.
[c]Significantly different from the control group, $p < 0.01$.
[d]Significantly different from the control group, $p < 0.02$.

REFERENCES

1. Kuratsune, M., T. Yoshimura, J. Matsuzaka and A. Yamaguchi. *Environ. Health Persp.* (1):119 (1972).
2. Kuratsune, M. *Halogenated Biphenyls, Terphenyls, Naphthalenes, and Dibenzodioxins and Related Products* (Amsterdam: Elsevier/North-Holland Biomed. Press, 1980), p. 287.
3. Nagayama, K., M. Kuratsune and Y. Masuda. *Bull. Environ. Contam. Toxicol.* 15:9 (1976).
4. Kamps, L.R., W. J. Trotter, S.J. Young, L.J. Carson, J.A.G. Roach, J.A. Sphon, J.T. Tanner and B. McMahon. *Bull. Environ. Contam. Toxicol.* 20:589 (1978).
5. Miyata, H., T. Kashimoto and N. Kunita. *J. Food Hyg. Soc.* 19:364 (1978).
6. Yoshihara, S., K. Kawano, H. Yoshimura, H. Kuroki and Y. Masuda. *Chemosphere* 8:551 (1979).
7. Yoshihara, S., K. Nagata, H. Yoshimura, H. Kuroki and Y. Masuda. *Toxicol. Appl. Pharmacol.* 59:580 (1981).
8. Kashimoto, T., H. Miyata, S. Kunita, T.C. Tung, S.T. Hsu, K.J. Chang, S.Y. Tang, G. Ohi, J. Nakagawa and S. Yamamoto. *Arch. Environ. Health* 36:321 (1981).
9. Kende, A.S., J.J. Wade, M.R. DeCamp, D. Ridge and A. Poland. Paper presented at the 167th National Meeting of the American Chemical Society, Los Angeles, CA, 1974.
10. Kuroki, H., Y. Masuda, S. Yoshihara and H. Yoshimura. *Food Cosmet. Toxicol.* 18:387 (1980).
11. Kuroki, H, and Y. Masuda. *Chemosphere* 7:771 (1978),.
12. Rappe, C., H.R. Buser, H. Kuroki and Y. Masuda. *Chemosphere* 8:259 (1979).

23

Nonmutagenicity of Phenoxy Acid Herbicides 2,4-Dichlorophenoxyacetic Acid and 4-Methyl-2-Chlorophenoxyacetic Acid

K. Linnainmaa

Chlorinated phenoxyacetic acids have widespread use as foliage and weed controls in forestry and agriculture. Of phenoxy acid herbicides, 2,4,5-trichlorophenoxyacetic acid (2,4,5-T), 2,4-dichlorophenoxyacetic acid (2,4-D) and 4-methyl-2-chlorophenoxyacetic acid (MCPA) have dominated. In recent years, however, major concern has focused on the possible health hazards of occupational exposure to phenoxy acid herbicides, primarily to 2,4,5-T. Commercial preparations of 2,4,5-T have been shown to be contaminated with highly toxic 2,3,7,8-tetrachlorodibenzo-*p*-dioxin (2,3,7,8-tetra-CDD) [1,2], known to possess carcinogenic, mutagenic and teratogenic properties [3,4]. 2,4,5-T has been withdrawn from market in several countries, primarily on the basis of increases in birth defects [5] and certain types of cancers [6] among populations occupationally exposed to 2,4,5-T. Thus far, however, the overall data concerning carcinogenicity, teratogenicity and mutagenicity of 2,4,5-T are rather limited and conflicting [7].

Epidemiological studies [6] suggest an increased risk of soft-tissue sarcomas and malignant lymphomas not only with exposure to 2,4,5-T and its tetra-CDD contaminant, but also to 2,4-D and MCPA. Furthermore, increased incidence of tumors has been observed in rats receiving 2,4-D orally [8,9] and in mice after intragastric treatments with 2,4-D isooctylester [10]. These studies have several limitations; therefore, final judgment on the carcinogenicity of 2,4-D has not been made. Commercial products of 2,4-D and MCPA are not contaminated with 2,3,7,8-tetra-CDD due to absence of appropriate precursors in the manufacturing process [11,12]. However, Rappe et al. [2] have shown that Scandinavian 2,4-D products may be contaminated with 1,3,6,8- and 1,3,7,9-tetra-CDD, PCDD isomers that are known to be several orders of magnitude less toxic than 2,3,7,8-tetra-CDD [12]. Whether the increases in the occurrence of cancer among populations exposed to 2,4-D and MCPA are caused by phenoxyacetic acids, by their polychlorinated dibenzodioxin (PCDD) and dibenzofuran (PCDF) contaminants, or by the combined effects of both of these factors, it has not been possible to determine.

Carcinogenicity of a chemical is known to be closely correlated with its mutagenic properties [13]. Interestingly, however, phenoxy acid herbicides have been nonmutagenic in most of the commonly used in vitro bacterial mutagenicity assays [14,15]. Studies with eukaryotic test systems, on the other hand, have yielded both positive and negative results [14,15]. In evaluations of the genotoxic risk of a chemical for humans, the results from mammalian test systems are most valuable. The available mammalian mutagenicity data with phenoxy acid herbicides are, however, quite limited so far. The results from different studies are conflicting, as positive findings [14,15] are opposed by negative ones [14–17]. With regard to the genotoxicity of MCPA in mammalian cells, there is only one published study, the result of which is negative [14]. Three studies have been published concerning chromosomal changes in the lymphocytes of workers exposed to herbicides, two with positive [18,19], and one with negative data [20]. In all of these studies, the workers have been exposed to a great variety of pesticides, including 2,4,5-T and 2,4-D. Therefore, no evaluation of the role of phenoxy acid herbicides for the positive findings can be made.

MCPA and 2,4-D, or their mixtures, have been the most frequently used herbicides for defoliation and weed control in Finland since the withdrawal of 2,4,5-T from the market in 1979. Because of the evident need for further knowledge on the genotoxicity of phenoxy acid herbicides, we have launched cytogenetic studies on the effects of 2,4-D and MCPA. Up to the present, we have studied the induction of sister chromatid exchanges (SCE) with 2,4-D and MCPA, both in vivo and in vitro. SCE are chromosomal alterations induced by a variety of mutagenic carcinogens, and their induction is considered a sensitive indicator of DNA damage in proliferating mammalian cells [21,22]. Special emphasis has been given to studies among occupationally exposed populations spraying foliage in forestry. The general utility of the SCE analysis for monitoring occupationally exposed subjects is, however, still unclear [23]; thus, experimental studies have also been carried out to find out whether 2,4-D and MCPA are able to induce SCE in laboratory animals in vivo, or in cell cultures in vitro.

MATERIALS AND METHODS

Phenoxyacetic Acid Compounds

All the phenoxy acid herbicides considered in the study came from Kemira Oy (Finland), and consist of the following products: "Vesakontuho tasku," containing 550 g/L of 2,4-D as amine salt, "Vesakontuho MCPA," containing 500 g/L of MCPA as isooctyl ester, and "Vesakontuho DM," containing 333 g/L of 2,4-D and 167 g/L of MCPA, both as isooctyl esters. In addition to 2,4-D or MCPA, no other effective herbicide compounds are included in these products. In experimental Chinese hamster ovary (CHO) cell culture studies, pure 2,4-D and MCPA (purity $>99\%$) phenoxy acid standards provided by the State Institute of Agricultural Chemistry, Finland) were included in the experiments in addition to commercial phenoxy acid herbicides.

Test Procedures

In Vivo Studies among Occupationally Exposed Subjects

SCE induction was studied in lymphocytes of workers spraying foliage in forestry during July–October 1981. To follow possible exposure-related changes in SCE frequencies, three successive blood samples were taken from 50 male sprayers. The first sample was taken before, the second in the middle of (after 6–36 days of exposure), and the third within two days after the end of spraying season (after 15–70 days of exposure). All the subjects in the study were working for the same wood industrial company in Finland (Enso-Gutzeit OY), using primarily "Vesakontuho DM," and, to a lesser extent, "Vesakontuho tasku" for sprayings. Phenoxy acid herbicides were the only pesticides used by the sprayers. To get an idea of individual exposure levels, samples for the determination of the 2,4-D and MCPA concentrations in urine were taken simultaneously with the second blood sample. In addition, blood samples for SCE analysis were taken from 15 controls not working with herbicides. The chemical analysis of phenoxy acid concentrations in urine and the methodology for SCE testing protocol have been described in detail elsewhere [24,25].

In Vivo Studies with Rats

SCE induction was studied in cultured blood lymphocytes of inbred male Wistar rats exposed to phenoxy acid herbicides "Vesakontuho tasku" and "Vesakontuho MCPA." The test animals were treated intragastrically by a single daily dose of 100 or 150–200 mg/kg of 2,4-D and MCPA, for two weeks, 5 days per week. The control animals were given saline by gavage. Every treatment group consisted of five animals. Blood samples for SCE cultures were drawn with cardiac puncture from animals 24 hr after the last treatment. For culturing, the protocol of Kligerman et al. [26], with small modifications [25], was followed. Slides for SCE analysis were prepared according to conventional methods [22].

In Vitro Studies with CHO Cells

SCE induction in CHO cells was studied according to conventional methods [22]. Both commercial solutions ("Vesakontuho tasku" and "Vesakontuho MCPA") and purified compounds of 2,4-D and MCPA were used in the experiments. Each chemical was tested in three different subtoxic concentrations of phenoxyacetic acid (10^{-5}, 10^{-4} and 10^{-3} *M*), and all treatments were performed in the presence and absence of rat liver microsome activating system (S9). The metabolic activation capacity of CHO cells is poor; thus, chemicals that require metabolic activation to become effective SCE inducers cannot be detected with CHO assay, unless an exogenous metabolic activation system is present [27,28]. As a positive control agent, benzo[a]pyrene (B[a]P), a potent SCE-inducer requiring metabolic activation, was used at a concentration of 10^{-5} *M*. The treatment time in every culture was 1 hr, after which cells were grown for two rounds of replication (25 hr) in the presence of bromodeoxyuridine, collected, fixed and stained for microscope analyses according to conventional protocol [22]. A detailed description of CHO protocol can be found elsewhere [25].

RESULTS

SCE in Lymphocytes of Occupationally Exposed Workers

The concentrations of 2,4-D and MCPA in the urine samples of exposed workers varied from 0.00 to 10.99 mg/mL, indicating that herbicides were absorbed into the body during spraying. The mean phenoxy acid concentration in the group of 50 subjects was 1.83 mg/L, with a standard deviation of 3.18.

The results of SCE analysis are presented in Figure 23.1. No significant

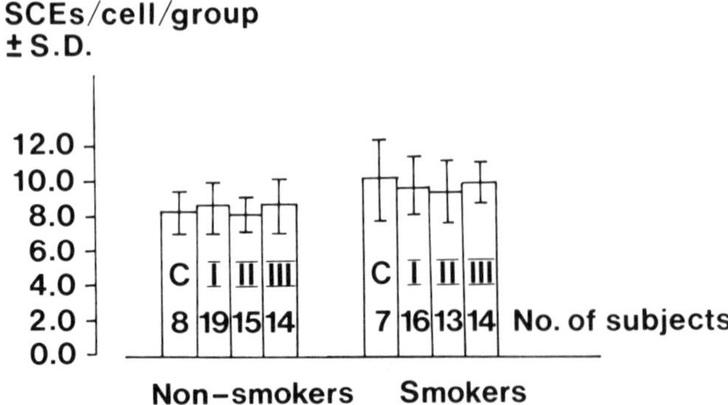

Figure 23.1. Frequencies of SCEs among workers exposed to 2,4-D and MCPA. The bars represent the group means of SCE in control subjects (C), and in herbicide sprayers; (I) samples taken before the spraying season; (II) samples taken in the middle of the spraying season (6–36 days of exposure); and (III) samples taken at the end of the spraying season (15–70 days of exposure).

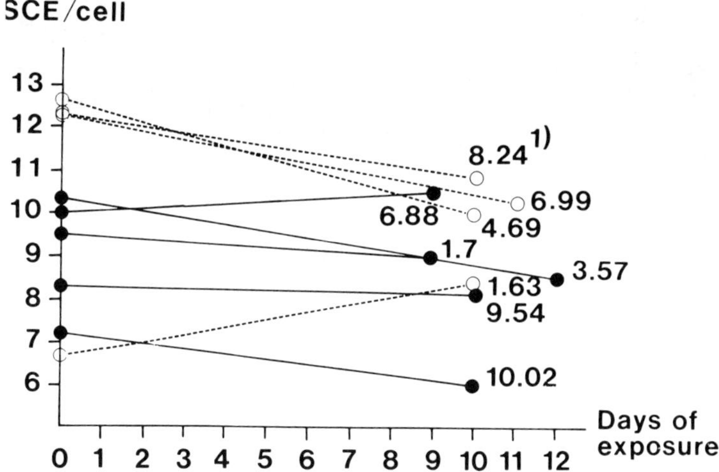

Figure 23.2. Frequencies of SCEs in blood samples taken before the spraying season (day 0), and in samples taken in the middle of the spraying season (sample II), among the subjects with the highest urinary phenoxy acid concentrations. The numbers represent the urinary 2,4-D (mg/mL) at the time when the second blood sample was taken from smokers (o- - - o) and nonsmokers (•——•).

exposure-related difference in the mean SCE frequency is to be found at the group level in samples taken before and after exposure. Further, the mean SCE in the control group of 15 subjects falls in the same range as those of the exposed subjects. A slight difference in SCE was observed, however, between smokers and nonsmokers; smokers showed higher mean values than did nonsmokers.

To study whether any exposure-related effects could be observed at the individual level, nine subjects with the highest phenoxy acid concentrations in urine were picked, and their SCE frequencies in lymphocytes were plotted against the number of days of exposure. As shown in Figure 23.2, there was no systematic change in individual SCE frequencies, either in relation to the level of urinary phenoxy acid concentrations, or in relation to the days of exposure.

SCE in Rat Lymphocytes

Results of SCE analysis from blood lymphocytes of exposed rats are presented in Figure 23.3. Sufficient data for SCE analysis could be obtained only from blood cultures of animals treated with the lower doses (100 mg/kg) of 2,4-D and MCPA, in addition to controls. In cultures from the animals treated with higher doses (150–200 mg/kg), only very few or no second-division metaphase cells suitable for SCE analysis were seen, indicating that the proliferation of lymphocytes was prevented by the toxic action of herbicides. In blood cultures from rats treated with the lower doses (100mg/kg), no toxic effects were observed, and there were no significant delays in cell cycle kinetics in treated cultures, based on

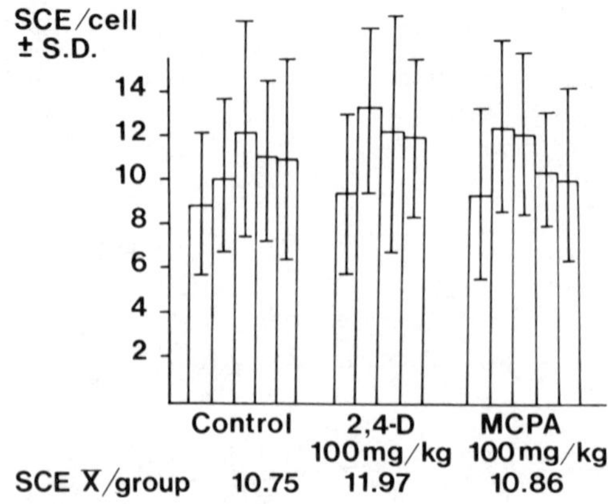

Figure 23.3. Average frequencies of SCE in blood lymphocytes of rats exposed to commercial 2,4-D and MCPA herbicides. Each bar represents an individual animal; the group means are seen below of each group.

Table 23.1. Percentages of First-, Second- and Third-Division Metaphase Cells in Blood Lymphocyte Cultures of Groups of Rats Exposed to Commercial 2,4-D and MCPA Herbicides

		Percentage of Cells ± SD		
Treatment	*No. of Animals*	*First Division*	*Second Division*	*Third Division*
Control	5	60.2 ± 5.0	29.0 ± 2.1	10.7 ± 3.7
2,4-D, 100 mg/kg	4	58.0 ± 5.5	32.1 ± 4.1	9.9 ± 5.2
MCPA, 100 mg/kg	5	64.5 ± 9.2	25.9 ± 5.2	9.6 ± 5.7

the distribution of first-, second- and third-division metaphase cells in cultures (Table 23.1).

As evident from Figure 23.3, there are large variations in average SCE frequencies between animals. These variations, however, are as characteristic within the control group as they are within the groups exposed to 2,4-D and MCPA. Furthermore, no statistical differences in mean SCE between the control group and either of the exposed groups can be observed.

SCE in CHO Cells

The results of the CHO SCE test with both commercial and pure 2,4-D and MCPA compounds are presented in Figure 23.4. No dose-related increase in the

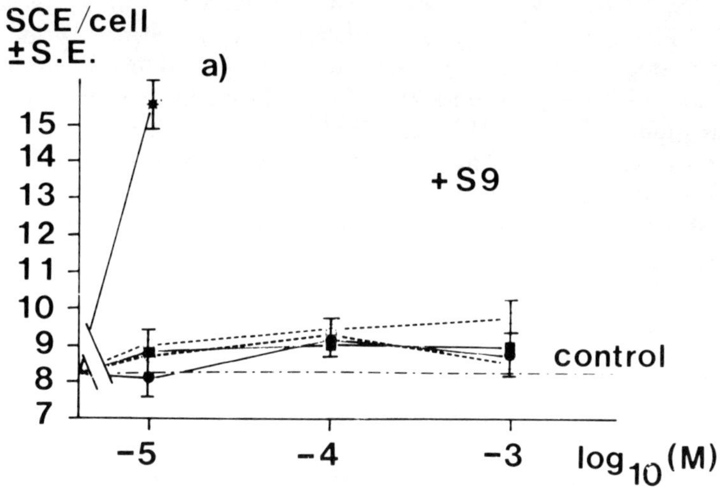

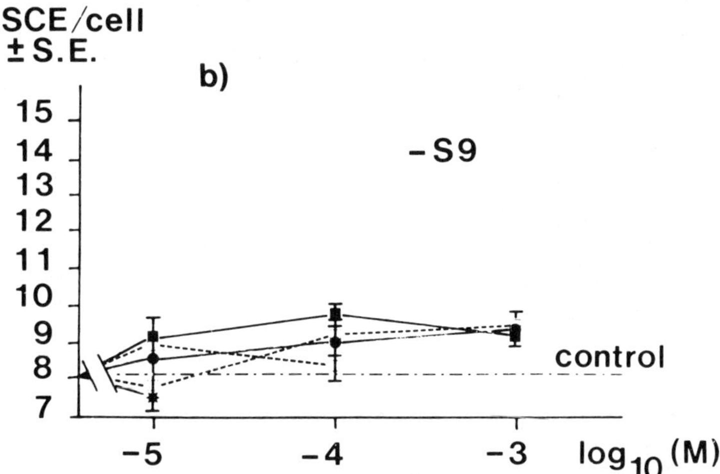

Figure 23.4. Frequencies of SCE in CHO cells after one-hour treatment with commercial and pure 2,4-D and MCPA compounds. Cells treated in (a) the presence and (b) the absence of an exogenous metabolic activation system (S9 mix). ——●—— = pure 2,4-D; - - - o - - - = commercial 2,4-D, "Vesakontuho Tasku"; ——■—— = pure MCPA; - - - □ - - - = commercial MCPA, "Vesakontuho MCPA"; ——*—— = B[a]P.

frequencies of SCE could be observed with any of the compounds in the study. However, a slight but statistically insignificant increase in SCE was characteristic for both commercial phenoxy acid herbicides as well as for the pure acids. Further, no differences can be observed in cultures treated in the presence of rat liver metabolic activation system (Figure 23.4a) compared to those treated in the absence of S9 (Figure 23.4b). Thus, neither commercial phenoxy acid herbicides containing amine salt of 2,4-D (Vesakontuho tasku), isooctyl ester of MCPA (Vesakontuho MCPA), nor pure 2,4-D and MCPA phenoxy acids were able to potentially induce SCE in CHO cells. The insignificance of the slight increases of SCE frequencies in cultures treated with phenoxy acids is clearly seen by comparing the results to those with a potent SCE inducer B[a]P (Figure 23.4a). B[a]P at 10^{-5} *M* (with S9) increased the average SCE frequency from base level (8.2 ± 2.7/cell) to 15.6 ± 5.5/cell.

DISCUSSION

The present results indicate that phenoxyacetic acid herbicides 2,4-D and MCPA do not induce SCE either in vivo or in vitro.

Crossen et al. [19] have studied the induction of SCE in lymphocytes of pesticide sprayers applying phenoxy acid herbicides 2,4-D and 2,4,5-T (among other pesticides). They found no overall difference between the control group and the sprayers. Five sprayers had, however, a SCE rate three standard deviations outside the mean of the control group. Because the subjects were exposed to a mixture of different pesticides during the same spraying season, a possible contribution by 2,4-D and 2,4,5-T to the elevated SCE frequencies remained obscure. The subjects in the present study were occupationally exposed only to the phenoxy acid herbicides 2,4-D and MCPA. The SCE frequencies of even the subjects with the highest urinary phenoxy acid concentrations (2–10 mg/L) did not deviate from the average SCE frequencies of the other subjects in the study. SCE induction in relation to an occupational exposure to phenoxy herbicides 2,4-D and MCPA seems most unlikely.

In accordance with the present results, studies of Lamb et al. [17] did not reveal significant increase in bone marrow cells of mice injected with various combinations of 2,4-D, 2,4,5-T and tetra-CDD. However, slightly higher SCE levels were characteristic in all the treated groups as compared to controls. This is quite similar to the present in vitro results with CHO cells.

As a whole, the available in vivo and in vitro data on SCE induction, together with the previous data from bacterial mutagenicity assays, indicate that phenoxy acid herbicides 2,4-D and MCPA do not act as direct DNA-damaging agents. Quite recently, however, data have accumulated which suggest that phenoxy acid herbicides may have genotoxic potential via an indirect mode of action. Reddy et al. [29] have introduced a novel class of chemical carcinogens possessing no mutagenic activity in bacterial assays. Characteristic of all these agents is that they cause a proliferation of peroxisomes in liver cells of rodents, together with an excessive production of hydrogen peroxide. The initiation of neoplastic transformation by these agents is suggested to be the result of the continued production of DNA-damaging oxygen radicals. Interestingly, Vainio et al. [30] have found that the chlorinated phenoxyacetic acids 2,4-D and MCPA also increase both the number and size of peroxisomes in liver cells of Chinese hamsters [30] and rats [31]. Special interest in studying peroxisome proliferation induced by phenoxy acid herbicides arose from the finding that esters of chlorinated phenoxyacetic acids are chemically closely related to clofibrate, one of the known peroxisome proliferators. The rats studied for peroxisome proliferation by Vainio et al. [31] were the same animals used in the present SCE study; thus, peroxisome proliferation in liver cells of rats was found not to be associated with simultaneous induction of SCE in their blood lymphocytes. On a dose basis, 2,4-D and MCPA were found to be equally potent peroxisome proliferators as clofibrate. Like 2,4-D and MCPA, clofibrate did not affect the frequency of SCE in the lymphocytes of exposed animals [25].

Based on the observations above, it has been suggested that phenoxy acid herbicides could act by a mechanism similar to the one proposed for tumor initiation by known peroxisome proliferators, e.g., via increased production of H_2O_2 and oxygen radicals [30]. If so, the carcinogenic potential would more likely be due to the phenoxy acid compound itself than due to the PCDD or PCDF contaminants, because of the structural analogy between phenoxyacetic acid and clofibrate molecules. Norstrom et al. [11] also suggested that phenoxy acids per se are the reason for the increase in cancer incidence, on the basis of their analysis of older Scandinavian formulations of 2,4-D and 2,4,5-T for contents of PCDD and PCDF.

Hydrogen peroxide, when tested in vitro on a V-70 Chinese hamster cell line, has been found to increase significantly the SCE frequency [32,33]. Thus, excessive production of H_2O_2 should be detectable by SCE induction. In vivo, proliferation of peroxisomes is, however, most prominent in liver cells; thus, the amounts of H_2O_2 formed in the target cells of the present experiments may have been too small for significant induction of SCE. Further, most of the H_2O_2 produced may rapidly be inactivated enzymatically (e.g., by catalase). Whether the slight increases in SCE frequencies observed in the present in vitro CHO experiment, and in the experimental in vivo studies of Lamb et al. [17] are due to the production of H_2O_2 in cells exposed to phenoxy acid herbicides remains to be elucidated.

ACKNOWLEDGMENTS

I wish to thank Dr. Harri Vainio for his critical reading of my manuscript, and for many valuable discussions. I am especially indebted to Dr. Pentti Vuojolahti, and his group at the Occupational Health Center for Enso-Gutzeit Oy (Imatra, Finland), for the excellent cooperation with the material for the study on SCE among the herbicide sprayers. The study has been supported by the Academy of Finland.

REFERENCES

1. Young, A.L., C.E. Thalken, E.L. Arnold and J.M. Cupello. U.S. FA-TR-76-18, Boulder, CO (1976).
2. Rappe, C., H.R. Buser and H.P. Bosshardt. *Chemosphere* 5:431 (1978).
3. IARC Monograph No. 15 (1977), p. 41.
4. IARC Internal Technical Report 78/001 (1978).
5. "The Effects of Herbicides in South Vietnam," National Academy of Sciences, Washington, DC (1974).
6. Hardell, L. Umeå University Medical Dissertations. New Series No. 65 (1981).
7. Grant, W.F. *Mutat. Res.* 65:83 (1979).
8. Hansen, W.H., M.L. Quaife, R.T. Haberman and O.G. Fitzhugh. *Toxicol. Appl. Pharmacol.* 20:122 (1971).
9. Haberman, T. "Pathological Changes in Rats Fed 2,4,-Dichlorophenoxy Acid for Two Years" (unpublished).
10. Innes, J.R.M., B.M. Ulland, M.G. Valerio, L. Petrucelli, L. Fishbein, E.R. Hart, A.J. Pallota, R.R. Bates, H.L. Falk, J.J. Gart, G.M. Klein, I. Mitchell and J. Peters. *J. Nat. Cancer Inst.* 42:1101 (1969).
11. Norström, Å., C. Rappe, R. Lindahl and H.R. Buser. *Scand. J. Work Environ. Health* 5:375 (1979).
12. Reggiani, G. In *Chlorinated Dioxins and Related Compounds: Impact on the Environment,* O. Hutzinger, R.W. Frei, E. Merian and F. Pocchiari, Eds. (Oxford: Pergamon Press, Ltd., 1982) pp. 463–493.
13. Ames, B.N., J. McCann, and E. Yamasaki. *Mutat. Res.* 31:347 (1975).
14. Seiler, J. *Mutat. Res.* 55:197 (1978).
15. Grant, W. *Mutat. Res.* 65:83 (1979).
16. Herbold, B.A., L. Machemer and G. Röhrborn. *Teratogen. Carcinog. Mutagen.* 2:91 (1982).
17. Lamb, J.C., T.A. Marks, B.C. Gladen. J.W. Allen and J.A. Moore. *J. Toxicol. Environ. Health.* 8:825 (1981).
18. Yoder, J., M. Watson and W.W. Benson. *Mutat. Res.* 21:335 (1973).
19. Crossen, P.E., W.F. Morgan, J.J. Horan and J. Stewart. *New Zealand Med. J.* 88: 192 (1978).
20. Högstedt, B., A.-M. Kolnig, F. Mitelman and S. Skerfving. *Hereditas* 92:177 (1980).
21. Perry, P.E. In: *Chemical Mutagens, Principles and Methods for Their Detection, Vol. 6,* F.J. deSerres and A. Hollaender, Eds. (New York: Plenum Press, 1980), pp. 1–39.
22. Latt, S.A., J. Allen, S.E. Bloom, A. Carrano, E. Falke, D. Kram, E. Schneider, R. Schreck, R. Tice, B. Whitfield and S. Wolff. *Mutag. Res.* 87:17 (1981).
23. Lambert, B., A. Lindblad, K. Holmberg and D. Francesconi. In: *Sister Chromatid Exchange,* S. Wolff, Ed. (New York: John Wiley & Sons, Inc., 1982), pp. 149–182.
24. Korhonen, R., and J. Kangas. Publications of University of Kuopio. Natural Sciences, Series Statistics and Reviews (1982), p. 26.
25. Linnainmaa, K. (in preparation).
26. Kligerman, A.D., J.L. Wilmer and G.L Erexson. *Environ. Mutagen.* 3:531 (1981).
27. Natarajan, A.T., A.D. Tates, P.P.W. van Buul, M. Meijers and N. de Vogel. *Mutat. Res.* 37:83 (1976).
28. Stetka, D.G., and S. Wolff. *Mutat. Res.* 41:343 (1976).
29. Reddy, J.K., and D.L. Azarnoff. *Nature* 283:397 (1980).

30. Vainio, H., J. Nickels and K. Linnainmaa. *Scand. J. Work Environ. Health.* 8:70 (1982).
31. Vainio, H., K. Linnainmaa, M. Kahonen, J. Nickels, E. Hietanen and J. Marniemi. *Biochem. Pharmacol.* (in press).
32. Bradley, M.O., I.C. Hsu and C.C. Harris. *Nature* 282:318 (1979).
33. Speit, G., W. Vogel and M. Wolf. *Environ. Mutagen.* 4:135 (1982).

24

Comparison of the Immunosuppressive Effects in Mice of 2,3,7,8-Tetrachlorodibenzo-*p*-dioxin and 2,3,7,8-Tetrachlorodibenzofuran

Annunciata Vecchi, Marina Sironi,
Maria Antonia Canegrati and
Silvio Garattini

Polychlorinated dibenzo-*p*-dioxins (PCDD) and dibenzofurans (PCDF) have been found to be persistent environmental contaminants. The 2,3,7,8-tetrachloro-isomers of both classes of compounds are highly toxic in laboratory animals [1,2]. Tetrachlorodibenzo-*p*-dioxin (tetra-CDD)can be formed during the manufacture of 2,4,5-tetrachlorophenol and may thus occur as a contaminant of herbicides. Accidental poisoning of animals and humans has occurred over the years [3,4], as a result of the production and use of chlorophenol (CP) and chlorobenzene compounds [5]. Tetrachlorodibenzofuran (tetra-CDF) is present in PCDF-mixtures detected in treated wood and polychlorinated biphenyls (PCB), including PCB contaminating the rice oil responsible for the human intoxication known as yusho sickness [6]. Tetra-CDD and tetra-CDF have also been found in fly ash and flue gases from municipal and industrial incinerators [7].

Tetra-CDF is closely related to tetra-CDD in structure and acute toxicity [8]. LD_{50} in the animal species investigated ranged from <10 $\mu g/kg$ for the guinea pig to >6000 $\mu g/kg$ for C57B1/6 mice [1]. As reported for tetra-CDD, signs of toxicity differ in different species, hepatic lesions being observed in rats and mice, but not in guinea pigs or monkeys [1]. Progressive weight loss and thymic atrophy are common findings. It is believed that tetra-CDF is as toxic as tetra-CDD at doses 15–30 times those of tetra-CDD [1].

The immunosuppressive effects of tetra-CDD have been studied extensively in animals; tetra-CDD is a potent inhibitor of humoral and cell-mediated responses [9,10], even depressing pluripotent bone marrow stem cells if given during the perinatal period [11]. The very limited data available on the immunosuppressive effects of tetra-CDF [12] suggest that they are similar to those of tetra-CDD.

We have previously shown [9] that in adult mice low doses of tetra-CDD depress humoral more than cellular responses and that strains of mice (DBA/2 and AKR) more resistant to tetra-CDD induction of hepatic aryl hydrocarbon hydroxylase (AHH) enzymes are also more resistant to the immunosuppressive effects than are very sensitive strains (C57B1/6 and C3H/He) [13,14]. The AHH enzyme system plays a central role in the biotransformation of aromatic xenobiotics; potentially carcinogenic metabolites are produced and/or degraded by this complex of enzymes [15,16].

In this study we compare the immunosuppressive activity of tetra-CDD and tetra-CDF in C57B1/6 mice after single exposure, evaluating humoral antibody production. Tetra-CDF is the more powerful in vitro inhibitor of tetra-CDD binding to the specific cytosolic receptor responsible for AHH enzyme induction; it is thus believed to act through a similar mechanism of action. For this reason tetra-CDF is studied here in C57B1/6 and DBA/2 mice, which differ in their level and/or affinity of tetra-CDD–specific receptor and are differently susceptible to the immunosuppressive effects of tetra-CDD [14,17].

MATERIALS AND METHODS

Mice

C57B1/6 and DBA/2 mice, hereafter called respectively B6 and D2, were obtained from Charles River, Calco, Italy. All animals were used at 10–12 weeks of age.

Chemicals

Tetra-CDD obtained from KOR Isotopes, Cambridge, Massachusetts, was given as a single intraperitoneal (i.p.) or oral (p.o.) injection in a volume of 0.1 ml/10 g body weight of an acetone:corn oil solution (1:6 v/v). Tetra-CDF was a gift from C. Rappe, University of Umeå, Sweden. It was dissolved and given as described for tetra-CDD.

Response to Sheep Red Blood Cells

Mice (7–8 per group) were injected i.p. with 4×10^8 sheep red blood cells (SRBC) 7, 21 or 42 days after drug treatment, and spleen hemolytic plaque-forming cells (PFC) were counted 5 days later [9].

Peritoneal Cells

Peritoneal exudate cells (PEC) were obtained by washing the peritoneal cavity with 5 ml of phosphate-buffered saline (PBS). Differential counts were made with Turk's solution [18].

Statistical Analysis

Results are presented as means ± S.E. Statistical significance was analyzed by Student's t test, or by Dunnett's test when more than two experimental groups were compared.

RESULTS

Body and thymus weights and spleen and peritoneal cell numbers were evaluated as signs of general toxicity. Results are shown in Table 24.1. No effects on body weight were observed at any of the doses tested, whereas doses higher than 180 μg/kg i.p. or p.o. significantly reduced thymus weights. This dose was also toxic for the spleen, as shown by the low number of splenocytes recovered after tetra-CDF treatment. No significant modifications were seen in the number of total peritoneal cells and macrophages. Peritoneal cell and macrophage counts after i.p. treatment were higher than those from p.o.-treated mice; it must be noted, however, that tetra-CDF was given in an acetone:oil solution that probably has some local irritant effect.

Immunosuppressive effect was then investigated, the toxic agent was given seven days before antigenic stimulation. A wide range of doses was tested, from 5 to 900 μg/kg. Figure 24.1 reports the dose-response curve of tetra-CDF inhibition of antibody production and the results with tetra-CDD for comparison. Humoral response was depressed similarly by both compounds; the major difference was in the effective doses: about 30 times more tetra-CDF than tetra-CDD, in terms of micrograms per kilogram, was needed to depress the immune response to the same level.

Since the i.p. route is an experimentally easy, but not realistic route of contamination, tetra-CDF was also given orally, and antibody production was evaluated. Results (Table 24.2) clearly show that the same degree of inhibition could be obtained with both treatments, in good agreement with previous reports for tetra-CDD [9].

The time-course of immunosuppression was investigated next. Doses of tetra-CDD (6 μg/kg) and tetra-CDF (180 μg/kg) found to be equally active on day 7 were used. Results (Figure 24.2) show that 42 days after tetra-CDD exposure, inhibition was still present and was as strong as that observed on day 7. In contrast, the depression induced by tetra-CDF had completely disappeared by day 42. In B6 mice the whole-body half-life of tetra-CDF is 2 days and metab-

Table 24.1. Effect of Tetra-CDF on Body Weight and Lymphoid Organs in B6 Mice

Route of Administration[a]	*Tetra-CDF* ($\mu g/kg$)	*Body Weight* (*g*)	*Thymus Weight* (*mg*)	*Spleen Cells* (10^6)	*PEC* (10^6)	*Macrophages* (10^6)
Intraperitoneal	0	24.1 ± 0.6	57.3 ± 5.4	136.9 ± 13.8		
	22.5	25.8 ± 1.0	57.0 ± 3.0	111.8 ± 6.5		
	45	24.6 ± 0.7	62.7 ± 2.8	107.7 ± 4.1[b]		
	90	24.1 ± 0.8	55.1 ± 5.1	120.7 ± 7.5		
	0	27.3 ± 0.7	42.8 ± 1.4	129.9 ± 20.3	7.6 ± 1.7	3.6 ± 0.9
	180	26.7 ± 1.3	30.5 ± 2.5[c]	74.9 ± 8.3[c]	11.0 ± 1.4	6.0 ± 1.1
	900	25.7 ± 1.7	13.5 ± 1.6[c]	64.4 ± 7.2[c]	11.5 ± 2.6	5.4 ± 0.6
Orally	0	28.8 ± 1.0	44.7 ± 1.4	143.1 ± 26.5	4.2 ± 0.6	1.9 ± 0.2
	180	26.2 ± 0.3	37.8 ± 2.2[c]	89.3 ± 9.6[c]	3.6 ± 0.4	1.8 ± 0.2
	900	25.1 ± 1.3	27.8 ± 2.9[c]	62.4 ± 9.0[c]	4.2 ± 0.7	1.9 ± 0.2

[a]Tetra-CDF was administered 12 days before testing.

[b]$p < 0.05$.

[c]$p < 0.01$.

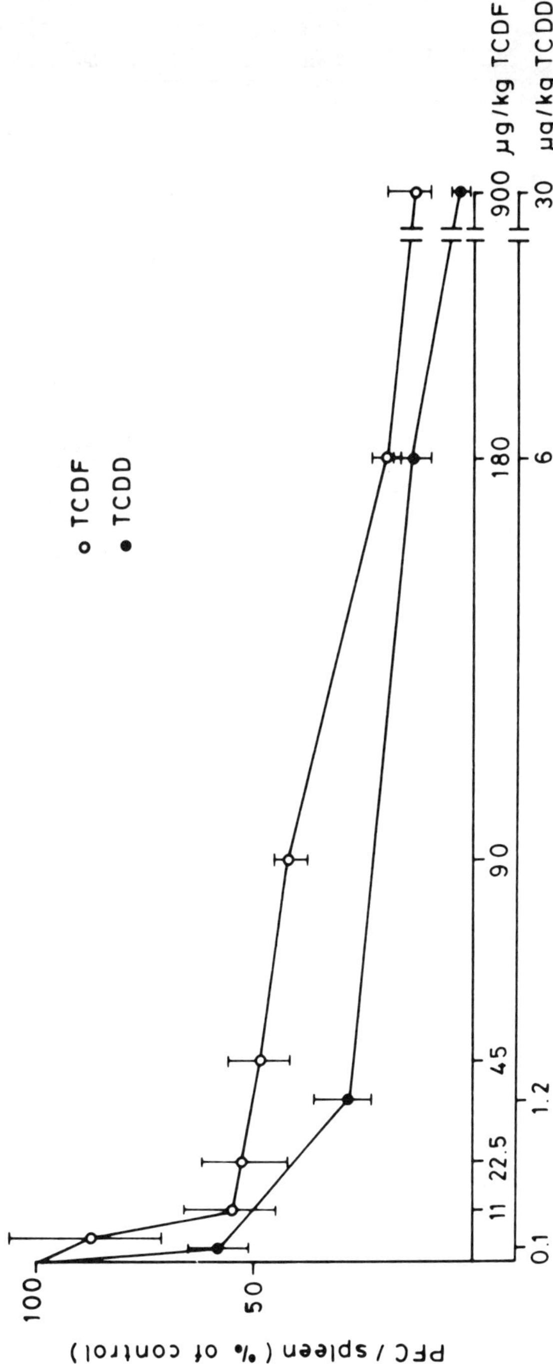

Figure 24.1. Dose-Response Effect on Humoral Antibody Production in B6 Mice of Tetra-CDD and Tetra-CDF Given i.p. Seven Days before the Antigen (4×10^8 SRBC). The test was performed five days after antigenic stimulation.

Table 24.2. Immunosuppressive Effect of Tetra-CDF Given by Different Routes[a]

Route of Administration[b]	*TCDF (μ g/kg)*	*PFC/10⁶ Splenocytes*	*PFC/Spleen*	*% of Control*
Intraperitoneal	0	206 181–234	24,795 20,890–29,428	100
	180	75[c] 52–107	5,692[c] 4,159–7,790	23
	900	62[c] 47–81	3,648[c] 1,419–4,037	15
Orally	0	336 308–365	48,334 35,778–65,297	100
	108	123[d] 84–181	10,323[c] 7,716–13,811	21
	900	111[d] 70–175	6,207[c] 4,120–9,352	13

[a]Logarithmic transformation of PFC values was done before statistical analysis. Results are antilogarithms of the means (± S.E.).

[b]Tetra-CDF was given on day −7; 4 × 10^8 SRBC were injected i.p. on day 0; test was performed on day +5.

[c]$p < 0.01$.

[d]$p < 0.05$.

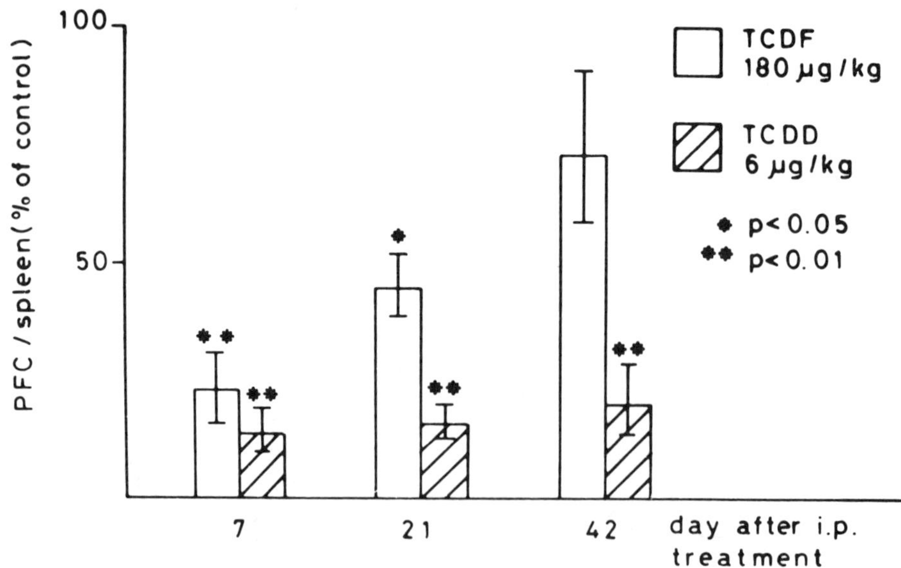

Figure 24.2. Time Course of the Immunosuppresive Effect of Tetra-CDD and Tetra-CDF Given i.p. at Days Indicated. Immunization and test as in Figure 24.1.

Table 24.3. Effect of TCDD and TCDF on Body and Thymus Weight in Different Mouse Strains

Strain	*Tetra-CDD*[a] *(μg/kg)*	*Tetra-CDF*[a] *(μg/kg)*	*Body Weight (g)*	*Thymus Weight (% of Control)*	
B6	0		25.9 ± 0.5	64.3 ± 2.9	(100)
	6		24.5 ± 0.4	45.0 ± 4.3[b]	(70)
		0	21.4 ± 0.9	31.2 ± 3.0	(100)
		180	22.1 ± 0.9	18.6 ± 4.6[b]	(60)
D2	0		21.1 ± 0.5	27.7 ± 2.2	(100)
	6		22.7 ± 0.8	23.4 ± 3.5	(84)
		0	25.8 ± 0.5	37.6 ± 3.0	(100)
		180	26.3 ± 0.5	32.0 ± 2.4	(85)

[a]Tetra-CDF and tetra-CDD were given i.p. 12 days before evaluation.
[b]$p < 0.05$.

olites were found in urine and feces [19], while tetra-CDD is not metabolized to an appreciable extent in the same animals, with a half-life of 17 days [20].

Because of the importance of the AHH enzyme system and the differences in tetra-CDD immunosuppression in B6 and D2 mice that differ in the level and/or affinity of the specific receptor responsible for AHH induction by tetra-CDD, tetra-CDF was also investigated in these two mouse strains. Table 24.3 shows the body and thymus weights. Thymus weight was reduced only in B6 mice. When humoral response was investigated, 85% inhibition of antibody production was observed in B6 mice after treatment with both compounds, while only 40% inhibition was induced in D2 mice (Figure 24.3). This difference between the two strains was confirmed by the number of PFC per million splenocytes (Figure 24.3, left).

DISCUSSION

Results reported here show that tetra-CDF has a definite immunosuppressive effect on humoral response in B6 mice. Similar degrees of immunosuppression can be obtained with tetra-CDF doses 30 times higher than those of tetra-CDD. However, because the LD_{50} for tetra-CDF is not yet well defined [1], it is difficult to compare equally suppressive doses of tetra-CDF and tetra-CDD and fractions of LD_{50}. Because the LD_{50} of tetra-CDF is >6000 μg/kg and that of tetra-CDD is 115–132 μg/kg [20,21], one can calculate that tetra-CDF (180 μg/kg) is equally active as tetra-CDD (6 μg/kg) at a dose less than 1/30 of the LD50, while that of tetra-CDD is 1/20. It might be concluded that tetra-CDF in B6 mice should be more suppressive than tetra-CDD, at least until definitive data on the tetra-CDF LD_{50} is available.

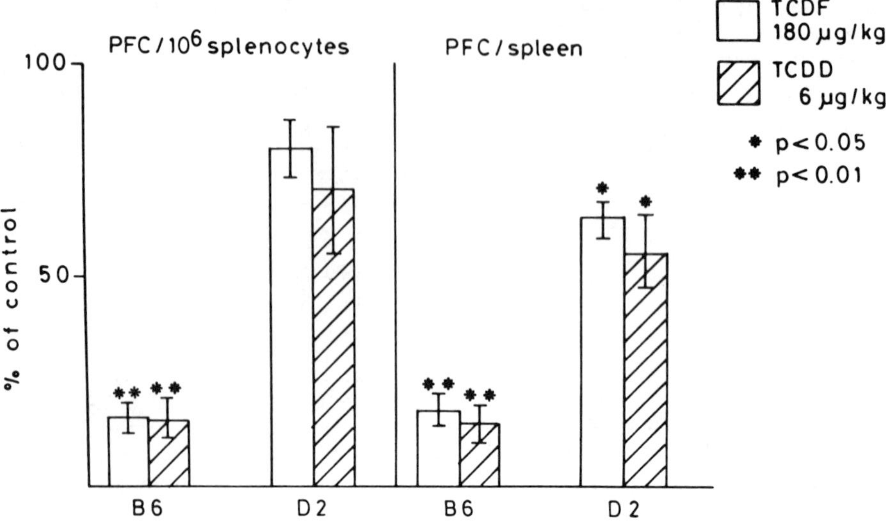

Figure 24.3. Effect of Tetra-CDD and Tetra-CDF on Humoral Antibody Production in B6 and D2 mice. Treatment, immunization and test as in Figure 24.1.

However, when we looked at the time course of the immunosuppressive effect, marked differences were observed between the two toxic agents. Tetra-CDD immunosuppression did not change from day 7 to day 42, whereas by day 42 tetra-CDF suppression had completely disappeared. The two agents reportedly have very different half-lives (2 days for tetra-CDF and 17 days for tetra-CDD); in addition, tetra-CDF is metabolized and, thus, excreted faster, while no relevant metabolism of tetra-CDD was observed in B6 mice [19,20]. Because the examination of immunological parameters in guinea pig studies stopped seven days after the last dose [12], it is difficult to say whether the time course of tetra-CDD and tetra-CDF effects are similar in species other than mice.

Results in B6 and D2 mice showing a good correlation between tetra-CDD and tetra-CDF effects may support the hypothesis that the tetra-CDD receptor plays a role, not only in vitro [22], but also in vivo for the biological acitivity of tetra-CDF [23]. This receptor was shown to be present not only in the liver, but also in other organs [24]; the highest concentration reported being in the thymus [24], the organ most affected by tetra-CDD in all the species investigated. In line with this view are the data of Greenlee and Poland [23], showing that tetra-CDD binding in vitro to thymic cytosol from D2 mice is far less than in B6. With the data on tetra-CDF half-life in B6 and D2 mice, it has been reported [19] that similar levels of radioactivity were found in thymus and spleen after administration of labeled tetra-CDF. This reasonably excludes that the differences observed in the two mouse strains were simply due to different levels of tetra-CDF in thymus and spleen. If tetra-CDF interacts in some way with the tetra-CDD receptor, including similar toxic effects, its binding should be shorter-lasting and/or weaker, as suggested by the half-life and by the time course of antibody production inhibition.

ACKNOWLEDGMENT

The generous contribution of the Gustav and Louise Pfeiffer Research Foundation, Los Angeles, California, is gratefully acknowledged.

REFERENCES

1. Moore, J.A., E.E. McConnell, D.W. Dalgard and M.W. Harris. *Ann. N.Y. Acad. Sci.* 320:151 (1979).
2. McConnell, E.E., J.A. Moore, J.K. Haseman and M.W. Harris. *Toxicol. Appl. Pharmacol.* 44:335 (1978).
3. Firestone, D. *Ecol. Bull.* 27:39 (1978).
4. Reggiani, G. *J. Toxicol. Environ. Health* 6:27 (1980).
5. Kimbrough, R.D. *CRD Crit. Rev. Toxicol.* 2:445 (1974).
6. Nagayama, J., Kuratsune and Y. Masuda. *Bull. Environ. Contam. Toxicol.* 15:9 (1976).
7. Rappe, C., H.R. Buser and H.P. Bosshardt. *Ann. N.Y. Acad. Sci.* 320:1 (1979).
8. McConnell, E.E., and J.A. Moore. *Ann. N.Y. Acad. Sci.* 320:138 (1979).
9. Vecchi, A., A. Mantovani, M. Sironi, W. Luini, M. Cairo and S. Garattini. *Chem.-Biol. Interact.* 30:337 (1980).
10. Vos, J.G. *CRC Crit. Rev. Toxicol.* 5:67 (1977).
11. Luster, M.I., G.A. Boorman, J.H. Dean, M.W. Harris, R.W. Luebke, M.L. Padarathsingh and J.A. Moore. *Int. J. Immunopharmacol.* 2:301 (1980).
12. Luster, M.I., R.E. Faith and L.D. Lawson. *Drug. Chem. Toxicol.* 2:49 (1979).
13. Poland, A., and E. Glover. *Molec. Pharmacol.* 11:389 (1975).
14. Garattini, S., A. Vecchi, M. Sironi and A. Mantovani. In: *Chlorinated Dioxins and Related Compounds,* O. Hutzinger, R.W. Frei, E. Merian and F. Pocchiari, Eds. (Oxford: Pergamon Press, 1982), p. 403.
15. Thorgeirsson, S.S., and D.W. Nebert. *Adv. Cancer Res.* 25:149 (1977).
16. Nebert, D.W. *Biochemie* 60:1019 (1978).
17. Okey, A.B., G.P. Bondi, M.E. Mason, G.F. Kahl, H.J. Eisen, T.M. Guenther and D.W. Nebert, *J. Biol. Chem.* 254:11636 (1979).
18. Mantovani, A., A. Vecchi, W. Luini, M. Sironi, G.P. Candiani, F. Spreafico and S. Garattini. *Biomedicine* 32:200 (1980).
19. Decad, G.M., L.S. Birnbaum and H.B. Matthews. *Toxicol. Appl. Pharmacol.* 59:564 (1981).
20. Neal, R.A., J.R. Olson, R.A. Gasiewicz and L.E. Geiger. *Drug Metab. Rev.* 13:355 (1982).
21. Vos, J.G., J.A. Moore and J.G. Zinkl. *Toxicol. Appl. Pharmacol.* 29:229 (1974).
22. Poland, A., E. Glover and A.S. Kende. *J. Biol. Chem.* 251:4936 (1976).
23. Greenlee, W.F., and A. Poland. *J. Biol. Chem.* 254:9814 (1979).
24. Carlstedt-Duke, J.M.B. *Cancer Res.* 39:3172 (1979).

Index